Electromagnetic Theory for Telecommunications

C.S. Liu

Professor of Physics
University of Maryland, College Park, MD, USA

and

V.K.Tripathi

Professor of Physics
Indian Institute of Technology Delhi, India

FOUNDATION®
BOOKS

Delhi • Bangalore • Mumbai • Kolkata • Chennai • Hyderabad • Pune

Published by
Cambridge University Press India Pvt. Ltd.
under the Foundation Books imprint
Cambridge House
4381/4, Ansari Road,
Daryaganj
New Delhi – 110 002

C-22, C-Block, Brigade M.M., K.R. Road, Jayanagar, **Bangalore** 560 070
Plot No. 80, Service Industries, Shirvane, Sector-1, Nerul, **Navi Mumbai** 400 706
10, Raja Subodh Mullick Square, 2nd Floor, **Kolkata** 700 013
21/1 (New No. 49), 1st Floor, Model School Road, Thousand Lights, **Chennai** 600 006
House No.3-5-874/6/4, (Near Apollo Hospital), Hyderguda, **Hyderabad** 500 029
Agarwal Pride, 'A' wing, 1308 Kasba Peth, Near Surya Hospital, **Pune** 411 011

First published 2007
Reprinted 2009

ISBN 9788175965447

Published by Manas Saikia for Cambridge University Press India Pvt. Ltd., and printed & bound at Raj Press, R-3 Inderpuri, New Delhi - 110 012.

Contents

Preface

Telecommunication has become an increasingly popular means of communication, enabling the rapid processes of globalisation. People are easily 'connected' through cell phones and the Internet, even though they are in different continents and thousands of kilometres apart. Today there are an estimated 1.3 billion people using the Internet daily, emailing, Googling, purchasing books on Amazon.com or buying merchandise on eBay.

The widespread use of means of telecommunication has been rendered possible by the rapid progress of the underlying science and technology, as well as the speedy development of requisite software and hardware. Discovery of electromagnetic waves has been the most important and fundamental breakthrough, initiating a revolution in the world of telecommunication. It all started with the discovery of laws of electromagnetic induction by *Faraday* in 1831. Soon after, in the 1850s, *Maxwell* wrote his masterful treatise on electromagnetic theory, culminating in the discovery of Maxwell's equations. Finally in 1887, *Hertz* experimentally observed the propagation of an electromagnetic wave emitted by an oscillatory dipole in free space. He said in 1989, 'Electricity has become a mighty kingdom. We perceive it in a thousand places where we had no proof of its existence before. The domain of electricity extends over the whole of nature.' Indeed in 1895, *Marconi* succeeded in transmitting the Hertzian wave over half a mile, eventually not only over the British channel (1897) but across the Atlantic ocean (1908), heralding us into the age of telegraph and wireless communication.

Electromagnetism, thus, has had profound effects on our lives and and has become an integral part of our global existence. While there are many excellent texts on electricity and magnetism we sensed a need, for students in engineering and applied sciences, for a textbook on the principles of electromagnetism with emphasis on the essentials of telecommunication. This book makes such an attempt and is intended for advanced undergraduate students (junior and senior) or beginning graduate students in engineering and physics, with prerequisites of only freshman physics and calculus courses.

In unit 1, the basic principles of electricity and magnetism are surveyed. In unit 2, electromagnetic waves in free space and various media such as dielectrics, conductors, semiconductors and plasmas are studied. Their propagation characteristics, phase and group velocities, reflection, refraction and transmission are also discussed. In unit 3, waveguides and electromagnetic wave propagation through various guiding structures have been dealt with. In particular, we introduce the surface wave–propagating along the boundary of free space and a medium such as metal or plasma.

In unit 4, radiating systems such as an oscillatory dipole antenna and the radiation properties of such sources are broached. In units 5-7, the physical principles for three primary ways of telecommunication – (i) radio, including cellular communication and the effects of ionosphere (unit 5), (ii) satellite communication including geostationary satellite (unit 6) and finally, (iii) optical communication including laser and optical fibre (unit 7) have been developed. In unit 8, the basic principle for remote sensing and geo-seismic exploration are introduced. Unit 9 introduces special relativity to develop a relativistic covariant formalism of electromagnetism. In unit 10 the radiation from accelerated charges has been covered.

This approach for a preliminary text has been used by one of us (VKT) for several years in teaching students at IIT Delhi, working for Bachelor of Technology and Master of Science degrees. We were gratified by the favourable response from our students, which encouraged us to write out this book. We hope it will help in meeting the growing needs of this field.

C.S. Liu
V.K. Tripathi

Unit 1

Electromagnetic Fields

Introduction

All the forces that we encounter in nature are manifestations of four forces : (i) gravitational force, arising between bodies by virtue of their masses, (ii) electromagnetic force, arising between bodies by virtue of their charges (stationary and moving), (iii) strong nuclear force, operative at short distances $\leq 10^{-15}$ metre, which holds nucleons *i.e.*, protons and neutrons together in a nucleus and (iv) weak nuclear force that accounts for certain kinds of radioactive decay. Of these four, the electromagnetic force appears in the most diverse forms and is encountered very frequently in our daily lives. For example, the push that we give to a block with our hands to make it move, is the electromagnetic force of repulsion between the charged constituents of molecules of our hands and those of the block. A book lying on a table experiences an electromagnetic force between the molecules of the surface of the table and those at the bottom of the book. The upward thrust exerted by a liquid on a body immersed in it (*Archimedes principle*) is the electromagnetic force between the liquid molecules and those of the body. The force that accounts for the stability of electrons rotating around the nucleus in an atom is again an electric force. The forces that bind the atoms in a solid are also electromagnetic in nature. Elastic forces too are electromagnetic forces. Our physiological functions, such as the beating of the heart and the transmission of neural signals, usually involve electromagnetic processes.

The basic constituents of matter in nature are electrons, protons and neutrons. All the three particles possess mass. The former two also possess an additional quantity called 'charge'. The electron has a negative charge $-e$ ($e = 1.6 \times 10^{-19}$ Coulomb) while the proton has a positive charge $+e$. The neutron does not have a net charge and is a neutral (hence its name!) particle. Most bodies that we see around us have exactly the same number of electrons and protons, and hence are electrically neutral overall. However, if we disturb this balance by removing a few electrons (say by rubbing), the body acquires a net positive charge. Similarly, if some electrons are added into it, the body acquires a net negative charge. One can easily sense the electric force in the presence of such charged bodies.

One important characteristic of electrons in all materials is their motion. In an atom, electrons revolve around the nucleus. They also spin about their own axis (Fig. 1.1(a)). In conductors and semiconductors, free electrons move with random velocities. When a conductor is connected to a battery, free electrons acquire an

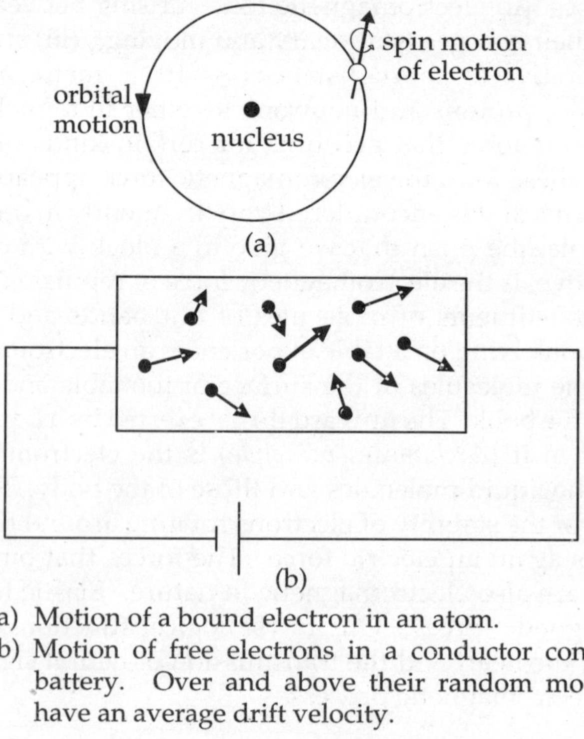

Figure 1.1: (a) Motion of a bound electron in an atom.
(b) Motion of free electrons in a conductor connected to a battery. Over and above their random motions, they have an average drift velocity.

average velocity constituting an electric current (Fig. 1.1(b)). This current produces a magnetic field. Oscillatory currents produce oscillatory electric and magnetic fields, called *electromagnetic waves*.

In this chapter, we shall study the fundamental laws governing the fields and develop a mathematical framework for studying electromagnetic waves.

1.1 Coulomb's Law

Coulomb observed that two static bodies (of small sizes), possessing like charges q_1, q_2 and separated from each other by a distance r, repel each other (along the line joining the two) by a force of magnitude

$$F = \frac{q_1 q_2}{4\pi \in_0 r^2}, \tag{1}$$

where $\in_0 = \frac{10^{-9}}{36\pi} \frac{\text{Coulomb}^2}{\text{Newton.m}^2}$ or $\frac{\text{Farad}}{\text{m}}$.

Bodies carrying opposite charges attract each other. If charges q_1 and q_2 are placed at distances $\mathbf{r_1}$ and $\mathbf{r_2}$ (from some common reference point) respectively, then the force on q_2 due to q_1 (in vector form) is

$$\mathbf{F} = \frac{q_1 q_2 (\mathbf{r_2} - \mathbf{r_1})}{4\pi \in_0 |\mathbf{r_2} - \mathbf{r_1}|^3}. \tag{2}$$

The force on q_1, due to q_2 is $-\mathbf{F}$ (Fig. 1.2). In case the charges are located in a uniform unbounded material medium, $\in_0$ in eq. (2) is replaced by $\in$–the permittivity of the medium. This may be used to define a quantity $\in_r = \in / \in_0$, called *relative permittivity* or *dielectric constant* of the medium.

It may be mentioned that Coulomb's law is not valid for moving charges, and hence has limited applicability.

1.2 Electric Field, E (r, t)

An electric field, at any point in space, is defined as the force experienced by a unit test charge placed at that point. In the case of

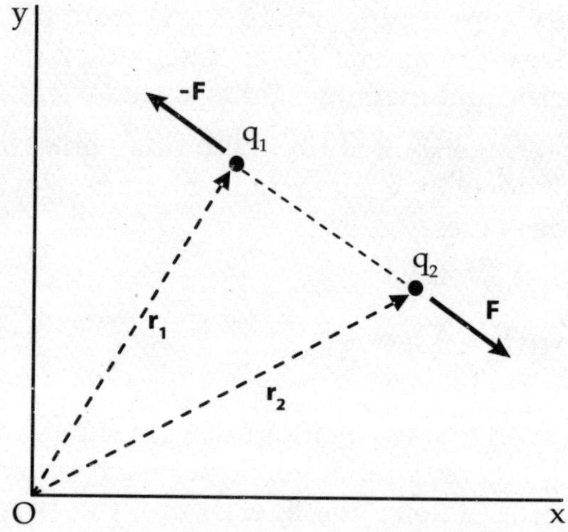

Figure 1.2: Coulomb force between charges q_1 and q_2.

a stationary charge q, located at $\mathbf{r_1}$, the electric field at $\mathbf{r}$ (according to Coulomb's law) is (Fig. 1.3) given by

$$\mathbf{E}\,(\mathbf{r}) = \frac{q\,(\mathbf{r} - \mathbf{r_1})}{4\pi\, \epsilon_0\, |\mathbf{r}-\mathbf{r_1}|^3}. \tag{3}$$

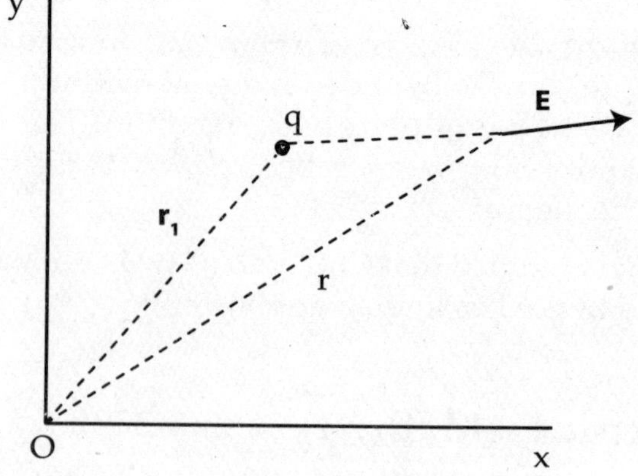

Figure 1.3: Electric field at $\mathbf{r}$ due to charge at $\mathbf{r_1}$.

This field (due to charge q) would exist in entire space, irrespective of whether the test charge to measure it is there or not. If charge q moves from $\mathbf{r_1}$ to $\mathbf{r_1}'$, its field does not change everywhere instantly. Change in $\mathbf{E}$ in close vicinity of the charge q is felt early whereas at farther distances, it is observed at a later time.

The electric field due to n charges– q_1 located at $\mathbf{r_1}$, q_2 at $\mathbf{r_2}$, and so on, is the linear sum of individual fields due to each of the charges

$$\mathbf{E}(\mathbf{r}) = \frac{1}{4\pi \in_0} \sum_{i=1}^{n} \frac{q_i (\mathbf{r} - \mathbf{r_i})}{|\mathbf{r} - \mathbf{r_i}|^3}. \tag{4}$$

If one has a continuous charge distribution with charge density (charge per unit volume) $\rho(r')$, then the above summation can be replaced by an integration over volume V', with q_i replaced by $\rho(\mathbf{r}')\, dV'$ and $\mathbf{r_i}$ by $\mathbf{r}'$ (where $\mathbf{r}'$ denotes the location of the charge element),

$$\mathbf{E}(\mathbf{r}) = \frac{1}{4\pi \in_0} \int \frac{\rho(\mathbf{r}')(\mathbf{r}-\mathbf{r}')}{|\mathbf{r} - \mathbf{r}'|^3} dV'. \tag{5}$$

As $\mathbf{r}$ and $\mathbf{r}'$ are independent variables – one denoting the position of the point of observation and the other the location of the charge element, we may write

$$\frac{(\mathbf{r} - \mathbf{r}')}{|\mathbf{r} - \mathbf{r}'|^3} = -\nabla \left(\frac{1}{|\mathbf{r} - \mathbf{r}'|} \right), \tag{6}$$

where ∇ operator refers to $\mathbf{r}$ variable and is given as $\nabla = \hat{x}\partial/\partial x + \hat{y}\partial/\partial y + \hat{z}\partial/\partial z$. Hence, electric field due to static charge distribution may be written as

$$\mathbf{E} = -\nabla\phi, \tag{7}$$

$$\phi = \frac{1}{4\pi \in_0} \int \frac{\rho(\mathbf{r}')\, dV'}{|\mathbf{r} - \mathbf{r}'|}. \tag{8}$$

ϕ is known as the *scalar potential* or *electrostatic potential*. If charge distribution is located in an infinite medium of permittivity $\in$, the electric field and potential given above are still valid with $\in_0$ replaced by $\in$.

PROBLEM 1.1. Show that $\nabla^2 (1/|\mathbf{r}-\mathbf{r}'|) = -4\pi\delta^3(\mathbf{r}-\mathbf{r}')$.

Solution: Before solving the problem, a remark about the one dimensional delta function $\delta(x-a)$ would be helpful. It is a function which is zero everywhere, except x=a. At x=a, its value is so large that

$$\int \delta(x-a)dx = 1$$

when limits of integration enclose the point x = a. In three dimensions

$$\delta^3(\mathbf{r}-\mathbf{r}') = \delta(x-x')\,\delta(y-y')\,\delta(z-z'),$$

$$\int \delta^3(\mathbf{r}-\mathbf{r}')\,dV = 1$$

when the volume of integration encloses the point $\mathbf{r} = \mathbf{r}'$. Now, let $\mathbf{r}-\mathbf{r}' = \mathbf{R}$ be a new variable. Then, ∇ operator takes the form $\nabla = \hat{x}\partial/\partial R_x + \hat{y}\partial/\partial R_y + \hat{z}\partial/\partial R_z$ and

$$\nabla^2 \frac{1}{|\mathbf{r}-\mathbf{r}'|} = \frac{1}{R^2}\frac{\partial}{\partial R}\left(R^2 \frac{\partial}{\partial R}\frac{1}{R}\right).$$

This function is zero for R≠0, just like a delta function. At R=0, its value is indeterminate. We obtain its volume integral. As the function vanishes at all $\mathbf{r}$ except $\mathbf{r} = \mathbf{r}'$, we choose (without loss of generality) the volume to be a sphere of radius R centered at $\mathbf{r} = \mathbf{r}'$,

$$\int \nabla^2 \frac{1}{|\mathbf{r}-\mathbf{r}'|} dV = \int \nabla \cdot \left(\nabla \frac{1}{|\mathbf{r}-\mathbf{r}'|}\right) dV = \oint \nabla \frac{1}{|\mathbf{r}-\mathbf{r}'|} \cdot \mathbf{dS}$$

$$= -\oint \frac{(\mathbf{r}-\mathbf{r}')}{|\mathbf{r}-\mathbf{r}'|^3} \cdot \mathbf{dS}$$

where we have used the Gauss's theorem ($\int \nabla \cdot \mathbf{A}\, dV = \oint \mathbf{A} \cdot \mathbf{dS}$) with volume integral being over volume V and surface integral over the surface enclosing the volume) to convert the surface integral into a volume integral. Since $\mathbf{dS}$ on the sphere is along $(\mathbf{r}-\mathbf{r}')$ and $|\mathbf{r}-\mathbf{r}'| = R$, the right-hand side becomes -4π. Thus $\nabla^2 (1/|\mathbf{r}-\mathbf{r}'|)$ behaves exactly as -4π times the delta function. One may also mention that due to the unique property of delta function

$$\int_V F(\mathbf{r})\, \delta^3 (\mathbf{r} - \mathbf{r}')\, dV = F(\mathbf{r}'),$$

where F(**r**) is any arbitrary function of **r** and the volume of integration encloses the point **r**′.

PROBLEM 1.2. For the electrostatic field, given by eqs. (7) and (8), show that $\nabla.\vec{E} = \rho/\epsilon_0$.

Solution: We have $\vec{\mathbf{E}} = -\nabla\phi$, hence

$$\nabla.\mathbf{E} = -\nabla^2\phi = -\frac{1}{4\pi\epsilon_0}\int \rho(\mathbf{r}')\, \nabla^2 \left(\frac{1}{|\mathbf{r} - \mathbf{r}'|}\right) dV'.$$

Since $\nabla^2 (1/|\mathbf{r} - \mathbf{r}'|) = -4\pi\delta^3 (\mathbf{r} - \mathbf{r}')$ (cf. problem 1), we obtain $\nabla.\mathbf{E} = \rho/\epsilon_0$.

This result, though obtained for static electric field, has been experimentally validated for all sorts of time dependent electric fields as well. Hence, it is a fundamental law of electrodynamics.

The electric field due to moving charges is more complex than that given by eq. (5). Before delving deeper into this issue we shall study the general nature of time dependent fields and the way they are transmitted, in this and next few units. A complementary quantity in this context is the magnetic field, that is produced by moving charges or currents.

1.3 Electric Current and Current Density J (r, t)

The charges in motion constitute electric current. One defines electric current in a wire, as the amount of charge passing through its cross-section per second. The charge flowing per unit area per second is called current density **J**. In case of conductors, the current is carried by free electrons. In semiconductors, it is carried by free electrons and holes. Consider the flow of charged particles, each of charge q, moving with average velocity **v**. Let there be a fictitious area of magnitude unity, perpendicular to $\mathbf{v}$ (Fig. 1.4). In one second, all the particles that are filled in a cylinder of length v and area unity will cross the area and the ones farther than v will not be able to reach the area. The number of such particles would be nv, where n is the number density of particles. The charge carried by these particles is

$$\mathbf{J} = nq\mathbf{v}. \tag{9}$$

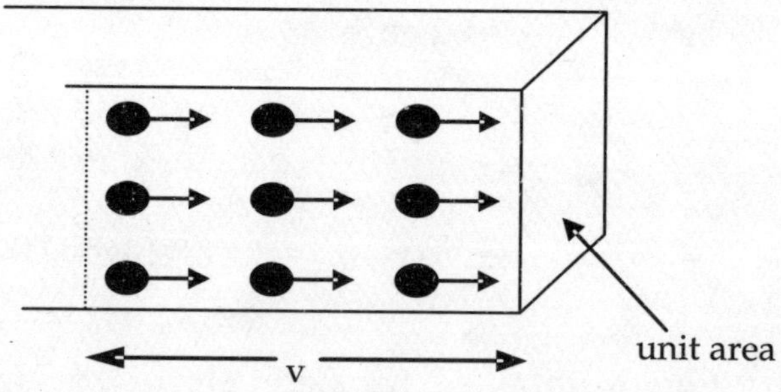

Figure 1.4: Electrons crossing unit area in one second.

In case the carriers are electrons, $q = -e$ and $\mathbf{J} = -ne\mathbf{v}$.

1.4 Equation of Continuity

In any region of space, if the incoming current is more than the outgoing current it leads to accumulation of charge. The imbalance of charge flow is denoted by a quantity $\nabla.\mathbf{J}$. In order to realise its

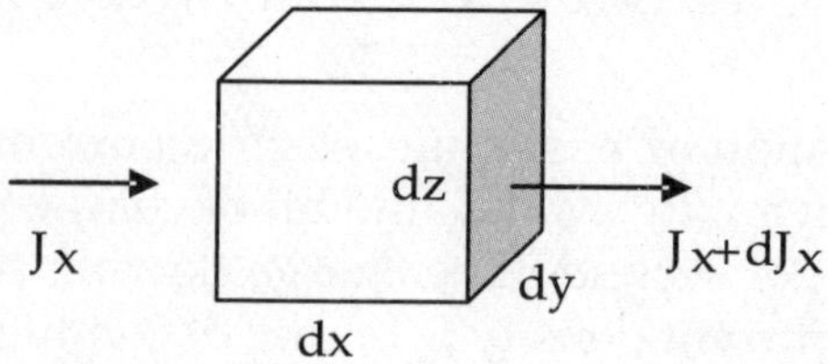

Figure 1.5: Charge accumulation in a box due to imbalance of inflow and outflow of current.

physical significance, consider a one-dimensional situation, where $\mathbf{J}||\hat{x}$ and $\mathbf{J}$ is a function of x. Let there be a fictitious box of length dx and cross-section dydz located between x and x + dx (Fig. 1.5). The charge entering the box per second is $j_x(x)dydz$ whereas the one leaving the box is $j_x(x + dx)dydz$. The net charge accumulated in the box per second is

$$J_x(x)dydz - J_x(x + dx)dydz.$$

Using Taylor's expansion, $J_x(x + dx) = J_x(x) + (\partial J_x/\partial x)\,dx$. Hence, the amount of accumulated charge becomes

$$-\frac{\partial J_x}{\partial x} dx\ dy\ dz$$

and the charge accumulated per unit volume per second is $-\partial J_x / \partial x$.

In case current density **J** has all the components, then the charge entering the box per second through face dydz located at x is $J_x(x, y, z)\,dy\,dz$, and that leaving the parallel face dydz at x + dx is $J_x(x + dx, y, z)dydz$. Similarly, the charges entering the dxdz face at y, per second, and those leaving at y+ dy are $J_y(x, y, z)dxdz$, $J_y(x, y + dy, z)dxdz$ respectively. The charges entering per second through dxdy face located at z and those leaving at z + dz are $J_z(x, y, z)dxdy$ and $J_z(x, y, z + dz)dxdy$ respectively. Thus, the difference of net charge entering and leaving can be found by using the Taylor expansions

$$J_x(x + dx, y, z) = J_x + \left(\frac{\partial J_x}{\partial x}\right) dx,\ \ J_y(x, y + dy, z) = J_y + \left(\frac{\partial J_y}{\partial y}\right) dy,$$

$$J_z(x, y, z + dz) = J_z + \left(\frac{\partial J_z}{\partial z}\right) dz,$$

where, J_x, J_y, J_z on the right sides of expansions refer to values at (x, y, z). The result is

$$-\left(\frac{\partial J_x}{\partial x} + \frac{\partial J_y}{\partial y} + \frac{\partial J_z}{\partial z}\right) dx\ dy\ dz = -\nabla.\mathbf{J}\ dx\ dy\ dz.$$

Thus, $-\nabla.\mathbf{J}$ represents net charge accumulation per unit volume per second. If we denote ρ as the charge density, then charge conservation implies $\partial\rho/\partial t = -\nabla.\mathbf{J}$ or

$$\frac{\partial\rho}{\partial t} + \nabla.\mathbf{J} = 0. \tag{10}$$

This is known as the *equation of continuity*. In the time independent situations, ρ does not change with time. Hence

$$\nabla.\mathbf{J} = 0. \tag{11}$$

1.5 Magnetic Field B (r, t)

A current carrying wire (or a combination of such wires) produces a magnetic field **B** around it. The magnetic field at a point can be defined in terms of the force **F**, that a current element of length **dl** and current I experiences when placed at the given point. Using the relation

$$\mathbf{F} = I\mathbf{dl} \times \mathbf{B} \tag{12}$$

where I is in ampere, dl in meter and **B** is in Tesla (1 Tesla = 10 kilogauss), we can calculate the force **F** experienced by element dl (in Newtons). In the case of a charged particle of charge q, moving with velocity **v**, the force on it due to the magnetic field is the 'Lorentz force',

$$\mathbf{F} = q\mathbf{v} \times \mathbf{B}. \tag{13}$$

Thus, the magnitude of magnetic field at a point can be measured by the force experienced by a unit charge, moving through the point with unit velocity and the direction of velocity is such that the resultant force is maximum.

1.6 Biot - Savart's Law

The magnetic field produced by steady currents is given by Biot-Savart's law. The magnetic field $\mathbf{dB}(\mathbf{r})$, due to a current element of volume dV' and static (dc) current density $\mathbf{J}(\mathbf{r}')$, placed at $\mathbf{r}'$, is given by (Fig. 1.6)

$$\mathbf{dB} = \frac{\mu_0}{4\pi} \frac{\mathbf{J}(\mathbf{r}') \times (\mathbf{r} - \mathbf{r}')}{|\mathbf{r} - \mathbf{r}'|^3} dV'$$

where $\mu_0 = 4\pi \times 10^{-7}$ Newton/Amp^2 or henry/meter is the free space magnetic permeability. By integrating the right-hand side over the volume of the current source (conductor), one gets the total magnetic field at $\mathbf{r}$ due to steady current distribution,

$$\mathbf{B}(\mathbf{r}) = \frac{\mu_0}{4\pi} \int \frac{\mathbf{J}(\mathbf{r}) \times (\mathbf{r} - \mathbf{r}')}{|\mathbf{r} - \mathbf{r}'|^3} dV'. \tag{14}$$

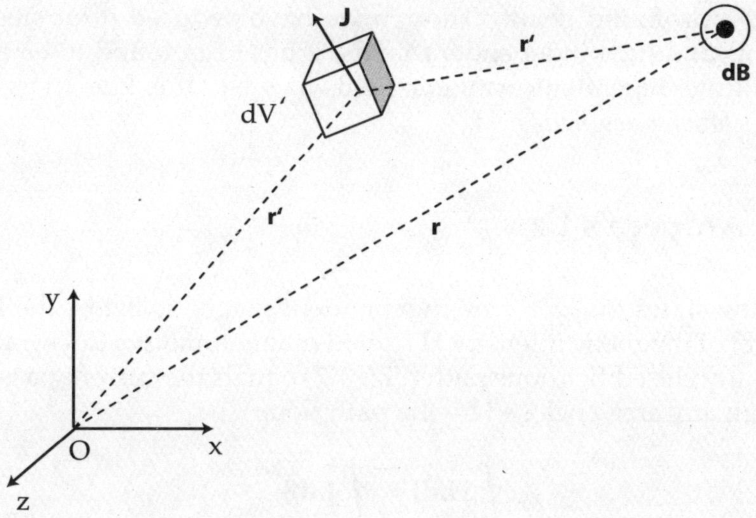

Figure 1.6: Magnetic field **dB** at **r** due to a current element at **r**′, **dB** is perpendicular to the plane containing **J** and **r**-**r**′.

If current distribution is located in an infinite medium of magnetic permeability μ, then μ_0 in the above expression is replaced by μ. Eq. (14) can be written in an alternate form using eq. (6),

$$\mathbf{B}(\mathbf{r}) = \frac{-\mu_0}{4\pi} \int \mathbf{J}(\mathbf{r}') \times \nabla \frac{1}{|\mathbf{r} - \mathbf{r}'|} dV' = \frac{\mu_0}{4\pi} \nabla \times \int \frac{\mathbf{J}(\mathbf{r}')}{|\mathbf{r} - \mathbf{r}'|} dV'$$

where we have employed the fact that $\mathbf{J}(\mathbf{r}')$ is a constant as far as ∇ operator is concerned. This is because ∇ refers to variable $\mathbf{r}$ and $\mathbf{r}'$ is independent of $\mathbf{r}$.

We may define a vector potential $\mathbf{A}\,(\mathbf{r})$ through

$$\mathbf{B}\,(\mathbf{r}) = \nabla \mathrm{x} \mathbf{A}\,(\mathbf{r}) \tag{15}$$

where

$$\mathbf{A}\,(\mathbf{r}) = \frac{\mu_0}{4\pi} \int \frac{\mathbf{J}\,(\mathbf{r}')}{|\mathbf{r} - \mathbf{r}'|} \mathrm{dV}'. \tag{16}$$

One may deduce an important property of magnetic field from eq. (15),

$$\nabla.\mathbf{B} = \nabla.\nabla \times \mathbf{A} = 0. \tag{17}$$

This is a profound result. Though we have deduced it for steady currents and time independent fields, it has been found to be true for all time-dependent magnetic fields as well. It is known as the *Gauss's law for magnetic field*.

1.7 Ampere's Law

This law states that for time independent magnetic fields, the line integral of magnetic intensity $\mathbf{H}$ (called magnetomotive force, mmf) along any closed fictitious path (Fig. 1.7) equals the current passing through any area enclosed by the path (loop), i.e.

$$\oint \mathbf{H.dl} = \int_{\mathrm{S}} \mathbf{J.dS} \tag{18}$$

where $\mathbf{J}$ is the electrical current density due to free charge carriers and $\mathbf{H}$ is related to magnetic field $\mathbf{B}$ as

$$\mathbf{B} = \mu_0(\mathbf{H} + \mathbf{M}).$$

Here, $\mathbf{M}$ is the magnetic dipole moment per unit volume, called magnetisation (defined later). In free space and non-magnetic materials, $\mathbf{M} = 0$ and $\mathbf{B} = \mu_0\mathbf{H}$. Using Stoke's theorem, the line integral in eq. (18) can be replaced by surface integral

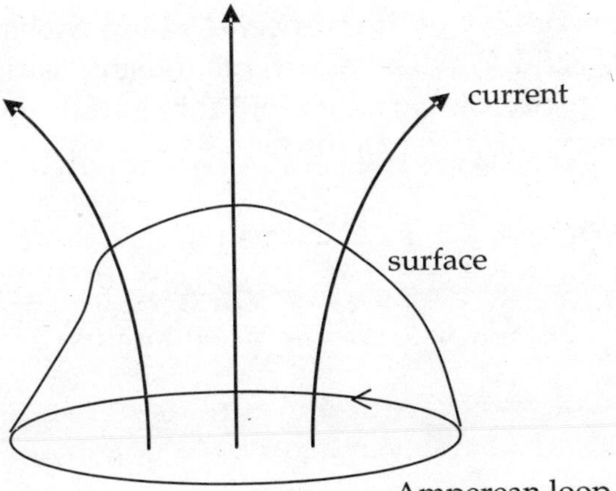

Figure 1.7: A fictitious loop on which line integral is evaluated.

$$\oint \mathbf{H.dl} = \int_S \nabla\text{x}\mathbf{H.dS}.$$

This leads to

$$\nabla \times \mathbf{H} = \mathbf{J}. \tag{19}$$

This is the differential form of Ampere's law.

PROBLEM 1.3. Deduce Ampere's law, $\nabla\times\mathbf{B} = \mu_0\mathbf{J}$, from Biot-Savart's law for steady state currents.

Solution: Since $\mathbf{B} = \nabla \times \mathbf{A}$, we have

$$\nabla \times \mathbf{B} = \nabla \times (\nabla \times \mathbf{A}) = -\nabla^2\mathbf{A} + \nabla(\nabla.\mathbf{A}). \qquad \ldots\text{(i)}$$

From eq. (16)

$$\nabla^2\mathbf{A} = \frac{\mu_0}{4\pi}\int \mathbf{J}(\mathbf{r}')\,\nabla^2\left(\frac{1}{|\mathbf{r}-\mathbf{r}'|}\right)\mathrm{dV}'$$

$$= -\frac{\mu_0}{4\pi}4\pi\int \mathbf{J}(\mathbf{r}')\,\delta^3(\mathbf{r}-\mathbf{r}')\,\mathrm{dV}' = -\mu_0\mathbf{J}(\mathbf{r}), \qquad \ldots\text{(ii)}$$

$$\nabla.\mathbf{A} = \frac{\mu_0}{4\pi}\int \nabla.\frac{\mathbf{J}(\mathbf{r}')}{|\mathbf{r}-\mathbf{r}'|}\mathrm{dV}' = -\frac{\mu_0}{4\pi}\int \nabla'.\frac{\mathbf{J}(\mathbf{r}')}{|\mathbf{r}-\mathbf{r}'|}\mathrm{dV}', \qquad \ldots\text{(iii)}$$

where ∇' is del operator with respect to $\mathbf{r}'$ and we have employed $\nabla'.\mathbf{J}(\mathbf{r}') = 0$, valid for steady currents. Using Gauss's theorem, the volume integral can be converted into surface integral over a surface enclosing all current sources.

$$\nabla.\mathbf{A} = \frac{\mu_0}{4\pi} \int \frac{\mathbf{J}(\mathbf{r_1})}{|\mathbf{r} - \mathbf{r_1}|} .d\mathbf{S_1} = 0 \quad \ldots(\text{iv})$$

as $\mathbf{J}$ vanishes on the surface. Using the equations (ii) and (iv) in (i), we obtain

$$\nabla \times \mathbf{B} = \mu_0 \mathbf{J}.$$

1.8 Faraday's Law of Electromagnetic Induction

One of the fundamental discoveries of electrodynamics was made by *Faraday*, when he observed that time-varying magnetic fields can produce electric field. If one considers a closed path, real or fictitious, then the line integral $\oint \mathbf{E}.d\mathbf{l}$ along the path is known as the electromotive force (emf). Faraday's law states that the emf induced in any closed path (Fig. 1.8) equals the time rate of change of

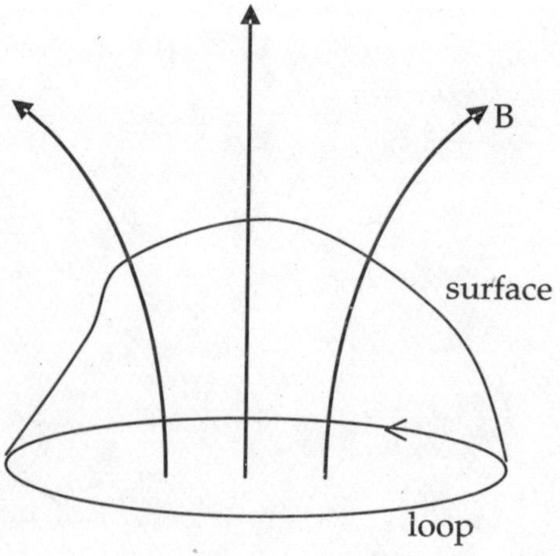

Figure 1.8: An open surface with open boundary as the loop for line integral evaluation.

magnetic flux through the area enclosed by the loop (the area could be any surface having the closed path as its open boundary),

$$\oint \mathbf{E.dl} = -\frac{\partial}{\partial t}\int_{s} \mathbf{B.dS}.$$

The negative sign arises due to the fact that the emf induced by the change of magnetic flux should be such that the current (if any) produced by it opposes the change in magnetic flux. This is known as *Lenz's law*. Using Stoke's theorem, the line integral can be converted into a surface integral leading to

$$\nabla \mathrm{x}\mathbf{E} = -\frac{\partial \mathbf{B}}{\partial t}.$$

1.9 Maxwell's Equations

Maxwell generalised the laws of electrodynamics to time-varying fields and put them in compact mathematical form, known as Maxwell's equations. These equations are *relativistically covariant*, i.e., they retain their form in every inertial frame of reference. In fact it is these equations, besides *Michelson Morley experiment*, that led Einstein to declare that the speed of light in vacuum is the same in all inertial frames of reference. All electric and magnetic fields, irrespective of how they are produced, must satisfy Maxwell's equations. These equations represent four fundamental laws of electrodynamics.

1.9.1 Gauss's Law for Electric Field

It states that the total outward normal electric flux, through any closed surface, is equal to the total charge enclosed by the surface (Fig. 1.9).

$$\oint \mathbf{D.dS} = \int_{V} \rho \, dV$$

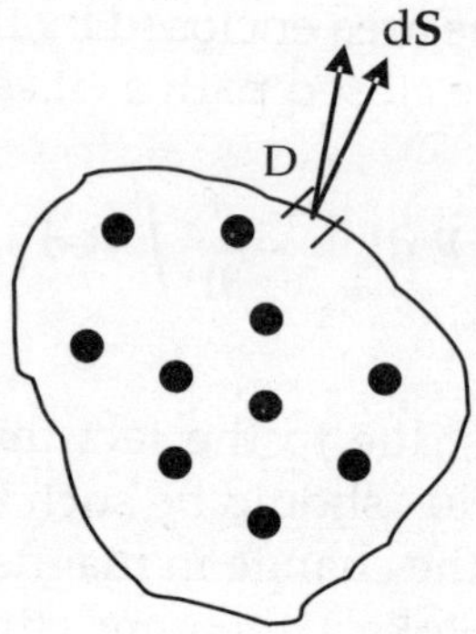

Figure 1.9: Displacement vector **D** at the surface element **dS** of a closed surface (called Gaussian surface).

where **D** is the displacement vector, defined as $\mathbf{D} = \epsilon_0 \mathbf{E} + \mathbf{P}$, $\epsilon_0 = 10^{-9}/36\pi$ Farad/m is the free space permittivity, **P** is the polarisation vector, i.e., the dipole moment per unit volume, ρ is the electric charge density and V represents the volume enclosed by the closed surface. Using Gauss's theorem, the surface integral can be converted to volume integral, leading to

$$\nabla.\mathbf{D} = \rho. \tag{20}$$

We have already seen the validity of this equation for static fields. However, it is valid for all time-varying fields as well.

1.9.2 Gauss's Law for Magnetic Field

It states that the total outward normal magnetic flux through any closed surface is zero as there are no magnetic monopoles,

$$\oint \mathbf{B}.\mathbf{dS} = 0.$$

It leads to

$$\nabla.\mathbf{B} = 0. \tag{21}$$

We had derived this equation for time independent fields, but it is valid for all time-varying fields too.

1.9.3 Faraday's Law of Electromagnetic Induction

We have already stated this in section 1.8. It connects the line integral of electric field (emf) over a closed path to the rate of change of magnetic flux linked with the area enclosed by the path and is written as

$$\int \mathbf{E.dl} = -\frac{\partial}{\partial t}\int \mathbf{B.dS}$$

or

$$\nabla \times \mathbf{E} = -\frac{\partial \mathbf{B}}{\partial t}. \tag{22}$$

1.9.4 Generalised Ampere's Law

Maxwell realised the incompatibility of Ampere's law, $\nabla \times \mathbf{H} = \mathbf{J}$, with the equation of continuity,

$$\frac{\partial \rho}{\partial t} + \nabla.\mathbf{J} = 0.$$

From Ampere's Law, $\nabla.\mathbf{J} = \nabla.(\nabla \times \mathbf{H}) = 0$, hence the continuity equation would give $\partial\,\rho/\partial\,t = 0$, i.e., accumulation or rarefaction of charges is never possible. This is contrary to numerous day-to-day observations, where charge density changes with time. Charging

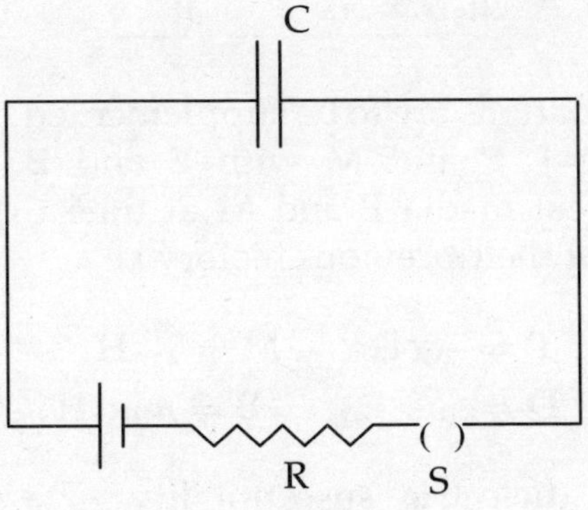

Figure 1.10: Charge accumulation in a capacitor plate begins, as soon as key S is closed.

of capacitor is the most striking example of charge accumulation with time (Fig. 1.10). Thus, Ampere's Law is valid only for time independent fields. Maxwell generalised it by adding a term $\mathbf{J_d}$, to it

$$\nabla \times \mathbf{H} = \mathbf{J} + \mathbf{J_d}.$$

Now $\nabla.\mathbf{J} = -\nabla.\mathbf{J_d}$. Using the equation of continuity and the first Maxwell's equation, one obtains

$$\nabla.\mathbf{J_d} = \frac{\partial \rho}{\partial t} = \nabla.\left(\frac{\partial \mathbf{D}}{\partial t}\right),$$

leading to $\mathbf{J_d} = \partial \mathbf{D}/\partial t$. This is known as the *displacement current density*. The generalised Ampere's law is thus

$$\nabla \times \mathbf{H} = \mathbf{J} + \frac{\partial \mathbf{D}}{\partial t}. \tag{23}$$

It would be useful to enlist all the Maxwell's equations once.

$$\boxed{\begin{aligned} \nabla.\mathbf{D} &= \rho, \\ \nabla.\mathbf{B} &= 0, \\ \nabla \times \mathbf{E} &= -\frac{\partial \mathbf{B}}{\partial t}, \\ \nabla \mathrm{x} \mathbf{H} &= \mathbf{J} + \frac{\partial \mathbf{D}}{\partial t}. \end{aligned}}$$

The Maxwell's equations are to be supplemented by the constituent relations, relating **J**, **P** and **M** with **E** and **B**. At microwave frequencies, in most media **P** and **M** at time t depend on **E** and **B** at time t, and not their previous history i.e.,

$$\begin{aligned} \mathbf{P} &= \varepsilon_0 \chi \mathbf{E}, \qquad & \mathbf{M} &= \chi_m \mathbf{H}, \\ \mathbf{D} &= \epsilon_0 \epsilon_r \mathbf{E}, \qquad & \mathbf{B} &= \mu_0 \mu_r \mathbf{H}, \end{aligned} \tag{24}$$

where χ is the dielectric susceptibility, χ_m is the magnetic permeability, $\epsilon_r = 1 + \chi$ is the relative permittivity (or dielectric constant) and $\mu_r = 1 + \chi_m$ is the relative permeability. For non-magnetic materials, $\mu_r = 1$. In the following discussion, we shall assume $\mu_r = 1$ unless stated otherwise. At optical frequencies, **P** and **M** (at time t) depend on **E** and **B** at all times preceding t. We shall return to this issue later. However, for harmonic fields with time dependence of the form $e^{-i\omega t}$, it is always possible to write **D** and **B** in terms of **E** and **H** in the form of eq. (24).

The current density **J** (t), in general, depends on the value of the electric field at time t as well as at earlier times. Let us obtain an expression for it.

1.10 Current Density and Electrical Conductivity

Consider a plasma, an ionised gas comprising of neutral atoms, ions and free electrons. When an electric field, **E** (t) is applied to the plasma, the free electrons and ions acquire drift velocities and constitute electric current. Since ions are much heavier than the electrons, their drift velocity is smaller and their contribution to current density can be ignored. The drift velocity of electrons **v**, is governed by the equation of momentum balance [a],

$$m\frac{d\mathbf{v}}{dt} = m\left(\frac{\partial \mathbf{v}}{\partial t} + \mathbf{v}.\nabla\mathbf{v}\right) = -\mathbf{E}t - \nu m\mathbf{v} \tag{25}$$

where –e and m are electronic charge and mass respectively, and ν is the momentum transfer (from electrons to ions and neutrals) collision frequency. (The connection between total and partial time derivatives of fluid velocity is derived in a note at the end of unsolved problems.) The last term in eq.(25) represents momentum lost by an electron per second, i.e., the product of momentum lost in a collision (m **v**) times the number of collisions suffered per second (ν). Here we approximate the total time derivative by partial time derivative and ignore the force due to the magnetic field of the wave. We shall justify these assumptions later. Multiplying the equation of motion by $-n_0e$ and dividing by m, we obtain the equation for current density $\mathbf{J} = -n_0\, e\, \mathbf{v}$ as,

$$\frac{\partial \mathbf{J}}{\partial t} + \nu\mathbf{J} = \frac{n_0 e^2}{m}\mathbf{E}, \tag{26}$$

where n_0 is the electron density. Eq. (26) gives

$$\mathbf{J}(t) = \frac{n_0 e^2}{m} e^{-\nu t} \int_{-\infty}^{t} e^{\nu t_1} \mathbf{E}(t_1)\, dt_1$$
$$= \frac{n_0 e^2}{m} \int_{0}^{\infty} e^{-\nu t_2} \mathbf{E}(t - t_2)\, dt_2. \tag{27}$$

Thus, the current density at time t depends on the electric field at t, as well as its previous history. In a special case when $\mathbf{E}(t) = \mathbf{A}e^{-i\omega t}$ (where **A** could be complex and have a space dependence), eq. (26) can be solved right away, by assuming steady state time dependence of **J** as $\exp(-i\,\omega\, t)$ and replacing $\partial/\partial t$ by $-i\,\omega$.

$$\mathbf{J} = \sigma \mathbf{E},$$
$$\sigma = \frac{n_0 e^2}{m\,(\nu - i\omega)}. \tag{28}$$

σ is called the *electrical conductivity*. It is a complex quantity. Expressing $\sigma = \sigma_r + i\sigma_i$ and assuming **A** to be real, the actual current density (real part of **J**) can be written as $\mathbf{J} = \sigma_r \mathbf{A} \cos\omega t + \sigma_i \mathbf{A} \sin\omega t$, while $\mathbf{E} = \mathbf{A}\cos\omega t$. σ_r gives the part of **J** which is in phase with **E** and is responsible for energy dissipation, i.e Ohmic heating of electrons. σ_i gives the inductive part. For $\nu << \omega$, $\sigma_r << \sigma_i$ and the inductive part dominates over the Ohmic part while the reverse happens for $\nu >> \omega$. Fig. 1.11 shows the time variation of electric field and current density when $\nu = 0$ and $\nu \neq 0$. In the former case **J** is $\pi/2$ out of phase with **E**, while in the latter phase difference is less.

Eq. (28) is valid in conductors also, where free electrons exist in the conduction band. However, the effective mass m of electrons could be different from the free space electron mass. The value of ν (the electron-phonon collision frequency) in most metals at room temperature is $\nu \approx 10^{12}\ s^{-1}$. At microwave frequencies, $\omega/2\pi \leq$ 30 GHz, $\omega << \nu$ and $\partial\mathbf{J}/\partial t$ term in eq. (26) can be neglected. Thus, **J** is an instantaneous function of **E** and σ is real.

Inside a semiconductor, two types of free charge carriers – electrons and holes exist. The electrical conductivity is the sum of electron conductivity and hole conductivity,

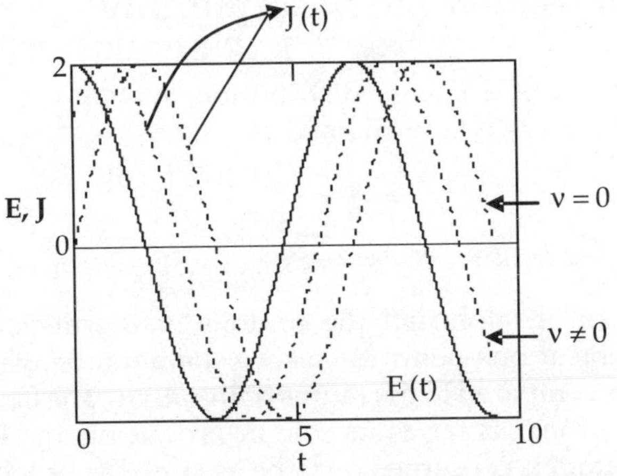

Figure 1.11: Temporal behavior of electric field (___) and current density (___) in collissionless and collissional plasmas.

$$\sigma = \frac{n_0 e^2}{m\,(\nu - i\omega)} + \frac{n_{0h} e^2}{m_h\,(\nu_h - i\omega)} \tag{29}$$

where quantities with subscript h refer to holes. At microwave frequencies, $\omega << \nu$, hence σ is largely real. In the following discussion, we shall consider only a single species of free charge carriers. Generalisation to multi-species media can be done easily.

PROBLEM 1.4. In the presence of an electric field of frequency $\omega = 3 \times 10^{12}$ rad / s, inside a conductor, the current density is $(\pi/3)$ out of phase with the electric field. Estimate the collision frequency.

Solution: From eq.(28), one may write $\sigma = \sigma_A \exp(i\phi)$, with

$$\sigma_A = \frac{n_0 e^2}{m(\nu^2 + \omega^2)^{1/2}}, \phi = \tan^{-1}\frac{\omega}{\nu}.$$

The current density due to the electric field $\mathbf{E} = \mathbf{A}\exp(-i\,\omega\,t)$ is

$$\mathbf{J} = \sigma_A \mathbf{A} \exp[-i(\omega t - \phi)]$$
$$= \sigma_A \mathbf{A} \cos(\omega t - \phi)\{\text{on taking the real part}\}.$$

As $\phi = \pi/3$, one obtains $\nu = \omega \cot\phi \cong 1.7 \times 10^{12}\,s^{-1}$.

1.11 Polarisation and Susceptibility

Dipole moment of a charge distribution, having charges q_1, q_2, q_3...located at $\mathbf{r_1}, \mathbf{r_2}, \mathbf{r_3} \ldots$ is defined as

$$\mathbf{p} = \sum_i q_i \mathbf{r_i}.$$

In the case of an atom, all the protons of the nucleus may be assumed to be at one point, as space separation between them is very small as compared to the radius of the atom. The fast revolving cloud of electrons, as far as its role in producing an electric field outside the atom is concerned, may be assumed to be located at the centre of the cloud. In the absence of any external electric field, the centre of negative charge of electrons coincides with the nucleus, giving zero dipole moment to the atom. When an electric field **E** is applied on the atom, the nucleus and electrons are pulled in opposite directions. The centre of negative charge thus gets shifted away from the nucleus and the atom acquires a net dipole moment. When the electric field is periodic in time, the induced dipole moment is also periodic.

Consider an atom with one electron rotating round the nucleus, with angular frequency ω_0. The atom may be located in a material. We apply a dc electric field. The field seen by the atom is $\mathbf{E}||\hat{\mathrm{x}}$. To obtain the shift in the centre of negative charge from the nucleus, we model the x-motion of the electron as if the electron is tied to the tip of a spring of spring constant $\omega_0^2 m$, where m is the mass of the electron and length of the spring is along x-axis (Fig. 1.12). The x-displacement of the electron from the equilibrium position is governed by the equation of motion,

$$m\frac{d^2x}{dt^2} + \omega_0^2 mx = -eE. \tag{30}$$

A general solution of this second order ordinary differential equation reveals that x at time t depends on E at t as well as its previous history. For a harmonic time dependence, $\mathbf{E} = \hat{\mathrm{x}} A \exp(-i\omega t)$ and the solution of eq.(30) can be trivially written by assuming x

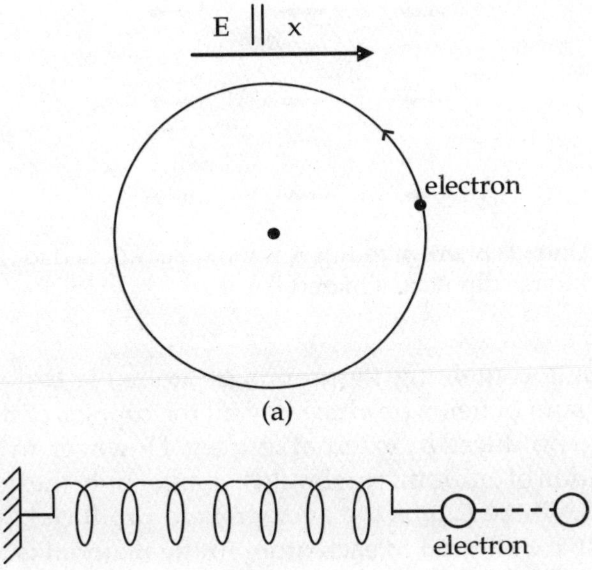

Figure 1.12: (a) A bound electron subjected to an electric field and (b) its equivalent spring model.

(in the steady state) to have the same time dependence as E, i.e., $x = x_0 e^{-i\omega t}$,

$$x = \frac{e\,E}{m(\omega^2 - \omega_0^2)}.$$

The induced dipole moment of the atom $p = -e\,x$ is

$$p = \frac{e^2\,E}{m(\omega_0^2 - \omega^2)} = \alpha E, \tag{31}$$

where $\alpha = e^2/m(\omega_0^2 - \omega^2)$ is called the ac polarisability of the atom. α shows resonance at $\omega = \omega_0$. For hydrogen, ω_0 falls in the ultraviolet region. For atoms with many electrons, α is the algebraic sum of susceptibilities due to all the orbiting electrons. Since the inner orbit electrons have higher ω_0 than the ones in outer orbits, the latter contribute more effectively to atomic polarisability.

Now let there be N atoms per unit volume, each with atomic polarisability α. The macroscopic field in the material (averaged

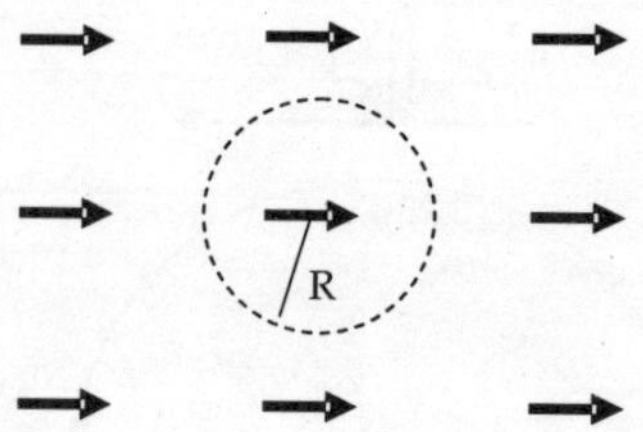

Figure 1.13: Dotted sphere of radius R is the space allotted to each induced atomic dipole in a material.

over a volume containing thousands of atoms) is $\mathbf{E} \parallel \hat{x}$. $\mathbf{E}$ is the cumulative sum of fields produced by all the dipoles of the material plus the field produced by external sources. However, in calculating the polarisation of an atom, we should consider only the field $\mathbf{E}-\mathbf{E_{self}}$ on the atom, where $\mathbf{E_{self}}$ is the average field produced by the atom itself. The space allotted to each atom in the material is a sphere of radius R, with atom at the centre (Fig. 1.13), where

$$\mathrm{N} = \frac{1}{\left(\frac{4\pi}{3}\right)\mathrm{R}^3} \quad \text{or} \quad \mathrm{R} = \left(\frac{3}{4\pi \mathrm{N}}\right)^{1/3}.$$

The average electric field produced by the atomic dipole in a sphere of radius R is (see problem 1.5)

$$\mathbf{E_{self}} = -\frac{1}{4\pi\,\epsilon_0}\frac{\mathbf{p}}{\mathrm{R}^3}.$$

Using this in $\mathbf{p} = \alpha\,(\mathbf{E} - \mathbf{E_{self}})$, one obtains the dipole moment of an atom as

$$\mathbf{p} = \frac{\alpha}{(1 - \mathrm{N}\alpha/3\,\epsilon_0)}\mathbf{E}. \tag{32}$$

Hence the polarisation vector $\mathbf{P}$ can be written as

$$\mathbf{P} = \epsilon_0\,\chi\,\mathbf{E}, \tag{33}$$

$$\text{where } \chi = \frac{\mathrm{N}\alpha/\,\epsilon_0}{(1 - \mathrm{N}\alpha/3\,\epsilon_0)}. \tag{34}$$

χ is the *dielectric susceptibility*. The *relative permittivity* is $\in_r = 1 + \chi$. In terms of $\in_r$, one may write the atomic polarisability α as

$$\alpha = \frac{3 \in_0}{N} \frac{\in_r - 1}{\in_r + 2}. \tag{35}$$

This is the **Clausius - Mossotti formula**. In optics, it is known as the *Lorentz–Lorenz equation*. Since $\in_r$ of a material can be measured easily, by filling it in a capacitor and measuring its capacitance, deduction of α is straight forward.

For molecules with finite permanent dipole moment, like water, the derivation of susceptibility is complex. These molecular dipoles, in the presence of an electric field, tend to align themselves along the field. However, this tendency is opposed by thermal motions and inertia of the molecules. At low frequencies, inertial effect can be neglected and the orientation polarisation can be obtained (as shown in problem 6). At high frequencies, inertial term dominates and dielectric susceptibility decreases.

PROBLEM 1.5. A charge q is placed at any point $S(\mathbf{r}_0)$ inside a sphere of radius R, centred at the origin. Show that the average field due to it, inside the sphere, is the same as the field at S due to a uniformly charged sphere with charge density $\rho = -q/\left(4\pi R^3/3\right)$. Hence show that the average field inside a sphere due to a dipole of dipole moment **p**, within the sphere, is

$$\mathbf{E}_{\mathbf{av}} = -\frac{\mathbf{p}}{4\pi \in_0 R^3}.$$

Solution: The electric field at **r** due to a charge q at $\mathbf{r}_0$ (Fig. 1.14) is

$$\mathbf{E} = \frac{q}{4\pi \in_0} \frac{\mathbf{r} - \mathbf{r}_0}{|\mathbf{r} - \mathbf{r}_0|^3}.$$

Average value of this field inside the sphere of radius R is

$$\mathbf{E}_{\mathbf{av}} = \int \frac{\mathbf{E} dV}{\left(\frac{4}{3}\pi R^3\right)} = \frac{q/\left(4\pi R^3/3\right)}{4\pi \in_0} \int \frac{\mathbf{r} - \mathbf{r}_0}{|\mathbf{r} - \mathbf{r}_0|^3} dV.$$

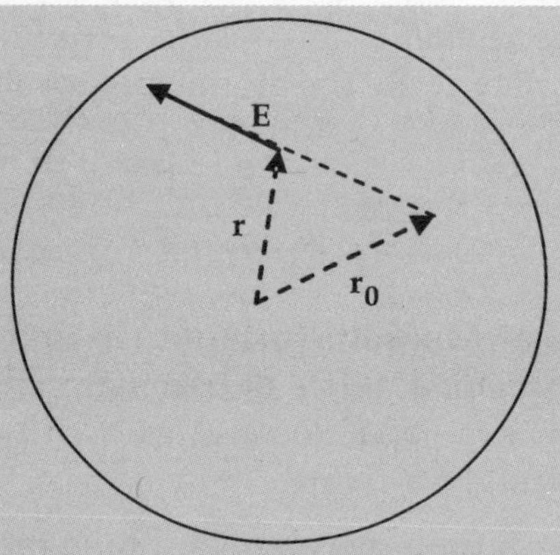

Figure 1.14: The field at point $\mathbf{r}$ due to a charge placed at $\mathbf{r}_0$.

If the sphere were charged with charge density ρ, then the electric field at $S(\mathbf{r}_0)$ would be

$$\mathbf{E}' = \frac{1}{4\pi \epsilon_0} \int \rho \frac{\mathbf{r}_0 - \mathbf{r}}{|\mathbf{r}_0 - \mathbf{r}|^3} dV.$$

For $\rho = -q/\left(4\pi R^3/3\right)$,

$$\mathbf{E}' = \mathbf{E}_{\mathbf{av}}.$$

The field at point $\mathbf{r}_0$ due to an uniformly charged sphere, considering a spherical Gaussian surface of radius r_0 and using Gauss's theorem, can be written as

$$\mathbf{E}' = \frac{\rho\left(4\pi r_0^3/3\right)}{4\pi \epsilon_0 r_0^3}\mathbf{r}_0 = -\frac{q\mathbf{r}_0}{4\pi \epsilon_0 R^3}.$$

If there are two charges $-q$ and $+q$, placed at $\mathbf{r}_0$ and $\mathbf{r}_0 + \mathbf{d}$ inside the sphere, then the average field inside the sphere due to both is

$$\mathbf{E}_{\mathbf{av}} = \frac{q\mathbf{r}_0}{4\pi \epsilon_0 R^3} - \frac{q\left(\mathbf{r}_0 + \mathbf{d}\right)}{4\pi \epsilon_0 R^3} = -\frac{\mathbf{p}}{4\pi \epsilon_0 R^3},$$

where $\mathbf{p} = q\mathbf{d}$.

PROBLEM 1.6. A dipole consisting of charges $-q$ and $+q$, separated by distance d, is placed in an electric field **E** (Fig.1.15). Obtain the potential energy of the dipole at an orientation θ with respect to the field, taking $\theta = \pi/2$ as the reference for potential energy. Then, consider a system with N dipoles per unit volume at temperature T. Assuming Boltzmann's law that the relative number of atoms with potential energy W is proportional to $\exp(-W / k_B T)$ [where k_B is the Boltzmann's constant and T is the temperature] obtain the net dipole moment per unit volume **P**.

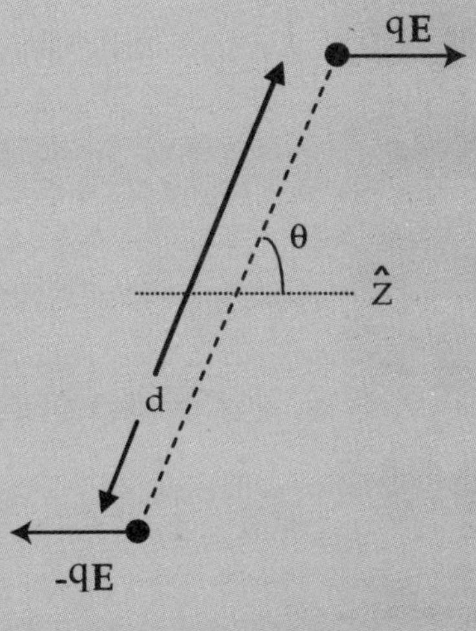

Figure 1.15: A dipole, placed in an electric field, experiencing a torque.

Solution: Let the electric field direction be z-axis. The dipole oriented at an angle θ with $\hat{z}$, experiences a torque q Ed sin θ. If one intends to increase θ by $d\theta$, he would have to do work $dW = q\,Ed \sin\theta\, d\theta$. Integrating over θ, one gets the potential energy of the dipole at θ as,

$$W = -qEd\,\cos\theta,$$

with $\theta = \pi/2$ as the reference for potential energy. From Boltzmann's law, the number of atoms per unit space volume per unit solid angle of dipole orientation is

$$n(\theta) = n_0 \exp(-W/k_B T),$$

where n_0 is related to the number of atoms per unit volume N as

$$N = \int_0^{2\pi} \int_0^{\pi} n(\theta) \sin\theta d\theta d\phi.$$

Here ϕ is the angle between the x-axis and the projection of dipole length on x-y plane and $\sin\theta d\theta \, d\phi$ is the solid angle in dipole moment space. The dipole moment per unit volume is

$$\mathbf{P} = \int_0^{2\pi} \int_0^{\pi} n\mathbf{p} \sin\theta d\theta \, d\phi.$$

Since $p_x = p\sin\theta \cos\phi$, $p_y = p\sin\theta \sin\phi$, $p_z = p\cos\theta$, on ϕ integration only the component along $\mathbf{E}$ survives,

$$P_z = \frac{Np \int_0^{\pi} \cos\theta \sin\theta e^{pE\cos\theta/k_B T} d\theta}{\int_0^{\pi} \sin\theta e^{pE\cos\theta/k_B T} d\theta}.$$

Using the substitution $(pE \cos\theta / k_B T) = y$, we integrate these integrals to obtain

$$P_z = Np\left(\coth a - \frac{1}{a}\right),$$

where $a = pE / k_B T$. For $a << 1$, this expression simplifies to

$$\mathbf{P} = \frac{Np^2\mathbf{E}}{3k_B T}.$$

Thus, the orientation polarisability is $\alpha = \dfrac{p^2}{3k_B T}$.

1.12 Magnetisation

In a material, every electron revolving around a nucleus of an atom can be viewed as a current loop of current I = e/T, where e is the magnitude of electron charge and T is the time period of revolution. The magnetic dipole moment of such a loop (called magnetic dipole), defined as the product of current and the area enclosed by the loop, is

$$m' = e\, r^2 \, \frac{\omega}{2},$$

where $\omega = 2\pi/T$. From Bohr theory of hydrogen atom, the angular momentum of an orbiting electron is an integral multiple of $\hbar$ i.e., $mr^2\omega = n\hbar$, where m is electron mass, $\hbar$ is the Planck's constant divided by 2π and n is an integer. Hence

$$m' = n\frac{e\hbar}{2m}. \tag{36}$$

Besides orbital motion, electrons also possess a magnetic dipole moment due to spin motion

$$m'_s = \frac{e\hbar}{2m}. \tag{37}$$

In the absence of external magnetic field, these dipoles are randomly oriented and there is no net magnetic dipole moment on the macroscopic scale (except in permanent magnets). When a magnetic field is applied, these dipoles tend to align themselves along the magnetic field, giving a net dipole moment on macroscopic scale. Magnetic dipole moment per unit volume is known as magnetisation **M**.

Atoms that have an odd number of electrons, have a finite magnetic moment. Sodium, for instance, has one electron in its unfilled shell. This electron gives the atom a spin and a magnetic moment. However, when compounds are formed, the extra electrons in the outer shells are coupled together with other electrons whose spin directions are exactly opposite. Due to this, the magnetic dipole moments of the valence electrons usually cancel out. That is why molecules, in general, do not have a magnetic moment.

In most bulk materials, there is a net magnetic moment when the inner electron shell is not filled, e.g. transition elements like chromium, manganese, iron, nickel, cobalt, palladium, and platinum. All the rare earth elements also have unfilled inner shells and permanent magnetic moments.

In the absence of a magnetic field, these atomic or molecular magnetic dipoles in a material are randomly oriented due to thermal motions. When a magnetic field **B** is applied, they tend to align themselves along the direction of magnetic field. The energy of a dipole of magnetic moment $\mathbf{m}'$, oriented at angle θ with respect to **B**, is $-m'B\cos\theta$. Following the procedure adopted for dielectric polarisation of polar molecules in problem 1.6 and assuming $m'B << k_B T$, we obtain the magnetisation

$$\mathbf{M} \cong \frac{Nm'^2}{3k_BT}\mathbf{B}.$$

In certain materials, called ferromagnets, the magnetic moments induced by applied magnetic field are quite enormous. The energy of orientation of a magnetic dipole **m** in a given applied field is $-\mathbf{m}.\mathbf{B}$, but the energy of the spinning electrons depends on the neighbouring spin alignments. In iron, if the moment of a nearby atom is pointing upwards, there is a very strong tendency of the neighbouring moment to also point upwards. This is why iron, cobalt and nickel are so strongly magnetic.

In contrast to ferromagnetic materials, *ferrimagnetic* compounds have high resistivity and a significant amount of anisotropy induced by the dc magnetic field. This field aligns the magnetic dipoles in the ferrite material to produce a net dipole moment and causes the dipole to precess at a frequency, controlled by the strength of the bias field. A microwave signal, circularly polarised in the same direction as this precession, strongly interacts with the dipoles as compared to an oppositely polarised field. This effect is utilised in fabricating directional devices like isolators, circulators, and gyrators. However, in this book we restrict ourselves to wave propagation in nonmagnetic media and not delve into the magnetisation of magnetic materials.

PROBLEM 1.7. An electron is orbiting round a nucleus with radius r and angular momentum $\mathbf{L} \| \hat{z}$ (Fig. 1.16). We gradually build a magnetic field **B**, normal to the plane of the orbit.

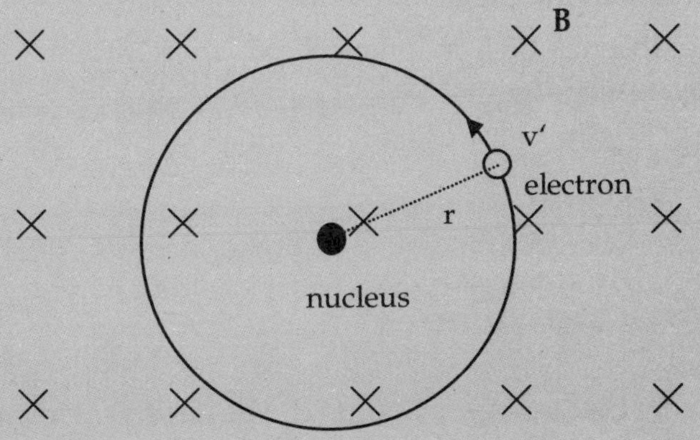

Figure 1.16: An atom in a magnetic field.

Show that the change in the magnetic dipole moment is $\Delta \mathbf{m}' = -\frac{e}{2m}\Delta \mathbf{L} = -\frac{e^2 r^2}{4m}\mathbf{B}$.

Solution: From Faraday's law of electromagnetic induction, an electric field **E** is induced tangential to the orbit, such that

$$E_\phi 2\pi r = -\frac{d}{dt}\left(B\pi r^2\right) \quad \text{or} \quad E_\phi = -\frac{r}{2}\frac{dB}{dt}.$$

The azimuthal equation of motion, on realising $L = m\, r^2 \dot{\phi}$, gives

$$\frac{1}{r}\frac{dL}{dt} = \frac{e\, r}{2}\frac{dB}{dt} \qquad \text{or} \qquad \Delta L = \frac{er^2}{2}B.$$

Corresponding change in magnetic dipole moment is

$$\Delta \mathbf{m}' = -\frac{e}{2m}\Delta \mathbf{L} = -\frac{e^2 r^2}{4m}\mathbf{B}.$$

The change in magnetic moment is opposite to the direction of **B**. In ordinary unmagnetised matter, the electron orbits are randomly oriented and the orbital magnetic dipole moment cancels out. In the presence of a magnetic field, each atom picks up an extra dipole moment and all these increments are antiparallel to the magnetic field. This is the mechanism which gives rise to *diamagnetism*. Diamagnetic property is universal to all atoms. However, it is typically weaker than paramagnetism and hence is observable mainly in atoms with even number of electrons, where paramagnetism is absent.

1.13 Wave Equation

In nonrelaxing media, where **D** and **J** can be written as instantaneous functions of **E**, (viz. $\mathbf{D} = \epsilon_0 \epsilon_r \mathbf{E}$, $\mathbf{J} = \sigma \mathbf{E}$) and **M** is an instantaneous function of **B**, one can deduce the wave equation for **E** or **H** from the Maxwell's equations for arbitrary dependence on time. In the case of relaxing media, where **D** and **J** depend on the time history of **E** and/or **M** depends on the time history of **B**, one can do so only when fields have sinusoidal time dependence. This, however, is not a serious limitation as fields with arbitrary time dependence can be Fourier analysed (Appendix A), and the propagation of each Fourier component studied independently. At the point of reception, different Fourier components can be superimposed to deduce the net signal. However, in this chapter, we restrict ourselves to nonrelaxing, nonmagnetic ($\mu_r = 1$) media.

Taking curl of eq.(22), $\nabla \times (\nabla \times \mathbf{E}) = \mu_0 \partial/\partial t\, (\nabla \times \mathbf{H})$ and employing eq.(23) along with the vector identity $\nabla \times \nabla \times \mathbf{E} = -\nabla^2 \mathbf{E} + \nabla\,(\nabla . \mathbf{E})$, we obtain

$$\nabla^2 \mathbf{E} - \nabla\,(\nabla . \mathbf{E}) - \frac{\epsilon_r}{c^2}\frac{\partial^2 \mathbf{E}}{\partial t^2} - \frac{\sigma}{\epsilon_0\, c^2}\frac{\partial \mathbf{E}}{\partial t} = 0, \tag{38}$$

where $c = \sqrt{\mu_0\, \epsilon_0} = 3 \times 10^8$ m/s. A similar wave equation can be deduced for **H**, by taking the curl of eq.(23) and employing eq.(22),

$$\nabla^2 \mathbf{H} - \frac{\epsilon_r}{c^2}\frac{\partial^2 \mathbf{H}}{\partial t^2} - \frac{\sigma}{\epsilon_0\, c^2}\frac{\partial \mathbf{H}}{\partial t} = 0, \tag{39}$$

where we have used $\nabla.\mathbf{H} = 0$ from eq.(21). Had we considered a magnetic material with $\mathbf{B} = \mu_0\mu_r\mathbf{H}$, eqs.(38) and (39) would still be valid with c^2 replaced by c^2/μ_r. We discuss some general properties of eq.(38) for different media. The properties of eq.(39) are similar.

1.13.1 Free space

In free space $\varepsilon_r = 1$, $\sigma = 0$ and $\nabla.\mathbf{E} = 0$ (from eq.20), hence eq.(38) takes the form

$$\nabla^2\mathbf{E} - \frac{1}{c^2}\frac{\partial^2\mathbf{E}}{\partial t^2} = 0. \tag{40}$$

For propagation in one dimension, say along $\hat{z}$, we take $\partial/\partial x = 0$, $\partial/\partial y = 0$ and write eq.(40) as

$$\frac{\partial^2\mathbf{E}}{\partial z^2} - \frac{1}{c^2}\frac{\partial^2\mathbf{E}}{\partial t^2} = 0. \tag{41}$$

A general solution of this equation is

$$\mathbf{E}(z, t) = \mathbf{F_1}\left(t - \frac{z}{c}\right) + \mathbf{F_2}\left(t + \frac{z}{c}\right). \tag{42}$$

If F_2 were zero, then $\mathbf{E}(z,t) = \mathbf{F_1}(t - z/c)$. Using this expression, if we evaluate electric field at $z = 0$, at time $(t - z/c)$, we obtain $\mathbf{E}(z,t) = \mathbf{E}(0, t - z/c)$. The field at z, at time t, is the same as it was at $z = 0$ at an earlier time $t - z/c$. This may be viewed as the propagation of disturbance from $z = 0$ to z in time z/c with velocity c. Thus, $\mathbf{F_1}(t - z/c)$ represents the propagation of wave along $+\hat{z}$. Similarly, $\mathbf{F_2}$ represents a wave propagating along $-\hat{z}$ with velocity c. c is thus known as the *velocity of light in vacuum*. For wave propagation along $+\hat{z}$, $\mathbf{F_2} = 0$. The wave field retains its temporal shape at all points z, as it was at $z = 0$, with the only difference that time gets delayed by z/c. Since $\partial E_z/\partial z = 0$ for these waves, E_z must be zero or a constant (independent of z and t). E_z = constant does not represent a propagating disturbance and hence can be discarded. Thus, an electromagnetic wave propagating along z has only E_x and E_y components of electric field, i.e., its electric field is transverse to the direction of wave propagation. From eq.(22) one may see that

$$\frac{\partial H_z}{\partial t} = -\frac{1}{\mu_0}(\nabla \times \mathbf{E})_z = 0.$$

Hence $H_z = 0$, i.e., the magnetic field is also transverse to the direction of propagation. Thus, electromagnetic waves in free space have transverse electric and magnetic fields. The direction of propagation of the one-dimensional wave could easily be chosen along x or y or any other direction. As long as the fields vary only in one direction, the above discussion holds. Further, eq. (40) governs the propagation of waves in two and three dimensions as well.

1.13.2 Dielectric

In a dielectric there are no free charges. Hence $\rho = 0$, $\sigma = 0$, $\nabla.\mathbf{E} = 0$ and eq.(38) takes the form

$$\nabla^2 \mathbf{E} - \frac{\varepsilon_r}{c^2}\frac{\partial^2 \mathbf{E}}{\partial t^2} = 0. \tag{43}$$

This equation is the same as eq.(44) with c replaced by $c/\sqrt{\varepsilon_r}$. These waves have same characteristics as those in free space but the velocity of propagation is $c/\sqrt{\varepsilon_r}$. There is no distortion of the signal on propagation in one direction in a nonrelaxing dielectric. Any real dielectric behaves like this at low frequencies ($\leq$ 30 GHz). At higher frequencies, relaxation effects become important and they give rise to phenomena like dispersion.

1.13.3 Conductors

In the case of conductors, let us first examine the nature of $\nabla.\mathbf{E}$. It is finite when $\rho \neq 0$. From equation of continuity, on using $\mathbf{J} = \sigma\,\mathbf{E}$, we have

$$\frac{\partial \rho}{\partial t} + \sigma\,\nabla.\mathbf{E} = 0.$$

Using eq. (20), we get

$$\frac{\partial \rho}{\partial t} + \frac{\sigma}{\varepsilon_0 \varepsilon_r}\rho = 0,$$

giving $\rho = \rho_0 \exp\left(-\left(\sigma/\varepsilon_0\varepsilon_r\right) t\right)$. Charge density cannot have any other time dependence than this. Hence, whenever **E** has any other time dependence $\rho = 0$, $\nabla.\mathbf{E} = 0$ and the wave equation can be written as

$$\nabla^2\mathbf{E} - \frac{\epsilon_r}{c^2}\frac{\partial^2\mathbf{E}}{\partial t^2} - \frac{\sigma}{\epsilon_0\, c^2}\frac{\partial\mathbf{E}}{\partial t} = 0. \tag{44}$$

One may see that a solution of the form $\mathbf{E} = \mathbf{F}\left(t - z/v\right)$ does not satisfy this equation, implying that a wave propagating in one dimension does not maintain its shape. In fact, conductivity leads to energy dissipation in the conductor and attenuation of the wave.

1.14 Energy Density of Fields and Poynting's Theorem

The electric and magnetic fields are produced by the charge and current distributions. In order to establish certain fields, one would have to establish certain charge and current distributions. The work done in doing so may be viewed as the energy stored in the electric and magnetic fields. We will derive the work done in establishing a static electric field and also that done in establishing a static magnetic field. We then generalise the result intuitively to time varying fields.

1.14.1 Energy Density of the Electric Field

The field is not merely a mathematical concept but a profound physical entity. It carries momentum and energy. Let there exist a charge distribution $\rho(\mathbf{r})$ at an instant, giving electric field $\mathbf{E}(\mathbf{r})$ and potential $\phi(\mathbf{r})$. Now, if we bring the charges from infinity to increase the charge density by $\delta\rho(\mathbf{r})$, then we will have to do work

$$\delta W = \int \phi\left(\mathbf{r}'\right)\delta\rho\left(\mathbf{r}'\right)dV',$$

where integration is over a large volume containing all the charges.

Using eq. (20) we may write $\delta\rho = \nabla.\delta\mathbf{D}$. Hence

$$\delta W = \int \phi\nabla.\delta\mathbf{D}dV' = \int \nabla(\phi.\delta\mathbf{D})dV' - \int \delta\mathbf{D}.\nabla\phi\, dV'.$$

Using Gauss's theorem, the first integral can be converted to a surface integral,

$$\oint \phi(\delta \mathbf{D}.\hat{\mathrm{n}})\mathrm{dS}'.$$

At the surface of a large volume (say a sphere of radius R) the area goes as R^2 while the integrand goes as R^{-3}. Hence, for $R \rightarrow \infty$, it vanishes giving

$$\delta \mathrm{W} = \int \mathbf{E}.\delta \mathbf{D} \mathrm{dV}' = \frac{1}{2}\delta \int \mathbf{E}.\mathbf{D}\mathrm{dV}'.$$

Thus, the work done in establishing the field per unit volume is

$$\mathrm{W_E} = \frac{1}{2}\mathbf{E}.\mathbf{D}. \tag{45}$$

This is known as the energy density of the electric field.

1.14.2 Energy Density of the Magnetic Field

Let there be a loop carrying static current I (Fig. 1.17). The cross-section of the tube may be taken as $\Delta\sigma$ and the magnitude of current density $J = I / \Delta\sigma$. This current produces a magnetic field in the entire space. The magnetic field **B** also exists in the area enclosed by the current loop. Now we wish to increase current dI in the loop, in time Δt.

The flux linked with the loop will change by

$$\Delta\phi = \int \delta \mathrm{B}.\mathbf{dS}.$$

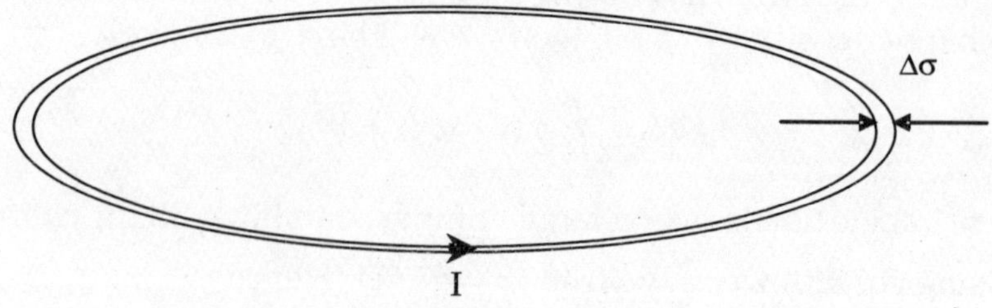

Figure 1.17: A current loop of cross-section $\Delta\sigma$.

The induced emf in the loop is $\Delta\phi/\Delta t$. Hence, the work done in moving the charge I + dI (per second) against the emf

$$\frac{dW}{dt} = (I + dI)\frac{\Delta\phi}{\Delta t} \approx I\frac{\Delta\phi}{\Delta t},$$

$$dW = I\Delta\phi = I\int \delta\mathbf{B}.d\mathbf{S}.$$

Using $\delta\mathbf{B} = \nabla \times \delta\mathbf{A}$ and employing the Stokes theorem, one obtains

$$dW = Jd\sigma\int \delta\mathbf{A}.\mathbf{dl} = \int \mathbf{J}.\delta\mathbf{A}dV = \frac{1}{\mu}\int \nabla \times \mathbf{B}.\delta\mathbf{A}dV,$$

where we have used $\mathbf{J} = \nabla \times \mathbf{B}/\mu$ from Ampere's law. We extend the volume integral over a large volume containing all the current sources. Using the vector identity

$$\nabla.(\mathbf{P} \times \mathbf{Q}) = \mathbf{Q}.\nabla \times \mathbf{P} - \mathbf{P}.\nabla \times \mathbf{Q}$$

one may write

$$dW = \frac{1}{\mu}\int \mathbf{B}.\nabla \times \delta\mathbf{A}dV + \frac{1}{\mu}\int \nabla.(\mathbf{B} \times \delta\mathbf{A})\,dV.$$

The last integral can be converted into a surface integral using Gauss's theorem. For localised current sources, the product $\mathbf{B} \times \delta\mathbf{A}$ goes as $1/r^3$ whereas the surface area goes as r^2. Hence over a large surface, this integral vanishes. Thus, one obtains

$$dW = \left(\frac{1}{\mu}\right)\int \mathbf{B}.\delta\mathbf{B}dV = \left(\frac{1}{2}\right)\delta\int \mathbf{B}.\mathbf{H}dV,$$

giving energy stored in the magnetic field per unit volume as

$$W_B = \mathbf{B}.\mathbf{H}/2. \qquad (46)$$

PROBLEM 1.8. A long solenoid of length l, radius R, number of turns per unit length N, $\mu = \mu_0$ and the axis along $\hat{z}$, carries a dc current I (Fig. 1.18). Ignoring end effects, show that (i) B_z=0 outside the solenoid, (ii) $B_z = \mu_0$ NI in the interior region, (iii) inductance of the solenoid L =μ_0 $N^2\pi$ R^2l and (iv) magnetic energy inside the solenoid is $(1/2)LI^2$.

Solution: If one assumes that the current flows in the azimuthal direction, then by symmetry r, ϕ, z components of the magnetic field and vector potential in the interior as well as exterior regions are independent of ϕ and z. Such a vector potential gives $B_r = 0$. Further, if one considers a circular Amperean loop of radius r in the x-y plane with centre on z-axis and applies Ampere's law, he obtains $B_\phi = 0$ for r < R.

Now consider an amperean loop in the x-z plane outside the solenoid, as shown in figure.1.18.

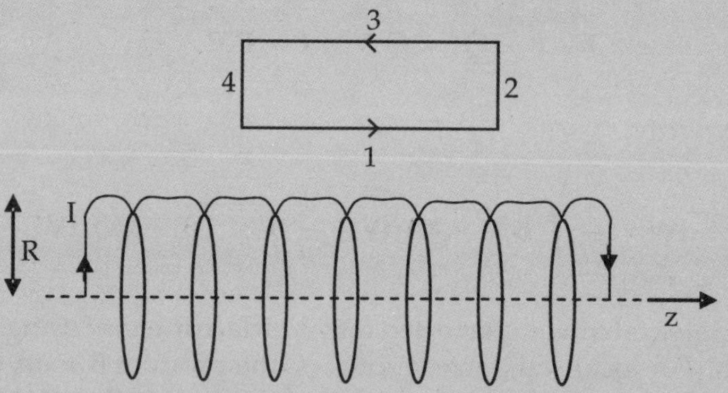

Figure 1.18: A solenoid carrying current I. Amperean loop has arms 1, 2, 3, and 4.

$\oint$ **B. dl** = 0 for the loop gives

$$B_z(1) = B_z(3)\,.$$

If one takes arm 3 to $x \to \infty$ and take $B_z \to 0$ in compliance with Biot-Savart's law, we get $B_z(1) = 0$, i.e., $B_z=0$ everywhere for $r>R$. Next push the arm1 of the above loop inside the solenoid and apply Ampere's law, we get

$$B_z = \mu_0 N I.$$

The magnetic flux Φ linked with each turn, is $\pi R^2 B_z$. Hence the flux linked with the solenoid is

$$\Phi = LI \text{ with } L = \mu_0 N^2 \pi R^2 l.$$

The energy stored in the magnetic field is

$$\varepsilon_m = \left(\frac{{B_z}^2}{2\mu_0}\right) \pi R^2 l = \left(\frac{1}{2}\right) L I^2.$$

1.14.3 Poynting's Theorem

Having established the energy stored in electric and magnetic fields, let us investigate how the energy flows from one region of space to another when fields vary with time.

From the generalised Ampere's law.

$$\mathbf{J} = \nabla \times \mathbf{H} - \frac{\partial \mathbf{D}}{\partial t},$$

$$\mathbf{J}.\mathbf{E} = \mathbf{E}.\nabla \times \mathbf{H} - \mathbf{E}.\frac{\partial \mathbf{D}}{\partial t}.$$

Using vector identity, one may write

$$\mathbf{E}.\nabla \times \mathbf{H} = -\nabla.(\mathbf{E} \times \mathbf{H}) + \mathbf{H}.\nabla \times \mathbf{E} = -\nabla.(\mathbf{E} \times \mathbf{H}) - \mathbf{H}.\frac{\partial \mathbf{B}}{\partial t}$$

$$= -\nabla.(\mathbf{E} \times \mathbf{H}) - \frac{1}{2}\frac{\partial}{\partial t}(\mathbf{B}.\mathbf{H}),$$

$$\mathbf{J}.\mathbf{E} = -\nabla.(\mathbf{E} \times \mathbf{H}) - \frac{\partial}{\partial t}(W_E + W_B),$$

$$\frac{\partial W_{em}}{\partial t} + \mathbf{J}.\mathbf{E} = -\nabla.(\mathbf{E} \times \mathbf{H}) \tag{47}$$

where $W_{em} = W_E + W_B$. One may interpret the left hand side of eq. (47) as the sum of the rate of increase of field energy density and the energy absorbed per unit volume per second. Hence, the right-hand side must represent energy entering unit volume per second. From the definition of divergence of a vector, $\mathbf{E} \times \mathbf{H}$ represents the energy flow density, i.e., the amount of electromagnetic energy flowing per unit area per unit time.

$$\mathbf{S} = \mathbf{E} \times \mathbf{H} \tag{48}$$

is known as the *Poynting vector*.

1.15 Conclusions

We have laid the framework of four fundamental equations – the Maxwell's equations, that govern the behaviour of electromagnetic fields in every possible situation, ranging from nuclear dimensions to astrophysical dimensions. These equations are to be supplemented by constituent relations, relating $\mathbf{P}, \mathbf{M}$ and $\mathbf{J}$ to electromagnetic fields. In general, $\mathbf{P}, \mathbf{M}, \mathbf{J}$ depend on the instantaneous values of **E** and **B** as well as their previous history and the corresponding relations are nonlinear. However, for time harmonic fields of small amplitude, these relations are simple and linear. For most part of this book, we will be confining ourselves to linear regime and sinusoidal or near sinusoidal fields.

UNSOLVED PROBLEMS

1. A sphere of radius r_0 and relative permittivity ε_r has charge distribution with charge density $\rho = \rho_0\,(1 - r^2/r_0^2)$. Using Gauss's theorem, obtain the electric field for (i) $r < r_0$ and (ii) $r > r_0$.

   [**Ans.** (i) $E_r = A\left(5r - \frac{3r^3}{r_0^2}\right)$, (ii) $E_r = 2A\,\frac{r_0^3}{r^2}$, $A = \frac{\rho_0}{15\varepsilon_0\varepsilon_r}$].

2. A coaxial cable of inner radius a and outer radius b is filled with a dielectric of relative permittivity ε_r. Show

that the capacitance per unit length of the cable is $C = 2\pi\, \varepsilon_0 \varepsilon_r / \ln(b/a)$.

[**HINT:** *Assuming charge per unit length on the inner conductor* σ*, obtain* E_r *in the region* $a < r < b$ *using Gauss's theorem. Then deduce the potential difference* V*, hence* $C = \frac{\sigma}{V}$].

3. The inner conductor of a coaxial cable carries current I and the outer one carries a current –I. Show that the magnetic field in the region $a < r < b$ (where a and b are the radii of inner and outer conductors respectively) is $B_\phi = \mu_0 I/2\pi r$ and the inductance per unit length of the cable is $L = (\mu_0/2\pi)\ln(b/a)$.

4. In high-power laser plasma interaction experiments, one observes streaming electrons producing a magnetic field $\mathbf{B} = \hat{\phi}\,(r/r_0)\exp(-r^2/r_0^2)$, where r, ϕ are cylindrical coordinates. Obtain the current density that produces it.

   [**ANS.** $\mathbf{J} = \hat{z}\left(\frac{2B_0}{\mu_0 r_0}\right)\left(1 - \frac{r^2}{r_0^2}\right)\exp\left(\frac{-r^2}{r_0^2}\right)$.]

5. In a region of relative permittivity $\varepsilon_r = 1, \mathbf{E} = \hat{x} A \cos(\omega t - kx)$. Obtain the charge density and the magnetic field. Such a field represents a special kind of wave in a plasma (an ionised gas), called plasma wave.

   [**ANS.** $\rho = \varepsilon_0 A k \sin(\omega t - kx)$, $\mathbf{B} = 0$.]

6. Consider the propagation of an electron beam of density n_0, drift velocity $v_0\hat{z}$, and radius r_0. Obtain the static electric and magnetic fields for $r < r_0$.

   [**ANS.** $E_r = -\frac{n_0 e r}{2\varepsilon_0}$, $B_\phi = \frac{E_r v_0}{c^2}$.]

7. Two grids A and B, placed at $z = 0$ and $z = d$, are maintained at a constant potential difference $V_{AB} = V_1 \cos\omega t$. At $t = t_1$, an electron enters grid A with velocity

$v_0\hat{z}$. Obtain the electron velocity at the exit, assuming $eV_1 << mv_0^2/2$ and the time at the exit as $t_1 + d/v_0$.

$$\left[\text{ANS. } v_z = v_0 - \left(\frac{eV_1}{mv_0\theta}\right) \sin\theta \cos(\omega t_1 + \theta),\ \theta = \frac{\omega d}{2v_0}.\right]$$

NOTE: This scheme is employed in a *Klystron* to obtain velocity modulation of an electron beam.

8. A free electron laser has a wiggler magnetic field

$$\mathbf{B_w} = \left(\hat{x}\, A_1(x) + \hat{z}\, A_2(x)\right) \exp(i\, k_w z)$$

where $A_2(x) = A_2(-x)$. Using $\nabla \times \mathbf{B_w} = 0$, $\nabla.\mathbf{B_w} = 0$, deduce the x dependence of A_1 and A_2.

$$\left[\text{ANS. } A_1 = -iA_0 \sinh(k_w x), \quad A_2 = A_0 \cosh(k_w x).\right]$$

9. An n-type semiconductor, with $-e$, m and ν as electronic charge, effective mass and electron-phonon collision frequency respectively, is placed in an ac electric field $\mathbf{E} = \hat{x}\, A_0 e^{-i\omega t}$. It imparts an oscillatory velocity $\mathbf{v}$ to electrons and heats them at a rate

$$\text{Re}(-e\mathbf{E}).\text{Re}(\mathbf{v}) \equiv -(e/2)\,\text{Re}\left(\mathbf{E}^*.\mathbf{v} + \mathbf{E}\cdot\mathbf{v}\right),$$

where Re refers to the real part of the quantity. Assuming that the rate of energy loss rate per electron per second is $(3/2)\,\delta\nu\, k_B(T_e - T_0)$, where k_B is the Boltzmann's constant, T_0 is the lattice temperature and δ is a constant signifying mean fraction of excess energy lost by an electron per collision, obtain the field amplitude A_0 at which time average electron temperature $T_e = 2\, T_0$.

$$\left[\text{ANS. } A_0 = [3\delta\, k_B T_0 m\left(\omega^2 + \nu^2\right)]^{1/2}/e.\right]$$

10. Two identical circular coils of radius R, carrying current I each, are placed at $z = -L/2$ and $z = L/2$ with their centres on z axis and planes perpendicular to z axis.

Such a configuration is realised in a mirror machine. Obtain the magnetic field at a point on z axis for -L/2 < z < L/2. Further, using $\nabla \cdot \mathbf{B} = 0$, show that at point r, ϕ, z, which is slightly off-axis, $B_r = -(r/2)\partial B_z/\partial z$.

$$\left[\textbf{Ans: } B_z = \left(\frac{\mu_0 I R^2}{2}\right)\left[\left\{R^2+\left(z+\tfrac{L}{2}\right)^2\right\}^{-3/2} + \left\{R^2+\left(z-\tfrac{L}{2}\right)^2\right\}^{-3/2}\right].\right]$$

11. In a mirror machine, magnetic field near the machine axis ($\hat{z}$) can be written as $\mathbf{B} = \hat{z}B_z + \hat{r}B_r$, $B_r = -(r/2)\partial B_z/\partial z << B_r$. Consider an electron of charge $-e$ and mass m, rotating about z axis with velocity $\mathbf{v} = v_\phi\hat{\phi} + v_z\hat{z}$ and gyroradius v_ϕ/ω_c, where $\omega_c = eB_z/m$. Using the z component of equation of motion, deduce

$$\frac{d}{dt}(m{v_z}^2/2) = -\mu\frac{dB_z}{dt},$$

where, $\mu = mv_\phi^2/2B$ is the magnetic moment of the electron. Employing the fact that the energy $\varepsilon = mv^2/2$ is a constant of motion and using the above equation, show that μ is a constant of motion.

> **Note:** μ = constant forms the basis of plasma confinement in a mirror machine that has minimum magnetic field in the centre. As a gyrating particle (electron or ion) moves away axially from the central region, B increases, hence v_ϕ increases to keep μ constant. With increasing v_ϕ, v_z decreases (as ε = constant). At some point, v_z vanishes and the particle turns back towards the centre.

12. Consider two overlapping spheres of singly ionised ions and free electrons of density n_0 each and radius R. The electron sphere is given a small displacement $\Delta\hat{x}$, with respect to the ion sphere. Assuming that the ions are immobile, show that the electric field in the overlapping region of the spheres is $\mathbf{E_s} = \hat{x}(n_0 e\Delta/3\varepsilon_0)$. This field exerts a restoration force $-e\mathbf{E_s}$ on each electron to pull back the sphere, causing simple harmonic motion. Show that the frequency of oscillation of the electron sphere is $\omega_p/\sqrt{3}$, where $\omega_p = (n_0 e^2/m\varepsilon_0)^{1/2}$ and $-e$ and m are the electronic charge and mass respectively.

13. A small plasma sphere of singly ionised ions and free electrons of density n_0 each and radius R is subjected to an external electric field $\mathbf{E} = \hat{x}A_0 \exp(-i\omega t)$. Solve the equation of motion $md^2\Delta/dt^2 = -eE_x - n_0 e^2 \Delta/3\varepsilon_0$ to obtain the displacement Δ.

[**HINT:** *Presume* $\Delta = a\exp(-i\omega t)$ *and replace* d/dt *by* $-i\omega$ *in the equation.*]

$$\left[\text{ANS: } \Delta = eE_x/[m(\omega^2 - \omega_p^2/3)], \omega_p = \left(\frac{n_0 e^2}{m\varepsilon_0}\right)^{1/2}.\right]$$

14. An electron of charge $-e$ and mass m is initially at rest, at the origin. It is subjected to crossed electric and magnetic fields $\mathbf{E}||\hat{x}$ and $\mathbf{B}||\hat{z}$. Obtain the electron velocity $\mathbf{v}$ and displacement as a function of time.

[**ANS:** $v_z = v_d \sin\omega_c t$, $v_y = v_d(1 - \cos\omega_c t)$, $v_x = 0$, $v_d = -E/B$, $x = (v_d/\omega_c)(1 - \cos\omega_c t)$, $y = v_d t - (v_d/\omega_c)\sin\omega_c t$, $z = 0$, $\omega_c = eB/m$.]

15. In an n-type semiconductor, the phase difference between current density and electric field at 200 GHz is $\pi/4$. What would be the phase difference at 1000 GHz?

[**ANS:** $\tan^{-1}(5)$.]

16. An axially symmetric electromagnet (a solenoid with axis along $\hat{z}$) of radius 0.2 m is used to cause breakdown of air. The resultant magnetic field B_z is uniform upto r=0.2m and 0 beyond it. B_z rises from 0 to 10 Tesla linearly with time in time τ. If the breakdown field for air is 3×10^6 V/m, estimate the requisite value of c.

[HINT: *Use Faraday's Law of electromagnetic induction to obtain* E_ϕ *at* r = 0.2m.]

[ANS: $\tau \leq 1/3\mu s$]

Footnote for eq. (25)

a. The connection between the total and partial time derivatives of the velocity field $\mathbf{v}(\mathbf{r}, t) = \mathbf{v}(x, y, z, t)$ can be realised by the fact that

$$\frac{\partial \mathbf{v}}{\partial t} = [\mathbf{v}(\mathbf{r}, t + \Delta t) - \mathbf{v}(\mathbf{r}, t)]\frac{1}{\Delta t}$$

$$\frac{d\mathbf{v}}{dt} = [\mathbf{v}(\mathbf{r} + \mathbf{v}\Delta t, t + \Delta t) - \mathbf{v}(\mathbf{r}, t)]\frac{1}{\Delta t},$$

as $\Delta t \rightarrow 0$. Here $v\Delta t$ is the change in position coordinate of the fluid element in time Δt. Partial time derivation keeps an eye on a point and represents the rate of velocity change at that point. Total time derivative keeps an eye on fluid element and measures its velocities at t and t $+\Delta$ t. Carrying out the Taylor expansion

$$\begin{aligned}\mathbf{v}(\mathbf{r} + \mathbf{v}\Delta t, t + \Delta t) &= \mathbf{v}(\mathbf{r}, t) + \left(\frac{\partial \mathbf{v}}{\partial x}\right) v_x \Delta t + \left(\frac{\partial \mathbf{v}}{\partial y}\right) v_y \Delta t \\ &\quad + \left(\frac{\partial \mathbf{v}}{\partial z}\right) v_z \Delta t + \left(\frac{\partial \mathbf{v}}{\partial t}\right) \Delta t \\ &= \mathbf{v}(\mathbf{r}, t) + (\mathbf{v} \cdot \nabla)\mathbf{v}\Delta t + \left(\frac{\partial \mathbf{v}}{\partial t}\right) \Delta t\end{aligned}$$

we obtain

$$\frac{d\mathbf{v}}{dt} = \frac{\partial \mathbf{v}}{\partial t} + (\mathbf{v} \cdot \nabla)\mathbf{v}.$$

REFERENCES AND SUGGESTED READING

[1] Griffiths D. J., *Introduction to electrodynamics*, New Delhi, Prentice Hall of India Pvt. Ltd., 2000.

[2] Jackson J. D., *Classical electrodynamics*, New York, John Wiley & Sons, 2005.

[3] Landau L. D. and Lifshitz E. M., *Electrodynamics of continuous media*, MA, Addison-Wesley, 1984.

[4] Ramo S., Whinnery J. R. and Duzer T. V., *Fields and waves in communication electronics*, New York, John Wiley & Sons, 1984.

[5] Johnk Carl T. A., *Engineering electromagnetic fields and waves*, New York, John Wiley & Sons, 1988.

[6] Chen F. F., *Introduction to plasma physics and controlled fusion*, Second Edition, New York and London, Plenum Press, 1984.

Unit 2

Plane Waves

Introduction

A wave is defined as a propagating disturbance. Sound wave, for instance, is a propagating pressure disturbance. When a person speaks, the air pressure in front of his mouth changes from equilibrium pressure P_0 to $P_0 + p$. p is called the *pressure disturbance.* It is first felt (*i.e.*, the speech is heard) near the mouth and later at farther distances, as it moves with finite velocity, say $\mathbf{v}$. If the person speaks through a horn, with its length along $\hat{z}$, the pressure disturbance may be considered to be propagating along $\hat{z}$. The pressure disturbance p at position z, at time t, is the same as that at z=0 at time $t-z/v$, shifted by the time of travel from 0 to z (Fig. 2.1).

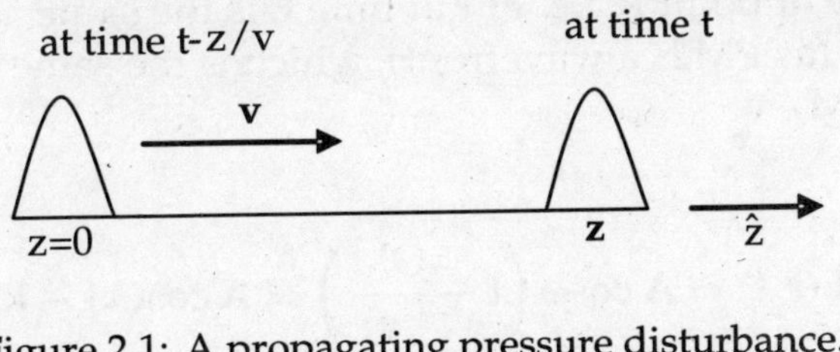

Figure 2.1: A propagating pressure disturbance.

If p (0, t) = f (t), then

$$p(z,t) = p\left(0, t - \frac{z}{v}\right) = f\left(t - \frac{z}{v}\right). \tag{1}$$

In the simplest form, f(t) can be taken to be f(t) = A cos ($\omega t + \theta_0$), where $\omega/2\pi$ is the frequency in cycles per second (Hertz), θ_0 is the initial phase and A is the amplitude. Then

$$p(z,t) = A\cos(\omega t - kz + \theta_0), \tag{2}$$

where $k = \omega/v$. k is called the *propagation constant.*

For an electromagnetic wave the propagating disturbances i.e. electric and magnetic fields (**E** and **B**) are vector quantities. Hence

$$\mathbf{E}(z,t) = \mathbf{A}\cos(\omega t - kz). \tag{3}$$

Similarly, one could write the magnetic field. Here, we have suppressed θ_0 by suitably defining the reference of time. The argument of the cosine function, (ωt - kz), is called the *phase of the wave* and ω/ k is defined as the **phase velocity** of the wave,

$$v_{ph} = \frac{\omega}{k}. \tag{4}$$

A plane of constant phase is defined as a wavefront. In this case, z = constant plane (i.e. the x-y plane) is a wavefront. It is perpendicular to the direction of propagation. One may note that

$$\mathbf{E}(z + \lambda, t) = \mathbf{E}(z,t),$$

where $\lambda = 2\pi / k$. The disturbance has the same value at wavefronts separated by λ. λ is identified as the *wavelength.*

For a wave propagating along an arbitrary direction $\hat{n}$, the wavefront would be a plane perpendicular to $\hat{n}$. This can be represented mathematically as $\hat{n}.\ \mathbf{r}$ = constant (Fig. 2.2). The field of the wave at point P, i.e. at **r** at time t, is the same as that at point M at time t (as PM is a wavefront), which is the same as that at O at time t – OM/ v.

Hence,

$$\mathbf{E}(\mathbf{r},t) = \mathbf{A}\cos\omega\left(t - \frac{OM}{v}\right) = \mathbf{A}\cos(\omega t - \mathbf{k}.\mathbf{r}), \tag{5}$$

where OM = $\mathbf{r}.\hat{n}$ and $\mathbf{k} = \hat{n}\omega/v$ is called the propagation constant. A more convenient representation of an electromagnetic wave is

$$\mathbf{E}(\mathbf{r},t) = \mathbf{A}e^{-i(\omega t - \mathbf{k}.\mathbf{r})}, \tag{6}$$

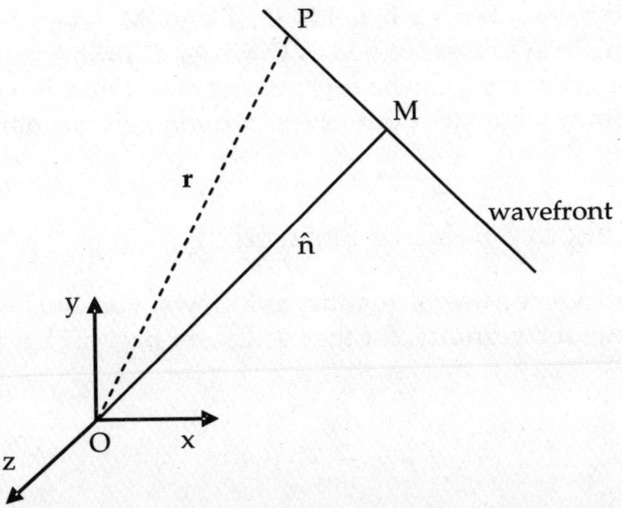

Figure 2.2: Propagation of a wave along $\hat{n}$.

where real part of the right-hand side is implied. The wave represented by eq. (6) is a plane wave as its wavefront $\mathbf{k}.\mathbf{r}$ = constant is a plane surface. The amplitude, frequency and velocity of propagation of the wave are determined by the source producing them and the properties of the medium through which it propagates. In this chapter, we deduce the properties of plane electromagnetic waves in different media from Maxwell's equations and study the phenomena of reflection and refraction from boundaries between media.

2.1 Maxwell's Equations for Sinusoidal Fields

Electromagnetic signals of any arbitrary time dependence can be Fourier analysed (Appendix A). If we learn the propagation of a frequency component ω of the signal through a medium, then the signal at the receiving end can be constructed by summing over the entire Fourier spectrum. Thus, we consider the propagation of a monochromatic wave with field varying with time as exp $(-i\omega t)$. In this case, one can write $\partial \mathbf{E}/\partial t = -i\omega \mathbf{E}$, i.e. the $\partial/\partial t$ operator operating over any oscillatory quantity **E**, **B**, **D**, **H**, **P** or **J** can be replaced by -iω. For such fields **D**, **J** and **M** (even in relaxing media) can always be expressed as $\mathbf{D} = \epsilon_0 \epsilon_r \mathbf{E}$, $\mathbf{J} = \sigma \mathbf{E}$, and $\mathbf{M} = \chi_m \mathbf{H}$, where ϵ_r, σ and χ_m can be functions of ω and may be complex in general. Since we restrict ourselves to nonmagnetic materials, we take $\mathbf{M} = 0$.

2.1.1 Effective Relative Permittivity

We may introduce a useful quantity as follows. For sinusoidal fields, the equation of continuity, for the oscillatory part of charge density ρ, gives

$$\rho = \frac{1}{\mathrm{i}\omega} \nabla \cdot (\sigma \mathbf{E}).$$

Using this equation, the first Maxwell's equation can be written as

$$\nabla \cdot (\epsilon_{\mathrm{eff}} \, \mathbf{E}) = 0, \tag{7}$$

where

$$\boxed{\epsilon_{\mathrm{eff}} = \epsilon_r + \frac{\mathrm{i}\sigma}{\omega \, \epsilon_0}} \tag{8}$$

is known as the *effective relative permittivity* of the medium. The other Maxwell's equations, eqs. (21) - (23) of unit 1, can be written as

$$\nabla \cdot \mathbf{H} = 0, \tag{9}$$

$$\nabla \times \mathbf{E} = \mathrm{i}\omega \, \mu_0 \mathbf{H}, \tag{10}$$

$$\nabla \times \mathbf{H} = -\mathrm{i}\omega \, \epsilon_0 \epsilon_{\mathrm{eff}} \, \mathbf{E}. \tag{11}$$

If the medium were a pure dielectric ($\sigma = 0$), then these equations would remain the same with ϵ_{eff} replaced by ϵ_r. Conversely, the Maxwell's equations in any conducting medium are the same as those in a dielectric with ϵ_r replaced by ϵ_{eff}. Usually, ϵ_r is real while ϵ_{eff} is complex. Hence nature of wave propagation is different. The entire information about the medium, as far as electromagnetic wave propagation is concerned, is contained in this single quantity ϵ_{eff}.

On using $\sigma = \dfrac{n_0 e^2}{m(\nu - i\omega)}$ from eq. (28) of unit 1, it may be written as

$$\epsilon_{eff} = \epsilon_r - \frac{\omega_p^2}{\omega^2\,(1 + i\nu/\omega)}, \tag{12}$$

where $\omega_p = (n_0\, e^2 /\, m\, \epsilon_0)^{1/2}$ is the *plasma frequency* and -e, m and n_0 are the charge, mass and equilibrium density of free electrons respectively.

2.1.2 Wave Equation

Taking curl of eq. (10) and using eq. (11), we get the wave equation, (see unit 1) as

$$\nabla^2 \mathbf{E} - \nabla(\nabla \cdot \mathbf{E}) + \frac{\omega^2}{c^2}\,\epsilon_{eff}\,\mathbf{E} = 0, \tag{13}$$

where $c = 1\sqrt{\mu_0\ \epsilon_0}$. Operating eq. (11) with $\nabla.$ and assuming ϵ_{eff} to be uniform (i.e., the medium to be homogeneous), we obtain

$$\epsilon_{eff}\ \nabla \cdot \mathbf{E} = 0. \tag{14}$$

There are two possibilities to satisfy eq.(14)

a. $\epsilon_{eff} = 0$, and
b. $\nabla \cdot \mathbf{E} = 0$.

The condition a is satisfied at a specific frequency given by $\omega\,(\omega + i\nu) = \omega_p^2 /\ \epsilon_r$. If collision frequency ν is ignored, this gives

$$\omega = \frac{\omega_p}{\sqrt{\epsilon_r}}. \tag{15}$$

This turns out to be the frequency of a space charge wave called *electrostatic wave*. We shall return to this wave later.

At all other frequencies $\epsilon_{eff} \neq 0$ and $\nabla.\mathbf{E} = 0$ is a must. The wave equation takes the form,

$$\nabla^2 \mathbf{E} + \frac{\omega^2}{c^2}\ \epsilon_{eff}\ \mathbf{E} = 0. \tag{16}$$

The waves governed by this equation are called *electromagnetic waves*. One may note that $\nabla \cdot \mathbf{E} = 0$ implies $\rho = 0$ (first Maxwell's equation), i.e. electromagnetic waves do not have space charge oscillations.

2.2 Electromagnetic Waves

The wave equation for electromagnetic waves could straightaway be solved by the separation of variables. However, we adopt a different approach. We presume a plane wave solution, given by eq. (4), and find the condition under which it satisfies the wave equation. In fact, whenever fields have space-time dependence of the form exp $[-i(\omega t - \mathbf{k}.\mathbf{r})]$, we can replace $\partial/\partial x$ operating over them by ik_x and ∇ by $i\mathbf{k}$. Replacing ∇^2 by $-k^2$ in eq. (16), we get

$$k^2 = \frac{\omega^2}{c^2} \epsilon_{eff} \,. \tag{17}$$

This is the dispersion relation for electromagnetic waves. In general k is complex, $\mathbf{k} = \mathbf{k}_r + i\mathbf{k}_i$ and the electric field of the wave can be written as

$$\mathbf{E} = \mathbf{A}\, e^{-\mathbf{k}_i \cdot \mathbf{r}} \cos(\omega t - \mathbf{k}_r \cdot \mathbf{r}), \tag{18}$$

where we have taken the real part of the complex expression (assuming $\mathbf{A}$ to be real). The amplitude of the wave, $\mathbf{A} \exp(-\mathbf{k}_i \cdot \mathbf{r})$, varies with space. The plane of constant amplitude is

$$\mathbf{k}_i \cdot \mathbf{r} = \text{constant}. \tag{19}$$

It is a plane perpendicular to $\mathbf{k}_i$. The argument of cosine function is called the *phase* of the wave. The plane of constant phase, i.e., wavefront is given by

$$\mathbf{k}_r \cdot \mathbf{r} = \text{constant}. \tag{20}$$

When $k_r || k_i$ the plane of constant amplitude and the wavefront are coincident. Such waves are called *homogeneous waves*. k_i is known as the *attenuation constant*.

2.2.1 Refractive Index

In a medium without absorption, *i.e.* where ϵ_{eff} is real and $k_i = 0$, the refractive index is defined as the ratio of velocity of light in free space to the phase velocity of the wave in the medium, $\eta = c/v_{ph}$. Since $v_{ph} = \omega/k$, one may write

$$\eta = \sqrt{\epsilon_{eff}}. \tag{21}$$

In media with finite absorption, *i.e.*, where ε_{eff} is complex, we still define refractive index through eq. (21). However, in this case η is complex, $\eta = \eta_r + i\eta_i$. The real part η_r characterises the phase velocity of the wave $v_{ph} = c/\eta_r$ and η_i characterises the attenuation constant $k_i = \omega\eta_i/c$.

The magnetic field of the wave can be obtained from eq. (10) by replacing ∇ by i**k**,

$$\mathbf{B} = \frac{\mathbf{k} \times \mathbf{E}}{\omega}. \tag{22}$$

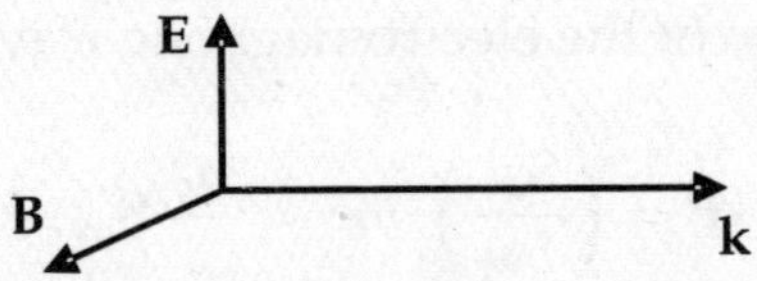

Figure 2.3: Electric field, magnetic field and propagation vector of a homogeneous electromagnetic wave.

B is perpendicular to **k** and **E**. Further, from $\nabla.\mathbf{E} = 0$, $\mathbf{k}.\mathbf{E} = 0$. Thus **E**, **B** and **k** are mutually perpendicular to each other (Fig. 2.3).

2.2.2 Energy and Momentum Flow

As an electromagnetic wave travels, it carries energy and momentum with it. The energy flow per unit area per second is $\mathbf{S} = \mathbf{E} \times \mathbf{H}$, where real parts of **E** and **H** are implied. Thus

$$\mathbf{S} = \mathrm{Re}\mathbf{E} \times \mathrm{Re}\mathbf{H} = \frac{1}{2}\,\mathrm{Re}[\mathbf{E} \times \mathbf{H} + \mathbf{E}^* \times \mathbf{H}], \tag{23}$$

where Re denotes the real part of the quantity, * denotes the complex conjugate, and we have used the complex number identity

$$\text{Re}\mathbf{A} \times \text{Re}\mathbf{B} = \frac{1}{2}\,\text{Re}[\mathbf{A} \times \mathbf{B} + \mathbf{A}^* \times \mathbf{B}]. \tag{24}$$

Using the values of **E** and **H**, employing the vector identity $\mathbf{A} \times (\mathbf{B} \times \mathbf{C}) = \mathbf{B}\,(\mathbf{A} \cdot \mathbf{C}) - \mathbf{C}\,(\mathbf{A} \cdot \mathbf{B})$ and using the fact that $\mathbf{k} \cdot \mathbf{E} = 0$, we have

$$\mathbf{S} = \frac{1}{2\mu_0}\,\text{Re}\,[\mathbf{k}A^2 e^{-2i(\omega t - \mathbf{k}\,.\,\mathbf{r})} + \mathbf{k}\,A^2 e^{-i(\mathbf{k}-\mathbf{k}^*).\mathbf{r}}]. \tag{25}$$

The time average of the first term, over a wave period, is zero. Expressing amplitude of the wave A exp $(-\mathbf{k}_i \,.\, \mathbf{r})$ as A′, the time average Poynting vector is

$$\mathbf{S}_{av} = \frac{A'^2}{2\mu_0}\mathbf{k}_r, \tag{26}$$

where $\mathbf{k}_r$ is the real part of **k**. S_{av} is called the *intensity of the wave.*

An electromagnetic wave can also be viewed as a stream of photons, each photon having energy $\hbar\omega$ and momentum $\hbar\mathbf{k}_r$. The number of photons crossing per unit area per second is $S_{av} / \hbar\omega$ and the momentum flow of the electromagnetic wave per unit area per second is

$$\mathbf{P} = \left(\frac{\mathbf{S}_{av}}{\hbar\omega}\right)\hbar\mathbf{k}_r = \frac{\eta_r}{c}\,\mathbf{S}_{av}. \tag{27}$$

The momentum carried by electromagnetic waves can be used to impart an impulse to a distant object or to drive an electrical current in a system. One of its major applications is in the current drive of a *tokamak,* a thermonuclear fusion device.

Now we discuss electromagnetic wave propagation in different media.

2.3 Wave Propagation in Free Space

In free space $\epsilon_r = 1$, $\sigma = 0$, $\epsilon_{eff} = 1$, hence the dispersion relation takes the form

$$k^2 = \frac{\omega^2}{c^2}, \tag{28}$$

and the wave fields can be written as

$$\mathbf{E} = \mathbf{A}e^{-i(\omega t - \mathbf{k} \,.\, \mathbf{r})}, \; \mathbf{B} = \left(\frac{\mathbf{k} \times \mathbf{A}}{\omega}\right) e^{-i(\omega - \mathbf{k} \,.\, \mathbf{r})}. \tag{29}$$

The phase velocity of the wave is $c = 3 \times 10^8$ m/s. The electric and magnetic fields are in the same phase and E/ H ratio, known as the *free space impedance* Z_0, turns out to be

$$Z_0 = \frac{E}{H} = \sqrt{\frac{\mu_0}{\epsilon_0}} = 377 \text{ ohms}. \tag{30}$$

The intensity of the homogeneous wave is S_{av}, given by

$$\mathbf{S}_{av} = \left(\frac{A^2}{2Z_0}\right) \hat{n}, \tag{31}$$

where $\hat{n}$ is the unit vector along **k**.

2.3.1 Inhomogeneous Wave

One may note from eq. (28), that though ω is real, **k** could still be complex. Let $\mathbf{k} = \mathbf{k}_r + i\mathbf{k}_i$. Substituting this in eq. (28) and equating the imaginary parts on both sides we get

$$\mathbf{k}_r . \mathbf{k}_i = 0 \tag{32}$$

while the electric field of the wave is given by eq. (18). Equation (32) reveals that $\mathbf{k}_r \perp \mathbf{k}_i$, i.e., the planes of constant phase and constant amplitude in free space (or any non-absorbing medium in general) are mutually perpendicular to each other (Fig. 2.4). Such waves are called *inhomogeneous* waves and occur in situations of total internal reflection, surface wave propagation and optical fibres.

PROBLEM 2.1. An inhomogeneous electromagnetic wave has $E = \hat{y}\, A \exp(-\sqrt{3}\omega\, x/c) \exp[-i\,(\omega\, t - 2\omega\, z/c)]$. Obtain **B** and $\mathbf{S}_{av}$.

Solution: Comparing the field with $\mathbf{E} = \mathbf{A} \exp[-i\,(\omega t - \mathbf{k.r})]$, we get $\mathbf{k} = (2\,\omega/\,c)\,\hat{z} + i\,(\sqrt{3}\omega/c)\,\hat{x}$. From the third Maxwell's equation,

$$\mathbf{B} = \mathbf{k} \times \mathbf{E}/\omega = (A/c)(-2\hat{x} + i\sqrt{3}\hat{z})\; e^{-\sqrt{3}\omega x/c} . e^{-i(\omega t - 2\omega z/c)}$$

$$\mathbf{S}_{av} = \text{Re}[\mathbf{E}^* \times \mathbf{H}/2] = \hat{z}(A^2/\mu_0 c) \,.\, \exp(-2\sqrt{3}\omega x/c).$$

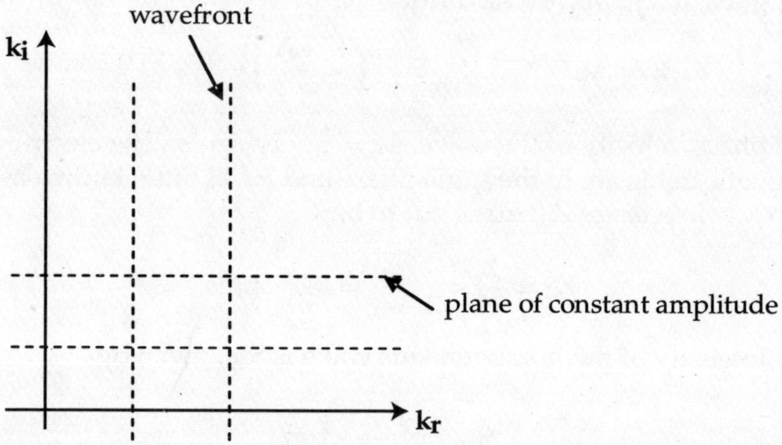

Figure 2.4: Planes of constant amplitude and constant phase of an inhomogeneous wave.

2.4 Wave Propagation in Dielectrics

In a dielectric $\sigma = 0, \mu_r = 1$, hence

$$k^2 = \frac{\omega^2}{c^2} \in_r, \eta = \in_r^{1/2} . \tag{33}$$

Since $\in_r > 1$, the wave travels with a phase velocity less than the velocity of light in vacuum. The *impedance of the medium* is

$$Z = \frac{E}{H} = Z_0/\eta. \tag{34}$$

In lossy dielectrics, $\in_r$ has finite imaginary part ($\in_r = \in' + i \in''$). Hence k is complex ($k = k_r + ik_i$) and wave suffers damping along the direction of propagation.

For $\in'' << \in'$,

$$k_r \approx \frac{\omega}{c}\sqrt{\in'}, \quad k_i \approx \frac{\omega}{2c}\frac{\in''}{\sqrt{\in'}} \tag{35}$$

and the time average Poynting vector is

$$\mathbf{S}_{av} = \left(\frac{c\mathbf{k}_r}{\omega}\right)\left(\frac{A^2}{2Z_0}\right). \tag{36}$$

The imaginary part in ϵ_r arises due to the resistance caused to the motion of electron cloud or reorientation of atomic dipoles by their neighbours. We have already seen in unit 1 that the inertia term, in the motion of electron cloud in an atom, gives rise to frequency dependence of ϵ_r. This makes the medium dispersive, where different frequency components of an electromagnetic signal propagate with different velocities and consequently the signal at the output end gets distorted.

The dielectrics could be gas, liquid or solid. In a gas, the dielectric susceptibility χ varies almost linearly with the number density of molecules, nevertheless $\chi << 1$. In the case of a liquid, it could be larger. It is also frequency sensitive at frequencies > 10 GHz. For instance, water has a dielectric constant of 80 at very low frequencies that reduces to 1.8 at optical frequencies.

The solid dielectrics, especially the crystalline ones, are characterised by large **band gap** ε_g between the valance and conduction bands ($\varepsilon_g \geq 3$ ev). At ultraviolet frequencies, when photon energy $\hbar\omega > \varepsilon_g$, the response of dielectrics to electromagnetic waves is characterised by the production of electron-hole pairs and this leads to strong absorption of the wave. With free carrier generation, dielectric becomes a semiconductor. This phenomenon is employed in making fast switching devices. However, we shall limit ourselves to propagation of waves with $\hbar\omega < \varepsilon_g$.

One of the most exciting applications of dielectrics is in the form of optical fibres, that can guide optical waves to long distances with very little attenuation. Silica (glass) is used for this purpose.

PROBLEM 2.2. In the damped harmonic oscillator model of atomic polarisation (Fig. 2.5), the displacement x of the centre of electron cloud, with respect to nucleus, is governed by the equation

$$\frac{d^2x}{dt^2} + \nu\frac{dx}{dt} + \omega_0^2 x = -\frac{eE}{m}.$$

Solve this equation for the harmonic field $\mathbf{E} = \hat{x}\, A \exp(-i\omega t)$ and deduce an expression for atomic polarisability.

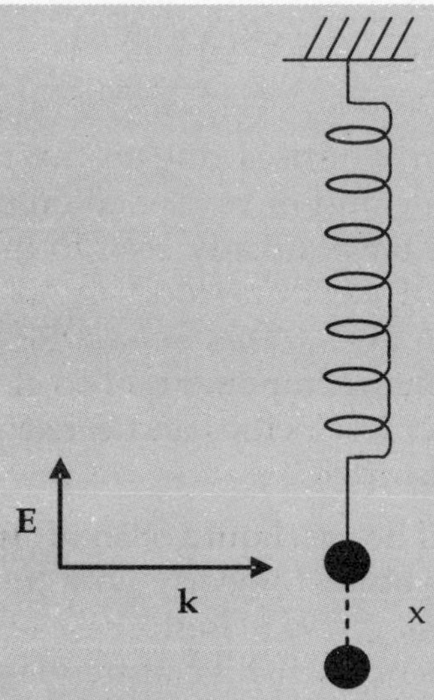

Figure 2.5: Motion of an electron attached to a spring of finite damping coefficient in the presence of an electromagnetic wave.

Solution: Taking $x = a \exp(-i\omega t)$ as the solution, we replace d/dt in the above equation by $-i\omega$ and obtain

$$x = eE/[m(\omega^2 - \omega_0^2 + i\nu\omega)].$$

Hence, the dipole moment $p = -ex$ of the atom is $p = \alpha E$ with

$$\alpha \quad = e^2/[m(\omega_0^2 - \omega^2 - i\nu\omega)]$$

as the atomic polarisability.

PROBLEM 2.3. Show that a relaxing medium is always dispersive.

Solution : In a relaxing medium, polarisation **P** at time t depends on **E** at t, as well as its previous history,

$$\mathbf{P}(t) = \int_0^\infty \beta(t_1)\mathbf{E}(t - t_1)dt_1.$$

For a harmonic field, $\mathbf{E}(t) = \mathbf{A} \exp(-i\omega t)$. This gives $\mathbf{P} = \epsilon_0 \chi \mathbf{E}$, with

$$\chi = \frac{1}{\epsilon_0} \int_0^\infty \beta(t_1) e^{i\omega t_1} dt_1.$$

Unless $\beta(t_1)$ is a delta function, χ is always a function of ω, i.e. the medium is dispersive. Conversely, a dispersive medium is always a relaxing medium.

2.5 Wave Propagation in Semiconductors and Conductors

Semiconductors are materials with a small energy gap between the valence and conduction bands, $\varepsilon_g \leq 2$ eV.

	Energy gap (eV)
Ge	0.66
Si	1.12
GaAs	1.42
GaP	2.26
InSb	0.17

In a pure semiconductor, called *intrinsic semiconductor*, the free carrier concentration is low and so is its electrical conductivity. However, carrier concentration can be greatly increased by doping the semiconductor with impurities, during the phase of its growth. The impurities can have energy levels in the forbidden band, either just below the conduction band or just above the valence band. The impurities whose energy levels are just below the conduction band are called *donors* as these energy levels are filled with electrons and even at low temperatures these electrons jump into the conduction band giving a high concentration of free electrons (Fig. 2.6). Such semiconductors, with excess free electrons, are called *n-type semiconductors*. The other type of impurities, that give vacant energy levels close to the valence band, are called *acceptors* as valence band electrons quickly jump into them creating holes in the valence band. The holes, or sites of electron deficiency, behave as free carriers while responding to electric and magnetic fields. Their charge is +e but mass m is significantly different from electron mass. A semiconductor, doped with acceptor type impurities, has excess of holes and is called *p-type semiconductor*. For group IV semiconductors (like Ge, Si), Group III dopants (e.g. B) are acceptors and group V dopants (e.g. P, As) are donors. For III-V compounds like GaAs, InSb, group II dopants (Zn, Cd) are acceptors and group VI dopants (Te, S, Se) are donors. In doped semiconductors, one can achieve free carrier concentration upto around 10^{19} cm^{-3}, which is five orders of magnitude higher than the carrier concentration in an intrinsic semiconductor at room temperature.

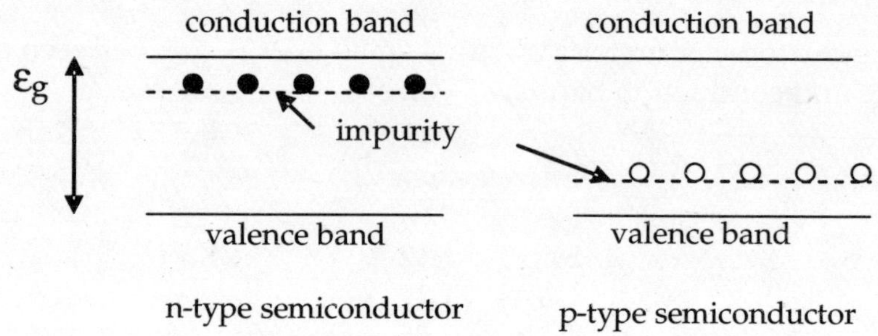

Figure 2.6: Impurity levels in n-type and p-type semiconductors.

When a semiconductor is subjected to an electromagnetic wave, the free carriers acquire a drift velocity. As they move, they suffer collisions with acoustical phonons, optical phonons, ionised impurities and crystal defects due to which they lose their drift momentum. The electron-ionised impurity collision frequency increases linearly with doping concentration and decreases as $T_e^{-3/2}$ with carrier temperature T_e. Thus, this class of collisions dominates at low temperatures. On the other hand, electron-phonon collision frequency increases with temperature and dominates at room temperature and above. Typically, the collision frequency ν varies in the range 10^{11} - 10^{13} s^{-1} depending on doping level, temperature and the nature of semiconductor.

One important feature of semiconductors is that the effective mass

$$m^{-1} = \hbar^{-2}\partial^2\varepsilon/\partial q^2$$

m, of free carriers, defined as (in the simplest of semiconductors having isotropic energy surfaces), where ε and q are energy and wave number of the carrier respectively. The effective mass of an electron could be very different from the electron mass in free space. For example, effective mass of electron in Ge and InSb (at low temperature) is respectively 0.3 and 0.17 times the electron mass in free space. In materials like InSb, where band gap is small, electron effective mass is a function of energy as ε-q relation is non-parabolic. In GaAs, there are many valleys, each valley having a specific ε-q relation with a specific energy minimum. Electron effective mass in the valley with lowest energy minimum is small ($m \simeq 0.06$ times the electron mass in free space) whereas in higher energy valleys, it is large ($m \sim 1.1$ times the free space mass). At low temperatures, electrons in n-type GaAs stay in the minimum energy valley. When one applies a high dc electric field, the electrons get heated and jump over to higher energy valleys where mass is high and dc electrical conductivity $\sigma = n_0 e^2 / m\nu$ falls off rapidly. The current density J, then behaves with dc electric field E as shown in fig. 2.7. As E increases J increases, then attains a maximum (at $E = E_1$) and falls to a minimum (at $E= E_2$) beyond which it increases again. The J-E curve between maximum and minimum has negative slope ($\partial J/\partial E < 0$), i.e., the semiconductor has *negative differential conductivity* (ndc). The semiconductor biased by a dc electric field, in this range, gives rise to space charge and current oscillations. Such a semiconductor, called **Gunn diode**, when placed in a resonator cavity gives rise to generation of microwaves and is widely used in radar and communication. (See problem 2.4).

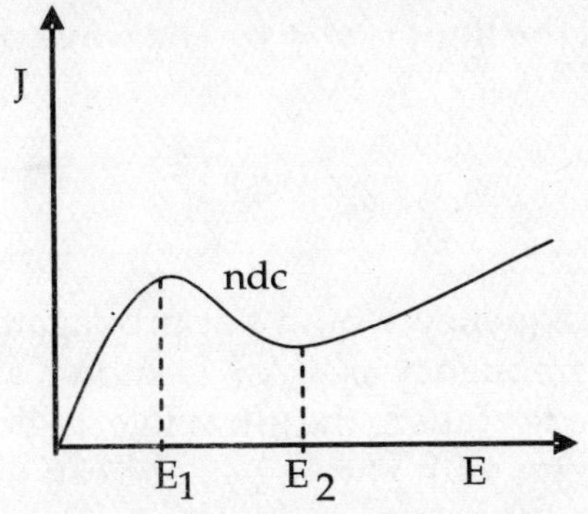

Figure 2.7: J-E characteristic of n-GaAs. At high dc field it shows negative differential conductivity.

Metals are characterised by overlapping energy bands and have high free electron density, $n_e \geq 10^{22}\text{cm}^{-3}$, even at low temperatures. Electrons suffer collisions primarily with phonons and collision frequency ν increases with temperature.

The effective relative permittivity of conductors and n-type semiconductors is given by eq. (12) and the dispersion relation for electromagnetic waves is

$$k^2 = \frac{\omega^2}{c^2}\left(\epsilon_L - \frac{\omega_p^2}{\omega^2\,(1 + i\nu/\omega)}\right), \tag{37}$$

where ϵ_L is the relative permittivity due to the lattice, $\omega_p = (n_0 e^2/m\ \epsilon_0)^{1/2}$, n_0 is the free electron density and m is the electron effective mass. The same dispersion relation holds for a p-type semiconductor with n_0, m and ν replaced by hole parameters. ω_p may be written as

$$\omega_p \approx 50(n_0/m_R)^{1/2}\ \text{rad/s}$$

where m_R is the ratio of electron effective mass to free space electron mass and n_0 is in m^{-3}. ϵ_L is typically in the range 4 to 20 (except for semimetals like bismuth where $\epsilon_L = 100$) and $\nu \sim 10^{12}\ s^{-1}$. At high frequencies, $\omega >> \nu$,

$$k^2 \approx \frac{\omega^2}{c^2}\left(\epsilon_L - \frac{\omega_p^2}{\omega^2}\right). \tag{38}$$

Fig. 2.8 gives ω as a function of k for electromagnetic wave in a semiconductor, in the limit $\nu/w = 0$. The wave propagates when k^2 is positive, i.e.

$$\omega > \frac{\omega_p}{\sqrt{\epsilon_L}}. \tag{39}$$

Waves of lower frequency would not propagate as k is imaginary and $\mathbf{S}_{av} = 0$. The frequency $\omega_p/\sqrt{\epsilon_L}$ is known as the **plasma edge**. In metals, plasma edge falls in the ultraviolet region. By scanning the reflectivity of a metal with varying ω, one can measure the plasma edge and hence deduce electron density or electron effective mass.

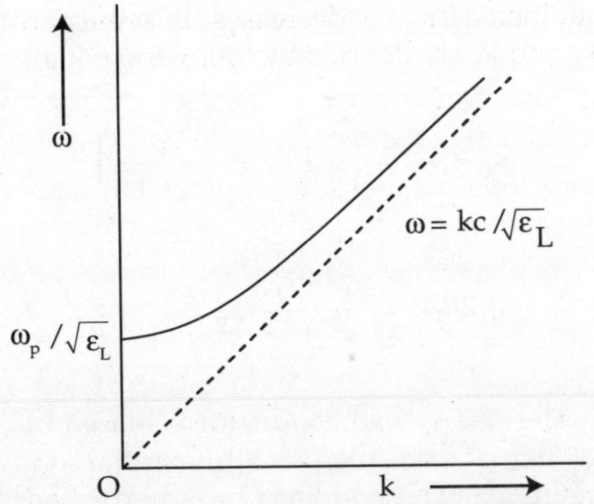

Figure 2.8: Dispersion relation for an electromagnetic wave in a semiconductor for $\upsilon = 0$.

The interaction of waves of $\omega < \omega_p/\sqrt{\epsilon_L}$ with conductors is also important. The electric field of these waves penetrates into the material upto a depth $\sim k_i^{-1}$, called *skin depth* δ,

$$\delta = \frac{c/\omega_p}{\sqrt{1 - \omega^2\, \epsilon_L\, /\omega_p^2}}. \tag{40}$$

At optical and infrared frequencies, $\delta \leq 0.1\ \mu$m in most metals. When the incident wave fluence, i.e. the product of intensity and pulse duration $S_{av}\tau \geq 0.3 J/cm^2$ (for a pulse duration $\tau \sim 1$ ps), it causes rapid heating and ablation of the skin layer, leaving the rest of the material intact. This technique, called *laser ablation*, is widely being used for pulsed laser deposition (PLD) and pulsed laser annealing (PLA) of thin films and deposition of super lattices (consisting of multiple thin layers of different materials).

At lower frequencies, comparable to or less than ν, the real and imaginary parts of k in a metal are comparable,

$$k_r \sim k_i \sim \frac{\omega_p}{c}\left(\frac{\omega}{2\nu}\right)^{\frac{1}{2}} \tag{41}$$

and skin depth increases as ω decreases. In semiconductors, where ω_p is lower, k_r and k_i are defined by following equations:

$$k_r^2 - k_i^2 = \frac{\omega^2}{c^2}\left(\epsilon_L - \frac{\omega_p^2}{\omega^2 + \nu^2}\right)$$

$$2k_r k_i = \frac{\omega^2}{c^2}\frac{\nu\omega}{\omega^2 + \nu^2}. \tag{42}$$

PROBLEM 2.4. An n-GaAs semiconductor is biased by a dc electric field $E_0\hat{x}$ (Fig.2.9) in the ndc region of differential conductivity σ_{ndc} (which is negative). The equilibrium is perturbed by a charge density $\rho_1 = \rho_A \cos(kx)$. Obtain its temporal evolution.

Solution : The charge density perturbation produces an electric field perturbation $E_1\hat{x}$. The net field in the semiconductor is $(E_0 + E_1)\hat{x}$. The current density $\hat{x}J$, which is a function of $(E_0 + E_1)$, on Taylor expansion around E_0 can be written as $\mathbf{J} = [J_0(E_0) + J_1]\,\hat{x}$, with

$$J_1 = \left.\frac{\partial J}{\partial E}\right|_{E_0} E_1 = \sigma_{ndc} E_1.$$

E_1, ρ_1 and J_1 satisfy the Poisson's equation and the equation of continuity

$$\epsilon_0 \epsilon_L \nabla \cdot \mathbf{E}_1 = \rho_1,$$

$$\frac{\partial \rho_1}{\partial t} + \nabla \cdot \mathbf{J}_1 = 0.$$

Using J_1 in the continuity equation and employing the Poisson's equation, we get

$$\frac{\partial \rho_1}{\partial t} + \frac{\sigma_{ndc}}{\epsilon_0 \epsilon_L}\rho_1 = 0.$$

Hence, ρ_1 goes as $\rho_A \exp\left[\left(|\sigma_{ndc}|/\ \epsilon_0\epsilon_L\right) t\right] \cos(kx)$ because σ_{ndc} is negative.

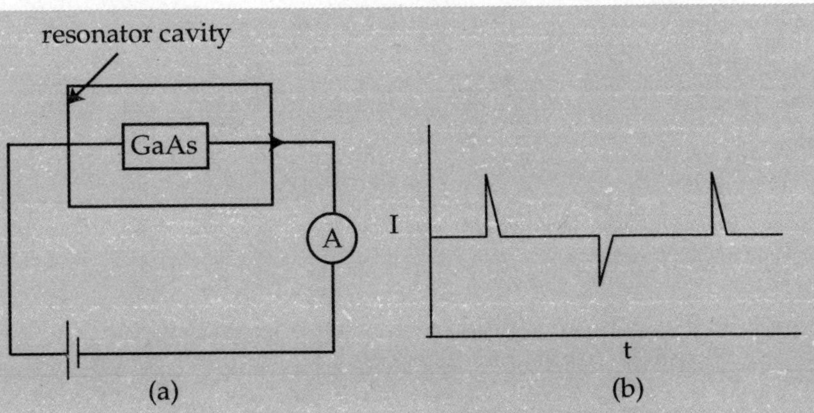

Figure 2.9: (a) A dc biased n-type GaAs semiconductor mounted inside a resonator cavity and (b) current flow in the circuit.

NOTE: In a sample of finite length, this gives rise to formation of domains of negative and positive charge. As they build up, the field in some regions increases beyond E_2 and elsewhere falls down below E_1. These regions, called domains, have different charge densities. As they move to the end of the sample, extra charge flows through the circuit, giving rise to spikes in current as shown in the fig. 2.9. When the diode is mounted inside a resonator cavity, the time dependent current gives rise to the growth of a suitable mode of microwave frequency.

2.5.1 Helicon Waves

The application of a static magnetic field to a semiconductor or metal changes the character of electromagnetic wave propagation dramatically. It permits the propagation of low frequency electromagnetic waves, with small attenuation and provides an important tool for material characterisation.

Consider a conductor immersed in a magnetic field $\mathbf{B}_s || \hat{z}$. An electromagnetic wave propagates through it, along $\hat{z}$ (Fig. 2.10) with electric field

$$\mathbf{E} = \mathbf{A}e^{-i(\omega t - kz)}. \tag{43}$$

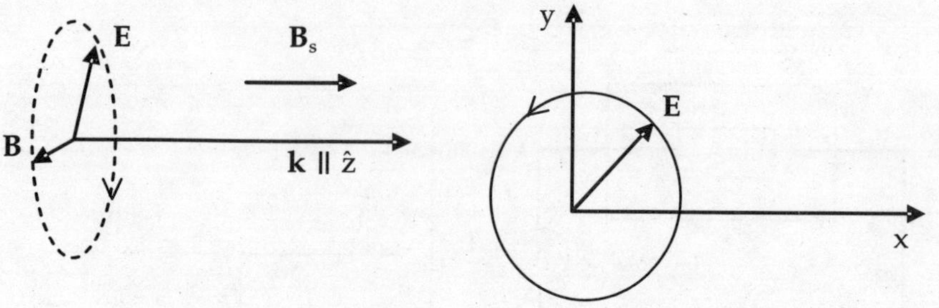

Figure 2.10: Helicon propagation in an n-type semiconductor. The tip of the electric field **E** rotates on a circle in the right-handed sense.

The magnetic field of the wave, from eq. (22) of unit 1, using $\partial/\partial t = -i\omega$ and $\nabla = i\mathbf{k}$ (where $\mathbf{k} = k\,\hat{z}$) is $\mathbf{B} = \mathbf{k} \times \mathbf{E}/\omega$. The equation of motion, governing the drift velocity of electron fluid, is

$$m\left[\left(\frac{\partial \mathbf{v}}{\partial t}\right) + \mathbf{v}.\nabla\mathbf{v}\right] = -e\mathbf{E} - e\mathbf{v} \times (\mathbf{B}_s + \mathbf{B}) - \nu m\mathbf{v} \qquad (44)$$

where the parenthesis on the left is basically $d\mathbf{v}/dt$ and the last term on the right denotes the rate of momentum loss via collisions. In equilibrium (i.e., in the absence of the wave), $\mathbf{v} = 0$. When wave is present, we treat **E** and **B** of the wave as small or perturbed quantities as well as the resultant electron drift **v** as a perturbed quantity. Then, we can linearise eq. (44) by neglecting the product of perturbed quantities, e.g., **v** with **v** and **v** with **B**, to obtain

$$\frac{\partial \mathbf{v}}{\partial t} = \frac{-e\mathbf{E}}{m} - \mathbf{v} \times \hat{z}\omega_c - \nu\mathbf{v} \qquad (45)$$

where $\omega_c = eB_s/m$ is the *electron cyclotron frequency*. In the steady state, **v** would have the same (**r**, t) dependence as **E**. Hence, we can replace $\partial/\partial t$ by $-i\omega$. Writing x, y, z components of eq. (45) and solving these, we obtain **v** and the current density $\mathbf{J} = -n_0 e\,\mathbf{v} = \underline{\underline{\sigma}}.\mathbf{E}$ as

$$v_x = -\frac{e\left[(\nu - i\omega)E_x - \omega_c E_y\right]}{m\left[(\nu - i\omega)^2 + \omega_c^2\right]}, v_y = -\frac{e\left[\omega_c E_x + (\nu - i\omega)E_y\right]}{m\left[(\nu - i\omega)^2 + \omega_c^2\right]},$$

$$v_z = -\frac{eE_z}{m(\nu - i\omega)}, \qquad (46)$$

$$J_x = \sigma_{xx}E_x + \sigma_{xy}E_y, J_y = \sigma_{yx}E_x + \sigma_{yy}E_y, J_z = \sigma_{zz}E_z, \qquad (47)$$

$$\sigma_{xx} = \sigma_{yy} = \frac{n_0 e^2(\nu - i\omega)}{m\left[(\nu - i\omega)^2 + \omega_c^2\right]},$$

$$\sigma_{xy} = \sigma_{yx} = -\frac{n_0 e^2 \omega_c}{m\left[(\nu - i\omega)^2 + \omega_c^2\right]},$$

$$\sigma_{zz} = \frac{n_0 e^2}{m(\nu - i\omega)}. \qquad (48)$$

The third and fourth Maxwell's equations, on replacing $\partial/\partial t$ by $-i\omega$ and ∇ by $i\mathbf{k}$, can be written as

$$\mathbf{k} \times \mathbf{E} = \omega\mu_0 \mathbf{H},$$
$$i\mathbf{k} \times \mathbf{H} = -i\omega \in_0 \in_L \mathbf{E} + \mathbf{J}. \qquad (49)$$

Taking $\mathbf{k}\,\times$ of the first equation above and using the second, we have

$$-k^2\mathbf{E} + \hat{z}k^2E_z + \frac{\omega^2}{c^2} \in_L \mathbf{E} + \frac{i\omega}{c^2 \in_0}\mathbf{J} = 0. \qquad (50)$$

Its z component gives $(\in_L + i\sigma_{zz}/\,\omega \in_0)\,E_z = 0$, giving $E_z = 0$. For x and y components, eq. (50) gives

$$\left[-k^2 + \frac{\omega^2}{c^2}\left(\in_L + i\frac{\sigma_{xx}}{\omega \in_0}\right)\right]E_x + i\frac{\omega^2}{c^2}\frac{\sigma_{xy}}{\omega \in_0}E_y = 0, \qquad (51)$$

$$\left[-k^2 + \frac{\omega^2}{c^2}\left(\in_L + i\frac{\sigma_{xx}}{\omega \in_0}\right)\right]E_y - i\frac{\omega^2}{c^2}\frac{\sigma_{xy}}{\omega \in_0}E_x = 0. \qquad (52)$$

A non-trivial solution of these equations demands that the determinant of coefficients of E_x and E_y must vanish,

$$\left(k^2 - \frac{\omega^2}{c^2} \in_+\right)\left(k^2 - \frac{\omega^2}{c^2} \in_-\right) = 0, \qquad (53)$$

$$\text{where } \epsilon_\pm = \epsilon_L + \frac{i(\sigma_{xx} \pm i\sigma_{xy})}{\omega\, \epsilon_0} = \epsilon_L - \frac{\omega_p^2}{\omega\,(\omega \mp \omega_c + i\nu)}. \tag{54}$$

Equation (53) gives two roots

$$k^2 = \frac{\omega^2}{c^2} \epsilon_+, \tag{55}$$

$$k^2 = \frac{\omega^2}{c^2} \epsilon_- \,. \tag{56}$$

For the first root, using eq. (51), we obtain $E_x = -iE_y$, whereas for the second root $E_x = iE_y$. Thus depending on the phase relationship between E_x and E_y, either the wave follows dispersion relation (55) or (56). The first root has striking features, hence we will discuss it in detail now.

Let us first understand the meaning of $E_x = -iE_y$. If we write E_x and E_y explicitly from eq. (43) and put $E_x = -iE_y$, we get $A_x = -iA_y$. We may choose (without loss of generality) $A_x = A_0$ as real. Then

$$A_y = iA_0.$$

Now, we write the actual E_x and E_y from eq. (43) by taking the real part of the complex right-hand side.

$$E_x = A_0 e^{-ik_i z} \cos(\omega t - k_r z)$$
$$E_y = A_0 e^{-ik_i z} \sin(\omega t - k_r z)\,, \tag{57}$$

where we have taken $k = k_r + ik_i$. One may note that $E_x^2 + E_y^2$ is independent of time, i.e., the magnitude of electric field of the wave does not change with time, only its direction changes. If we denote E_x on x axis and E_y on y axis of a graph paper, at a fixed value of z, then as time changes the tip of the resultant electric field describes a circle in the anticlockwise sense. This is called a **right circularly polarised mode**. If we take a right-handed screw and rotate it in the way electric field rotates, it advances in the direction of wave propagation. The propagation vector of this mode is denoted as k_R and can be explicitly written as

$$k_R^2 = \frac{\omega^2}{c^2}\left(\epsilon_L - \frac{\omega_p^2}{\omega\,(\omega - \omega_c + i\nu)}\right). \tag{58}$$

If we follow a similar procedure for the other root, we find that $E_x = iE_y$, representing a **left circularly polarised wave** with propagation constant k_L,

$$k_L^2 = \frac{\omega^2}{c^2}\left(\epsilon_L - \frac{\omega_p^2}{\omega\left(\omega + \omega_c + i\nu\right)}\right). \tag{59}$$

Thus there are two distinct modes of wave propagation in a magnetised semiconductor. For wave propagation along the static magnetic field, these modes are circularly polarised in the opposite sense. The phase velocity of the wave depends on its polarisation. A right circularly polarised wave propagates with $v_{ph} = \omega/k_R$ whereas the left circularly polarised wave has $v_{ph} = \omega/k_L$. Wave of any arbitrary polarisation can be decomposed into right and left circular polarisations and these components will travel with their respective phase velocities. For an initially linearly polarised wave, this leads to a change of orientation of the electric field with z, called *Faraday rotation*. However, here we focus primarily on the right circularly polarised mode.

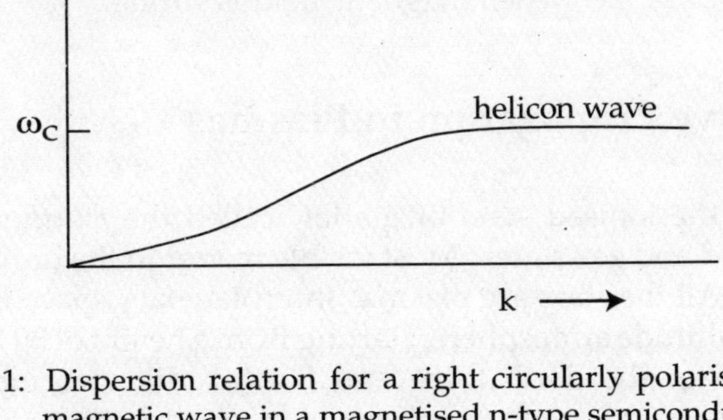

Figure 2.11: Dispersion relation for a right circularly polarised electromagnetic wave in a magnetised n-type semiconductor.

If one ignores ν, then k_R^2 is positive for all $\omega < \omega_c$ and all waves of frequencies below the electron cyclotron frequency propagate through the conductor, with velocity c/η_R, where

$$\eta_R = \left(\epsilon_L + \frac{\omega_p^2}{\omega(\omega_c - \omega)} \right)^{\frac{1}{2}}. \tag{60}$$

In fig. 2.11, we have plotted frequency versus wave number for a right circularly polarised wave. The low frequency branch, $\omega < \omega_c$, is called the **helicon wave**. At low frequencies $\omega << \omega_c << \omega_p$, η_R is huge and these waves travel with small phase velocities. Around $\omega \sim \omega_c$ (cyclotron resonance) η_R acquires large values. Beyond this frequency ($\omega > \omega_c$), there is a frequency interval in which η_R is imaginary and wave does not propagate. Propagation begins when $\omega > \omega_R$, where ω_R is given by

$$\omega_R(\omega_R - \omega_c) = \frac{\omega_p^2}{\epsilon_L}.$$

When ν is finite k_R becomes complex ($k_R = k_{Rr} + ik_{Ri}$). However if $(\omega - \omega_c)^2 >> \nu^2$, we have

$$k_{Rr} = \frac{\omega}{c} \left(\epsilon_L + \frac{\omega_p^2}{\omega(\omega_c - \omega)} \right)^{\frac{1}{2}},$$

$$k_{Ri} = \frac{\omega_p^2 \nu \omega}{2(\omega_c - \omega)^2 c^2 k_{Rr}}. \tag{61}$$

Thus, the helicon waves ($\omega << \omega_c$) travel with small attenuation when $\omega_c^2 >> \nu^2$, i.e., when magnetic field is strong.

2.6 Wave Propagation in Plasmas

Plasma is the ionised state of matter, called the *fourth state* after solid, liquid and gas states. Most visible matter in the universe is in this state. All the stars are plasma. Interplanetary space is plasma. Our high-altitude atmosphere, starting from a height of 90 km above the surface of the Earth to several Earth radii, viz., ionosphere, plasmasphere and magnetosphere, is in plasma state. Flame is a plasma. In fact flames, seeded with low work function material to increase the electron density, are used in magnetohydrodynamic (MHD) power generation yielding a higher efficiency than thermal power plants. In the laboratory, plasmas are produced by dc discharge, radio frequency discharge, photoionisation, and by lasers, microwaves, electron beams and ion beams. Their temperatures vary from 10^3 K to 10^8 K. Above a temperature of 10^5 K plasma is nearly fully ionised and electron-ion collision frequency scales with electron temperature T_e as $\nu \sim T_e^{-3/2}$. The dc conductivity of the plasma could be higher than that of metals.

Plasmas have long range Coulomb interactions between charged particles. If we consider a localised sphere of plasma, the electrons (due to their higher thermal velocity than ions) would move out of the plasma. However, as they come out, they leave behind a net positive charge on the sphere that pulls them back. The electron-ion separation can occur only upto a distance of the order of **Debye length** $\lambda_D = v_{th}/\omega_p$, where v_{th} is the electron thermal speed and ω_p is the electron plasma frequency. Plasmas of dimensions bigger than λ_D can be treated as quasi-neutral with

$$n_e = n_i Z_i, \tag{62}$$

where n_e is the electron density, n_i is the density of ions of charge $Z_i e$. In fact, ionised gases of dimension greater than λ_D only can be characterised as a state of matter and called plasma, when collective effects dominate over the individual particle effects. One simple demonstration of collective effects is *plasma oscillations* (Stix[1]). If one disturbs the electron density of the plasma from its equilibrium value, all the electrons undergo oscillations of frequency ω_p.

The propagation of electromagnetic waves in an unmagnetised plasma is similar to that in a conductor with $\epsilon_r = 1$. We may write the dispersion relation as

$$k^2 = \frac{\omega^2}{c^2}\left(1 - \frac{\omega_p^2}{\omega^2\,(1 + i\nu/\omega)}\right). \tag{63}$$

At high frequencies, $\omega >> \nu$,

$$k \approx \frac{\omega}{c}\sqrt{1 - \frac{\omega_p^2}{\omega^2}} \quad \text{or} \quad \omega^2 = \omega_p^2 + k^2c^2. \tag{64}$$

The wave can propagate only when $\omega > \omega_p$ or electron density $n_0 < n_c$, where

$$n_c = \frac{m\omega^2 \epsilon_0}{e^2} \tag{65}$$

is the critical density at which plasma frequency equals the wave frequency. Critical density scales as ω^2 i.e., higher frequency waves can propagate in plasmas of higher electron density. For $\omega > \omega_p$, the refractive index of the plasma is

$$\eta = \sqrt{1 - \frac{\omega_p^2}{\omega^2}} \tag{66}$$

which is less than 1, hence the phase velocity of an electromagnetic wave in a plasma is greater than the velocity of light in vacuum. The dispersion relation for electromagnetic waves in a plasma is the same as that in a semiconductor, (see fig. 2.8), with $\epsilon_L = 1$. The time average Poynting vector is

$$\mathbf{S}_{av} = \left(\frac{A^2}{2Z_0}\right) \quad \eta \quad \hat{n}. \tag{67}$$

When collisions are included, k becomes complex, $k = k_r + ik_i$. For $\omega^2 >> \nu^2$, one has

$$k_r \approx \frac{\omega}{c}\left(1 - \frac{\omega_p^2}{\omega^2}\right)^{\frac{1}{2}},$$

$$k_i \approx \frac{\omega_p^2}{c^2}\frac{\nu}{2\omega k_r}. \tag{68}$$

The attenuation constant decreases with electron temperature as $T_e^{-3/2}$ and increases with electron density faster than n_0^2 (as ν scales linearly with electron density n_0). In an underdense plasma, collisional absorption is the main source of energy dissipation. However, in an inhomogeneous plasma where critical layer exists (at which $\omega_p = \omega$), the laser may get converted into a plasma wave giving rise to much higher level of absorption (Liu and Tripathi)[2].

2.6.1 Whistler and Alfven Waves

Plasmas often have a static magnetic field. It helps their confinement and introduces a wide variety of modes of wave propagation. One such mode is whistler. *In semiconductors and conductors, this mode is called helicon.* When a whistler propagates along the direction of static magnetic field $B_s\,\hat{z}$, it is right circularly polarised and its dispersion relation (from eq. (58) on taking $\epsilon_L = 1$) is

$$k^2 = \frac{\omega^2}{c^2}\left(1 - \frac{\omega_p^2}{\omega\,(\omega - \omega_c + i\nu)}\right), \tag{69}$$

where ω_p is the electron plasma frequency, $\omega_c = eB/m$ is the electron cyclotron frequency and ν is the electron collision frequency. Usually, $\nu << \omega_c$. k is sensitive to cyclotron resonance. For ω slightly above ω_c, k^2 has large negative value and wave does not propagate unless

$$\omega(\omega - \omega_c) > \omega_p^2$$

or $\omega > \omega_R$, where

$$\omega_R = \frac{1}{2}\left[\omega_c + \sqrt{\omega_c^2 + 4\omega_p^2}\right].$$

For $\omega < \omega_c$, wave propagates at all plasma densities. Actually, it is at these low frequencies that the wave is called a whistler. Taking $\nu << (\omega_c - \omega)$, the real and imaginary parts of k can be written as

$$k_r = \frac{\omega}{c}\left(1 + \frac{\omega_p^2}{\omega\,(\omega_c - \omega)}\right)^{\frac{1}{2}},$$

$$k_i = \frac{\omega_p^2 \nu \omega}{2c^2\,(\omega_c - \omega)^2\,k_r}. \tag{70}$$

For $\omega << \omega_c < \omega_p^2$, k_r goes as $\sqrt{\omega}$. It is a dispersive wave with higher frequencies propagating with higher phase velocities. Whistlers are a common occurrence in the ionosphere with frequencies $\omega/2\pi \leq$ 30 KHz. They are produced as pulses in lightening discharges. However, as they travel towards the ground, the higher frequency components are received earlier and lower frequency components arrive later giving a whistling tone. The whistlers propagate with low attenuation. They can travel from the north pole to south pole, along the field lines of Earth's magnetic field. Since their wavelength in plasma is much smaller than the wavelength they would have in free space, they suffer less diffraction. Whistlers also propagate through the magnetosphere. In laboratory, they are used for plasma heating.

At low frequencies, one must consider the ion motion also. The reason is that the electron drift velocity **v**, due to the electromagnetic wave, has a denominator m $(\omega - \omega_c + i\nu) \sim - m\omega_c$. Had we considered ion drift due to the wave, we would get a similar denominator m_i $(\omega + \omega_{ci} + i\nu_i)$, where $\omega_{ci} = eB_s/m_i$ is the ion cyclotron frequency, m_i is the ion mass, ν_i is the ion collision frequency and e is the charge of an ion (assumed to be singly ionised). Since $m\omega_c = m_i\omega_{ci}$, the two denominators (for $\omega \leq \omega_{ci}$) become comparable and ion response becomes as important as the electron response. This leads to the modified dispersion relation,

$$k^2 = \frac{\omega^2}{c^2}\left(1 - \frac{\omega_p^2}{\omega(\omega - \omega_c + i\nu)} - \frac{\omega_{pi}^2}{\omega(\omega + \omega_{ci} + \nu_i)}\right)$$

$$\approx \frac{\omega^2}{c^2}\frac{\omega_{pi}^2}{\omega_{ci}(\omega + \omega_{ci})} \tag{71}$$

when $\omega \leq \omega_{ci}$ and collisions are ignored. Here, $\omega_{pi} = \left(n_0 e^2/\in_0 m_i\right)^{1/2}$ is the ion plasma frequency. At $\omega << \omega_{ci}$, the wave is called **Alfven wave** and its dispersion relation takes the form

$$k = \frac{\omega}{v_A}, \tag{72}$$

where $v_A = c\omega_{ci}/\omega_{pi}$ is the *Alfven velocity*. The left circularly polarised ($E_x = iE_y$) wave also gives the same dispersion relation at $\omega << \omega_{ci}$. Thus, Alfven waves propagate with Alfven velocity, irrespective of their polarisation. When they propagate at an angle to magnetic field, their propagation characteristics change. In fig. 2.12, we have plotted the dispersion relation for right circularly polarised mode in a plasma.

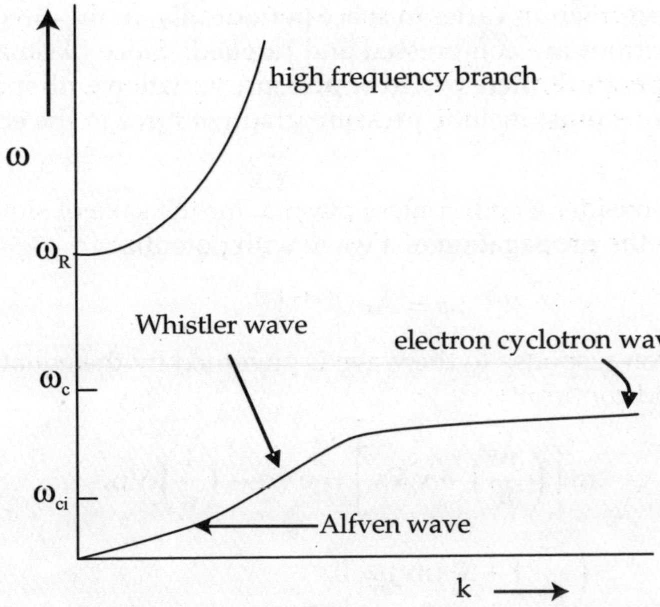

Figure 2.12: Dispersion curves for the right circularly polarised mode propagating along the dc magnetic field in a plasma.

Alfven waves are an important source of information about distant stars. They are believed to play a vital role in the dynamics of accretion discs and star formation, as they can carry momentum and energy across the magnetic field while charged particles cannot move (they only gyrate about the magnetic field lines).

2.7 Electrostatic Waves

The plane wave under the condition $\epsilon_{\text{eff}} = 0$, has $\mathbf{k} \times \mathbf{H} = 0$, $\mathbf{k}.\mathbf{H} = 0$ (second Maxwell's equation with ∇ replaced by i$\mathbf{k}$). Hence $\mathbf{H} = 0$, i.e., the wave is a purely electric wave called electrostatic wave. Since $\nabla \times \mathbf{E} = 0$, $\mathbf{E}$ can be expressed as $\mathbf{E} = -\nabla\phi$, where ϕ is called electric potential. Further, $\nabla . \mathbf{E} \neq 0$ for this wave. Hence, in compliance with the first Maxwell's equation, $\rho \neq 0$, i.e., the space charge density oscillates with time. At a given instant of time, density perturbation varies in space periodically, in the direction of **k**, i.e., electrons are compressed and rarefied. Since plasmas have finite temperature, there would be pressure variation with space and time and one must include pressure gradient force in the equation of motion.

Let us consider a collisionless plasma, for the sake of simplicity, and study the propagation of a wave with potential

$$\phi = Ae^{-i(\omega t - kz)}. \tag{73}$$

The electron response to the wave is governed by the equations of motion and continuity,

$$m\left[\left(\frac{\partial \mathbf{v}}{\partial t}\right) + \mathbf{v}.\nabla\mathbf{v}\right] = e\nabla\phi - \left(\frac{1}{n}\right)\nabla p, \tag{74}$$

$$\left(\frac{\partial n}{\partial t}\right) + \nabla.(n\mathbf{v}) = 0, \tag{75}$$

where $p = nk_BT$, n is the density of electrons, T is the electron temperature and k_B is the Boltzmann's constant. The last term in eq. (74) represents the pressure gradient force per electron. In equilibrium, when $\phi = 0$, $n = n_0$, $\mathbf{v} = 0$. When the equilibrium is perturbed by ϕ, we write

$$n = n_0 + n_1, \quad \mathbf{v} = 0 + \mathbf{v}_1,$$

where n_1 and $\mathbf{v}_1$ are perturbed quantities and temperature is a constant (isothermal approximation). In the adiabatic approximation, one may consider temperature perturbation using the relation $pn^{-\gamma}$ = constant, where γ is the ratio of specific heats of electron gas at constant pressure and constant volume. Using these n and **v** in eqs. (76) and (77) and ignoring the products of perturbed quantities, we have

$$\frac{\partial \mathbf{v}_1}{\partial t} = \frac{e}{m}\nabla\phi - \frac{k_BT}{n_0 m}\nabla n_1, \tag{76}$$

$$\frac{\partial n_1}{\partial t} + \nabla.(n_0\mathbf{v}_1) = 0. \tag{77}$$

All perturbed quantities vary with the same space-time dependence as ϕ, in the steady state. Hence, we can replace $\partial / \partial t$ by $-i\omega$ and ∇ by $i\mathbf{k}$. Then the above equations give

$$\mathbf{v}_1 = -\left(e\mathbf{k}\frac{\phi}{m\omega}\right) + \left(v_{th}^2\mathbf{k}\frac{n_1}{\omega n_0}\right), \tag{78}$$

$$n_1 = n_0\left(\mathbf{k}.\frac{\mathbf{v}_1}{\omega}\right) = -\frac{n_0 e k^2 \phi}{m\left(\omega^2 - k^2 v_{th}^2\right)} \tag{79}$$

where $v_{th} = (k_B T / m)^{1/2}$ is the *electron thermal velocity*. Poisson's equation $\epsilon_0 \nabla . \mathbf{E} = \rho$ for the wave takes the form

$$\nabla^2\phi = +\frac{n_1 e.}{\epsilon_0}$$

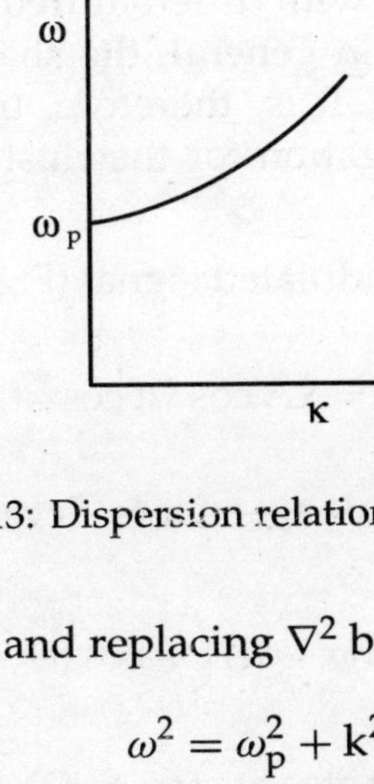

Figure 2.13: Dispersion relation for a plasma wave.

Substituting for n_1 and replacing ∇^2 by $-k^2$, we get

$$\omega^2 = \omega_p^2 + k^2 v_{th}^2 \tag{80}$$

as the dispersion relation for the electrostatic wave (Fig. 2.13). Had one employed the adiabatic approximation, one would get a factor γ multiplying the last term. This wave is called **Langmuir wave** or **plasma wave**. It can be easily excited by passing an electron beam through the plasma. It can also be excited by mode conversion of an electromagnetic wave of frequency close to ω_p. A plasma wave heats the plasma electrons via collisions or *Landau damping*. The latter is a process in which plasma electrons, moving with random velocities close to wave phase velocity resonantly exchange energy with the wave. To these electrons, the Doppler shifted wave frequency seems to be vanishingly small, i.e., the wave field appears as a dc field. Thus, they can get accelerated by the wave for a long time, efficiently exchanging energy with it (for details, see Stix[1] or Liu and Tripathi[2]). Large amplitude plasma waves are currently being excited by lasers in plasmas, for electron acceleration upto GeV energy.

2.8 Group Velocity

The radio signals employed in communication are either amplitude modulated, phase modulated or pulsed. Such signals, by Fourier integral representation, are a superposition of many frequencies. In a dispersive medium, waves with different frequencies travel with different velocities. Hence, in general, the shape of the signal at the output end is distorted. It is, therefore, useful to define the velocity of amplitude propagation for the cluster of waves, called group velocity.

Consider an amplitude-modulated signal (Fig. 2.14),

$$\mathbf{E}(z = 0, t) = \hat{x} A \cos \Omega t \cos \omega t, \tag{81}$$

where $\Omega << \omega$, propagating along +z axis. It can be expressed as

$$\mathbf{E}(0, t) = \hat{x}\frac{A}{2}\left[\cos(\omega + \Omega)\, t + \cos(\omega - \Omega)\, t\right].$$

Let the propagation constants at $(\omega + \Omega)$ and $(\omega - \Omega)$ be $k_+ = k(\omega + \Omega)$ and $k_- = k(\omega - \Omega)$. Then

$$\mathbf{E}(z, t) = \hat{x}\frac{A}{2}\left[\cos\left\{(\omega + \Omega)\, t - k_+ z\right\} + \cos\left\{(\omega + \Omega)\, t - k_- z\right\}\right]. \tag{82}$$

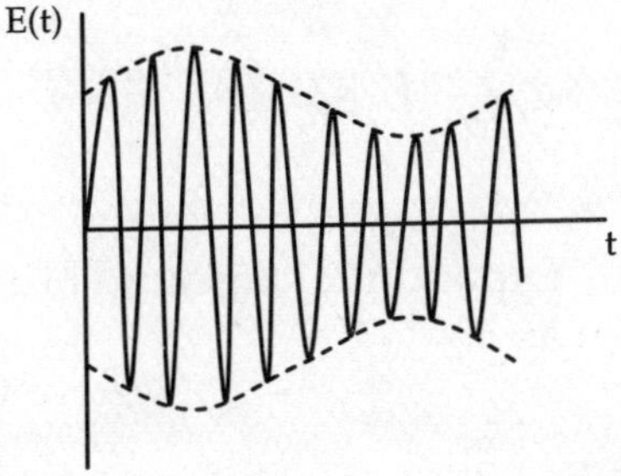

Figure 2.14: An amplitude modulated signal.

Expanding $k\pm$ around ω, $k_{\pm} = k(\omega) \pm \frac{\partial k}{\partial \omega}\Omega$,

$$\mathbf{E} = \hat{x}\frac{A}{2}\left[\cos\left(\omega t - kz + \Omega\left(t - \frac{z}{v_g}\right)\right) + \cos\left(\omega t - kz - \Omega\left(t - \frac{z}{v_g}\right)\right)\right]$$

$$= \hat{x}A\cos\left[\Omega\left(t - \frac{z}{v_g}\right)\right]\cos(\omega t - kz), \qquad (83)$$

where $v_g = \partial\omega / \partial k$, $k(\omega) \equiv k$. Thus, the amplitude of the wave propagates with velocity v_g, whereas its phase propagates with velocity $v_{ph} = \omega / k$.

In the general case, let $\mathbf{E}(0, t) = \hat{x}\, A(t)\, e^{-i\omega t}$, where A (t) is a slowly varying function of time, i.e. $\partial A / \partial t << \omega A$. We express A(t) in terms of a Fourier integral,

$$A(t) = \frac{1}{\sqrt{2\pi}}\int_{-\infty}^{\infty} g(\Omega)\, e^{-i\Omega t} d\Omega, \qquad (84)$$

where

$$g(\Omega) = \frac{1}{\sqrt{2\pi}}\int_{-\infty}^{\infty} A(t)\, e^{i\Omega t} dt \qquad (85)$$

is the Fourier transform. The field at (z, t) can be written from $\mathbf{E}(0, t)$ by delaying each frequency component by the time of flight,

$$\mathbf{E}(z,t) = \hat{x}\frac{1}{\sqrt{2\pi}}\int_{-\infty}^{\infty} g(\Omega)\, e^{-i[(\omega+\Omega)t-k'z]}d\Omega \tag{86}$$

where $k' \equiv k(\omega+\Omega)$. Expanding k′ around ω as $k' = k(\omega) + \frac{\partial k}{\partial \omega}\Omega$, eq. (86) can be written as

$$\mathbf{E}(z,t) = \hat{x}\left[\frac{1}{\sqrt{2\pi}}\int_{-\infty}^{\infty} g(\Omega)\, e^{-i\Omega\left(t-\frac{z}{v_g}\right)}d\Omega\right]\cdot e^{-i(\omega t-kz)}$$

$$= \hat{x}A\left(t-\frac{z}{v_g}\right)e^{-i(\omega t-kz)} \tag{87}$$

where we have employed eq. (84). Thus, the amplitude of any signal travels with group velocity $v_g = \partial\omega / \partial k$. In three dimensions,

$$\mathbf{v_g} = \partial\omega/\partial\mathbf{k}. \tag{88}$$

PROBLEM 2.5. A microwave pulse propagates in a collisionless plasma along z-axis. At z = 0, $\mathbf{E}(0, t) = \hat{x}\, A_0 \exp(-t^2/\tau^2).\cos(\omega t)$. Write $\mathbf{E}(z, t)$, hence deduce $E|_{z} = z_0 = 100\sqrt{2}\pi c/\omega$ and $t = t_0 = 200\pi/\omega$, when $\omega_p = \omega/\sqrt{2}$, $\omega\tau >> 1$.

Solution: $\eta = \left(1-\omega_p^2/\omega^2\right)^{1/2} = 1/\sqrt{2}, v_{ph} = c\sqrt{2}, v_g = c/\sqrt{2}$.

Hence, $\mathbf{E}(z,t) = \hat{x}A_0 e^{-\left(t-\sqrt{2}z/c\right)^2/\tau^2}\cos\left[\omega\left(t-z/c\sqrt{2}\right)\right]$,

$\mathbf{E}(z_0, t_0) = \hat{x}A_0$.

2.9 Boundary Conditions

When an electromagnetic wave falls on an interface between two media, it suffers partial reflection and partial transmission. In order to determine the values of reflection and transmission coefficients, let us deduce the boundary conditions that the fields must satisfy at the interface.

2.9.1 Continuity of Tangential Component of E

Consider the boundary between medium I and medium II, at $z = 0$ (Fig. 2.15). The quantities $\in_r$, σ and μ, characterising the two media, are denoted by subscripts 1 and 2 respectively. Consider the line integral $\oint \mathbf{E} \cdot \mathbf{dl}$ along the rectangular path ABCDA, when $BC \to 0$, $AB = 1$. From the Faraday's law of electromagnetic induction, this integral must equal the time derivative of magnetic flux linked with the loop. As the area of the loop vanishes, the flux linked with it also vanishes. Hence, $\oint \mathbf{E} \cdot \mathbf{dl} = 0$, or

$$E_{xI} = E_{xII}. \tag{89}$$

Similarly, $E_{yI} = E_{yII}$ at the interface. Thus, the tangential component of the electric field is continuous across the boundary.

In a special case, when medium II is a perfect conductor ($\sigma_2 = \infty$), electric field inside it is zero (otherwise infinite power dissipation will occur). Hence, the continuity condition demands that the tangential component of the electric field, E_{11}, in the first medium must also vanish at the boundary.

2.9.2 Continuity of Tangential Component of H

Apply the generalised Ampere's law

$$\oint \mathbf{H}.\mathbf{dl} = \int \left[\mathbf{J} + \left(\frac{\partial \mathbf{D}}{\partial t} \right) \right] .\mathbf{dS} \tag{90}$$

to the loop ABCDA of the fig. 2.15. In the limit $BC \to 0$, the surface integral on the right-hand side vanishes. Hence,

$$H_{xI} = H_{xII} \tag{91}$$

Similarly, one could show $H_{yI} = H_{yII}$. Thus, the tangential component of **H** is continuous across the boundary.

2.9.3 Continuity of Normal Component of D

Consider a Gaussian surface S of area of cross-section A and length d ($d \to 0$) at the interface between two media (Fig. 2.16). If the

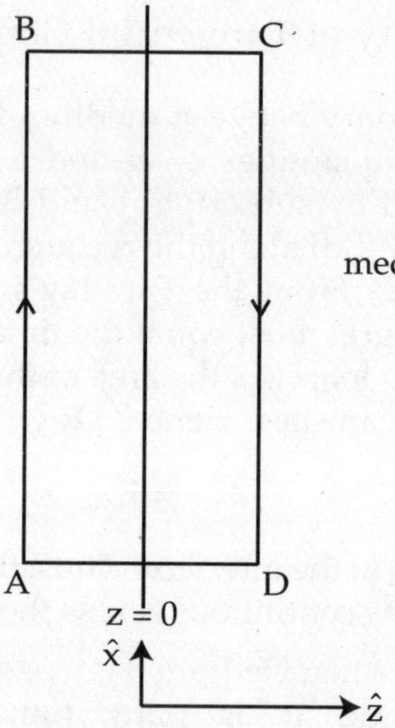

Figure 2.15: The path of the line integral.

two media are dielectric, then the charge enclosed inside the surface vanishes. Hence, from the Gauss's law $\oint \mathbf{D} \,.\, \mathbf{dS} = 0$, we obtain

$$D_{ZI} = D_{ZII}, \tag{92}$$

i.e., the normal component of the displacement vector is continuous across the boundary. In a special case, when medium II is a perfect conductor, finite surface charge density ρ_s is possible. In this case, charge enclosed inside the Gaussian surface would be ρ_s A. Hence

$$D_{ZII} - D_{ZI} = \rho_s. \tag{93}$$

In the case when conductivities of the two media are finite, the Gauss's law is $\oint \mathbf{D.dS} = \int \rho \, dV$. For electromagnetic field of frequency ω, the oscillatory space charge density (from the equation of continuity), is $\rho = \nabla.\, \mathbf{J}/i\,\omega$. Using this value of ρ and converting the volume integral into surface integral, we have

$$\oint (\mathbf{D} + i\mathbf{J}/\omega).\mathbf{dS} = 0 \text{ or } \oint \epsilon_{\rm eff}\, \mathbf{E.dS} = 0.$$

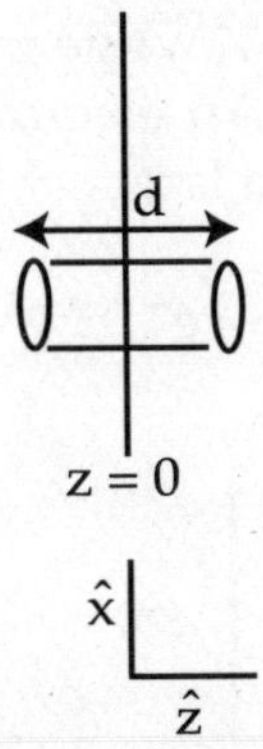

Figure 2.16: A Gaussian surface over which surface integral is to be evaluated.

It leads to

$$\epsilon_{eff1}\, \mathbf{E}_{ZI} = \epsilon_{eff2}\, \mathbf{E}_{ZII}, \tag{94}$$

i.e. the normal component of effective displacement vector, $\mathbf{D} - \mathbf{J}/i\,\omega$, is continuous across the boundary.

2.9.4 Continuity of Normal component of B

Applying the Gauss's law for magnetic field, $\oint \mathbf{B}.\mathbf{d}S = 0$, over the surface S (Fig. 2.16) we obtain

$$B_{ZI} = B_{ZII}. \tag{95}$$

i.e., the normal component of the magnetic field is continuous across the boundary.

2.10 Reflection and Transmission of Waves

2.10.1 Normal Incidence

Consider the interface (z = 0) between two media (Fig. 2.17). In both media, we assume $\mu_r = 1$. $\epsilon_r, \sigma, \epsilon_{eff}$ and η in medium I (z < 0) will be denoted by subscript 1, while in medium II ($z > 0$) they will be denoted by subscript 2. An electromagnetic wave is normally incident on the interface from left,

$$\mathbf{E}_i = \hat{x}e^{-i(\omega t - k_1 z)}, \tag{96}$$

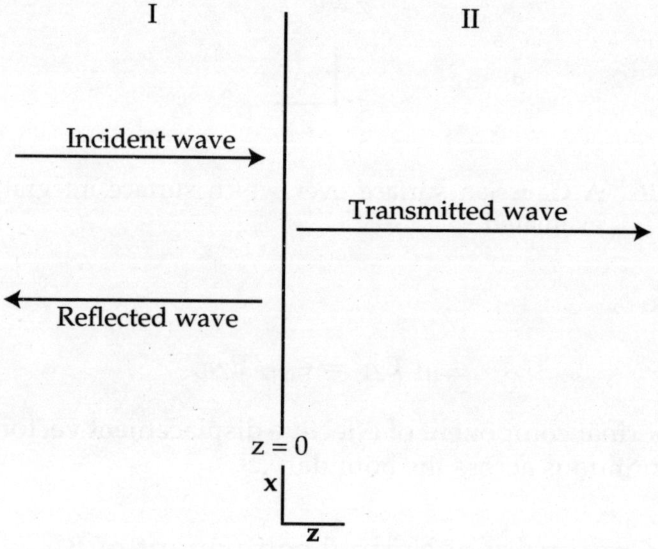

Figure 2.17: Reflection and transmission of normally incident electromagnetic wave on an interface.

where $k_1 = \omega\eta_1/c$, $\eta_1 = \sqrt{\epsilon_{eff1}}$, The fields of the reflected and transmitted waves can, in general, be written as

$$\mathbf{E}_r = \mathbf{A}_r \exp[-i(\omega_r t - \mathbf{k}_r.\mathbf{r})] \text{ and } \mathbf{E}_t = \mathbf{A}_t \exp[-i(\omega_t t - \mathbf{k}_t.\mathbf{r})].$$

However, at $z = 0$,

$$E_{ix} + E_{rx} = E_{tx}$$

must hold at all times and for all values of x and y. This demands that E_{rx} and E_{tx} must have the same t, x, y dependence as E_{ix}, i.e.,

$$\omega_r = \omega_t = \omega, \quad k_{rx} = k_{ry} = 0, \quad k_{tx} = k_{ty} = 0.$$

The continuity of E_y and H_x across z=0 gives $E_{ry} = E_{ty} = 0$. Since $\mathbf{k}_r.\mathbf{E}_r = 0, \mathbf{k}_t.\mathbf{E}_t = 0$, one has $E_{rz} = E_{tz} = 0$. Thus one may write

$$\mathbf{E}_r = \hat{x}\, R' A_0 e^{-i(\omega t + k_1 z)},$$

$$\mathbf{E}_t = \hat{x}\, T' A_0 e^{-i(\omega t - k_2 z)}, \tag{97}$$

where $k_2 = \omega\eta_2/c$. The magnetic intensities of the incident, reflected and transmitted waves can be obtained using the third Maxwell's equation,

$$\mathbf{H}_i = \mathbf{k}_1 \times \frac{\mathbf{E}_i}{\omega\mu_0} = \hat{y}\left(\frac{\eta_1}{c\mu_0}\right) E_i$$

$$\mathbf{H}_r = -\hat{y}\left(\frac{\eta_1}{c\mu_0}\right) E_r, \mathbf{H}_t = \hat{y}\left(\frac{\eta_2}{c\mu_0}\right) \mathbf{E}_t. \tag{98}$$

Demanding the continuity of Ex and Hy at $z = 0$, at all times, we have

$$1 + R' = T', 1 - R' = T'\frac{\eta_2}{\eta_1},$$

i.e.,

$$R' = \frac{\eta_1 - \eta_2}{\eta_1 + \eta_2}, \quad T' = \frac{2\eta_1}{\eta_1 + \eta_2}. \tag{99}$$

For $\eta_2 > \eta_1$, R′ is negative, i.e., there is a phase change of π on reflection from an optically denser medium. If medium II is a conductor or a semiconductor, then η_2 is complex and phase change on reflection is different from π.

The intensities or time average power flow densities of the incident, reflected and transmitted waves are :

$$I_i = \left(\frac{\eta_{1r}}{2c\mu_0}\right) A_0^2, \quad I_r = \left(\frac{\eta_{1r}}{2c\mu_0}\right) A_0^2\, |R'|^2, \quad I_t = \left(\frac{\eta_{2r}}{2c\mu_0}\right) A_0^2\, |T'|^2,$$

where η_{1r} and η_{2r} are the real parts of η_1 and η_2. The intensity reflection coefficient $R = I_r/I_i$ and intensity transmission coefficient $T = I_t/I_i$ are

$$\mathrm{R} = \left|\frac{\eta_1 - \eta_2}{\eta_1 + \eta_2}\right|^2, \quad \mathrm{T} = \frac{4(\eta_{2r}/\eta_{1r})|\eta_1|^2}{|\eta_1 + \eta_2|^2}. \tag{100}$$

As the contrast between η_1 and η_2 increases, R increases. We have plotted R as a function of ω/ω_p in fig. 2.18, for the case when medium 1 is free space and medium 2 is a plasma with $\eta_2 = \left(1 - \omega_p^2/\omega^2\right)^{1/2}$. As ω rises above ω_p, R decreases monotonically approaching zero for $\omega/\omega_p >> 1$.

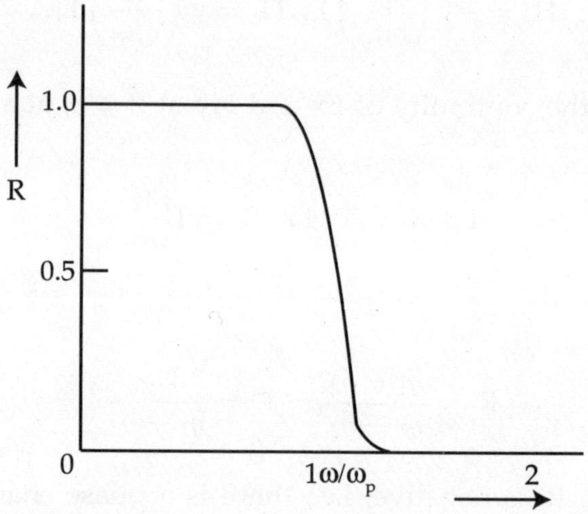

Figure 2.18: Reflectivity as a function of normalised frequency of the electromagnetic wave, normally incident on a vacuum plasma interface.

PROBLEM 2.6. A laser of $\omega = 2 \times 10^{15}$ rad/s and power density 10^{10} W/m^2 is normally incident from free space on a conductor of $\epsilon_r = 9$ and $\omega_p = 8 \times 10^{15}$ rad/ s. Estimate the field of the laser, 0.2 μm inside the conductor. Take $\nu = 0$.

Solution: The amplitude of the incident laser is $A_0 = (2\mu_0 c S_{av})^{1/2} = 2.7 \times 10^6$ V/m. For given parameters $\eta = \eta_r + i\eta_i = \sqrt{7i}$ in the conductor. Hence the field amplitude just inside the conductor is $A_t = |2/(1+\eta)|A_0 \approx 0.7A_0$. Amplitude at the depth of 0.2 μm is

$$A = 0.7A_0 \exp\left(\frac{-\omega\eta_i d}{c}\right) \approx 5 \times 10^4 \mathrm{V/m}$$

PROBLEM 2.7. A radio wave of amplitude 1 V/m and frequency 1 MHz is normally incident on the surface of the earth ($\epsilon_r \approx 4, \sigma \sim 1$ mho/ m). Obtain the wave amplitude at a depth of 2 m.

Solution: For the earth

$$\eta = \eta_r + i\eta_i = (4 + i1.8 \times 10^4)^{1/2} \approx 94(1+i).$$

Hence, the wave amplitude at the interface is $A_t \approx 1.5 \times 10^{-2}$ V/m and that at a depth of 2 m is $A \approx 3 \times 10^{-4}$ V/m.

2.10.2 Oblique Incidence: P-Polarisation

Consider an electromagnetic wave, obliquely incident on the interface (z = 0) between two media at an angle of incidence θ_i, $\mathbf{k_1} = k_1 \sin\theta_i \hat{x} + k_1 \cos\theta_i \hat{z}$, $k_1 = \omega\eta_1/c$ (Fig. 2.19). The electric field of the wave is polarised in the plane of incidence, which is defined as the plane containing the direction of wave propagation and the surface normal (i.e., x-z plane in the present case),

$$\mathbf{E_i} = \left(A_0\hat{x} + A_0'\hat{z}\right) e^{-i(\omega t - k_{1x}x - k_{1z}z)}. \qquad (101)$$

As $\mathbf{k_1} \cdot \mathbf{E_i} = 0$, we have $A_0' = -A_0 k_{1x}/k_{1z}$. The magnetic intensity of the incident wave, from the third Maxwell's equation, is

$$\mathbf{H_i} = \frac{\mathbf{k_1} \times \mathbf{E}_i}{\omega\mu_0} = \hat{y}\frac{k_1^2 A_0}{k_{1z}\omega\mu_0} e^{-i(\omega t - \mathbf{k_1}.\mathbf{r})}. \qquad (102)$$

In view of the continuity conditions on tangential components of **E** and **H** at z = 0, the reflected and transmitted waves must also be

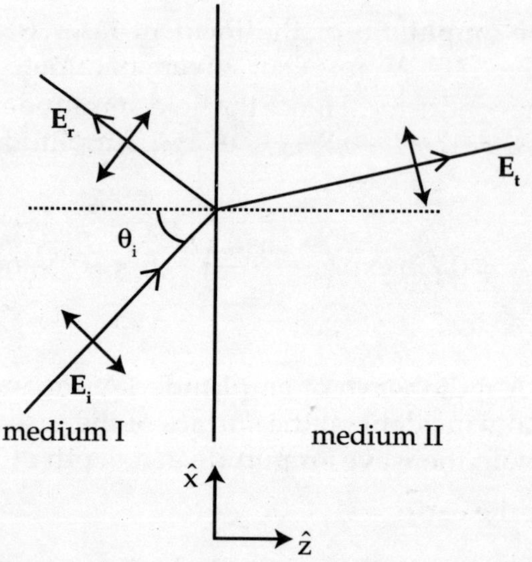

Figure 2.19: Reflection and transmission of a p-polarised electromagnetic wave obliquely incident at the interface, z=0, between two media.

p-polarised and have the same frequency as the incident wave. The electric fields of the reflected and transmitted waves, thus can be written as

$$\mathbf{E_r} = R'A_0 \left(\hat{x} - \frac{k'_{1x}}{k'_{1z}}\hat{z}\right) e^{-i(\omega t - k'_{1x}x - k'_{1z}z)} \tag{103}$$

$$\mathbf{E_t} = T'A_0 \left(\hat{x} - \frac{k_{2x}}{k_{2z}}\hat{z}\right) e^{-i(\omega t - k_{2x}x - k_{2z}z)} \tag{104}$$

where $k'_1 = \omega\eta_1/c$, $k_2 = \omega\eta_2/c$. Corresponding magnetic intensities are

$$\mathbf{H_r} = \hat{y}\frac{k_1^2 A_0}{k'_{1x}\omega\mu_0} e^{-i(\omega t - \mathbf{k'_1} \cdot \mathbf{r})},$$

$$\mathbf{H_t} = \hat{y}\frac{k_2^2 A_0}{k_{2x}\omega\mu_0} e^{-i(\omega t - \mathbf{k_2} \cdot \mathbf{r})}. \tag{105}$$

Demanding the continuity of E_x and H_y at $z = 0$, for all values of x and t, we obtain

$$k_{1x} = k'_{1x} = k_{2x}, \tag{106}$$

$$I + R' = T', \quad 1 - R' = \alpha T',$$

$$R' = \frac{1-\alpha}{1+\alpha}, \quad T' = \frac{2}{1+\alpha}, \tag{107}$$

where $\alpha = k_2^2\, k_{1z}/k_1^2\, k_{2z} = \eta_2^2\, k_{1z}/\eta_1^2\, k_{2z}$.

Equation (106) states that the tangential component of the wave vector remains unchanged on reflection or transmission. When η_1 and η_2 are real, one may write $k'_{1x} = k_1 \sin\theta_r$ and $k_{2x} = k_2 \sin\theta_t$, where θ_r and θ_t are the angles of reflection and transmission and respectively. Eq. (106) gives the Snell's laws :

$$\theta_r = \theta_i,$$

$$\eta_1 \sin\theta_i = \eta_2 \sin\theta_t. \tag{108}$$

Further, $k'_{1z} = -k_{1z}$ and

$$k_{2z} = \left(k_2^2 - k_{2x}^2\right)^{1/2} = \frac{\omega}{c}\left(\eta_2^2 - \eta_1^2 \sin^2\theta_i\right)^{1/2}. \tag{109}$$

For wave propagation in medium II, $(\eta_2^2 - \eta_1^2 \sin^2\theta)$ must be greater than zero. One may note from eq. (107), that for a specific angle of incidence $\theta = \theta_B$ (for which $\alpha = 1$), R′ vanishes. Using Snell's law, θ_B turns out to be

$$\theta_B = \tan^{-1}\left(\frac{\eta_2}{\eta_1}\right). \tag{110}$$

It is known as the *Brewster's angle.* In case of reflection from a denser medium ($\eta_2 > \eta_1$), $\theta_B > \pi\ /\ 4$. In the opposite case of $\eta_2 < \eta_1$, θ_B is less than $\pi\ /\ 4$. It can be shown that at $\theta_i = \theta_B$, the reflected and transmitted rays are perpendicular to each other ($\theta_r + \theta_t = \pi\ /\ 2$). At all other angles $\theta \neq \theta_B$, R′ is finite.

The time average Poynting vectors of the incident, reflected and transmitted waves, for the case when η_1, η_2 and k_{2z} are real, are (eq. 36):

$$\mathbf{S}_i = \frac{k_1^2 A_0^2}{2k_{1z}^2 \omega\mu_0}\mathbf{k}_1, \quad \mathbf{S}_r = \frac{k_1^2 A_0^2}{2k_{1z}^2 \omega\mu_0}|R'|^2\mathbf{k}_1', \quad \mathbf{S}_t = \frac{k_2^2 A_0^2}{2k_{2z}^2 \omega\mu_0}|T'|^2\mathbf{k}_2.$$

One may define intensity reflection and transmission coefficients, with respect to normal components of Poynting vector :

$$R = \frac{|S_{rz}|}{S_{iz}} = \left|\frac{1-\alpha}{1+\alpha}\right|^2, T = \frac{S_{tz}}{S_{iz}} = \frac{4\alpha}{|1+\alpha|^2}. \tag{111}$$

2.10.2.1 *Critical Reflection*

One may note from eq. (108) that for $\eta_2 < \eta_1$, $\theta_t > \theta_i$, i.e., a wave incident from a denser medium onto a rarer medium, moves away from the surface normal on transmission. For $\theta_i = \theta_c$, where

$$\theta_c = \sin^{-1}\left(\frac{\eta_2}{\eta_1}\right), \tag{112}$$

$\theta_t = \pi/2$, i.e. the transmitted wave propagates parallel to interface. θ_c is called the *critical angle*. For $\theta_i > \theta_c$, k_{2z} is imaginary. Writing $k_{2z} = i\,|k_{2z}|$ in eq. (104), we obtain

$$\mathbf{E}_t = T'A_0\left(\hat{x} + i\frac{k_{2x}}{|k_{2z}|}\hat{z}\right)e^{-|k_{2z}|z}e^{-i(\omega t - k_{1x}x)}. \tag{113}$$

The amplitude of the transmitted wave falls off exponentially away from the interface, with no phase variation along z (Fig. 2.20). The planes of constant amplitude (z = constant) and constant phase (x = constant) are perpendicular to each other. The time average Poynting vector turns out to be parallel to x-axis. Thus there is no energy propagation, normal to the interface, in the rarer medium.

2.10.3 Oblique Incidence: S-Polarisation

When the incident wave is polarised perpendicular to the plane of incidence (S-polarisation),

$$\mathbf{E}_i = \hat{y}A_0 e^{-i(\omega t - \mathbf{k}_1 \cdot \mathbf{r})}, \tag{114}$$

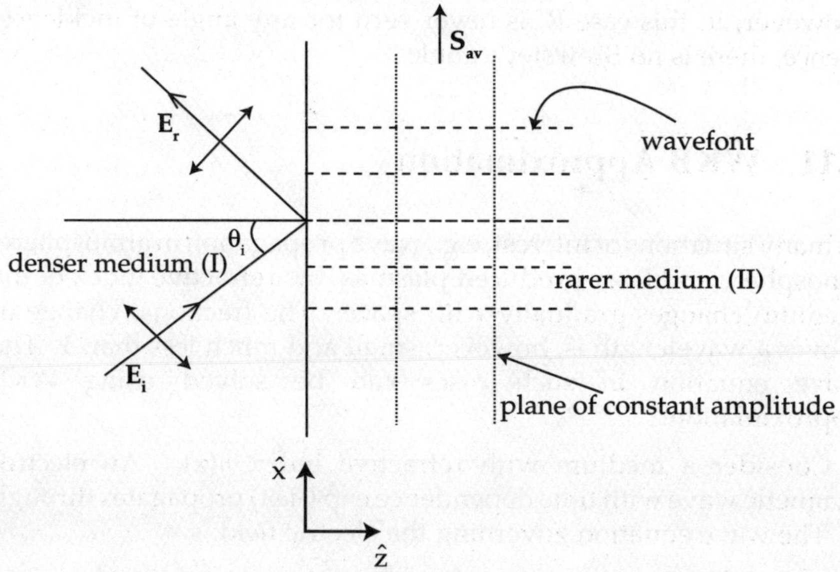

Figure 2.20: Total internal reflection of an electromagnetic wave; $\theta_i > \theta_c$.

the reflected and transmitted waves are also S-polarised,

$$\mathbf{E}_r = \hat{y}R'A_0e^{-i(\omega t - k_1' \cdot r)}$$
$$\mathbf{E}_t = \hat{y}T'A_0e^{-i(\omega t - k_2' \cdot r)}. \tag{115}$$

The magnetic intensities of the three waves are $\mathbf{H}_i = [(k_1 \times E_i)/\omega\ \mu_0]$, $\mathbf{H_r} = [(\mathbf{k}_1' \times \mathbf{E}_r)/\omega\mu_0], \mathbf{H}_t = [(\mathbf{k}_2 \times \mathbf{E}_t)/\omega\ \mu_0]$. Using the continuity of E_y and H_x at $z = 0$, we obtain

$$k_{1x} = k_{1x}' = k_{2x}, \tag{116}$$

$$1 + R' = T', 1 - R' = T'\frac{k_{2z}}{k_{1z}},$$

$$R' = \frac{1 - k_{2z}/k_{1z}}{1 + k_{2z}/k_{1z}}, T' = \frac{2}{1 + k_{2z}/k_{1z}}. \tag{117}$$

Eq. (116) is the same as that for the case of p-polarisation. Hence, Snell's laws are independent of the nature of polarisation of the wave. The critical angle θ_c, at which total internal reflection of the electromagnetic wave occurs from a rarer medium, is also same.

However, in this case R′ is never zero for any angle of incidence. Hence, there is no Brewster's angle.

2.11 WKB Approximation

In many situations of interest, e.g., wave propagation in atmosphere, ionosphere and laser produced plasmas, the refractive index of the medium changes gradually with space. The fractional change in it over a wavelength is, however, small and much less than 1. The wave equation in such cases can be solved using WKB approximation.

Consider a medium with refractive index $\eta(x)$. An electromagnetic wave with time dependence exp $(-i\omega t)$ propagates through it. The wave equation governing the electric field is

$$\nabla^2 \mathbf{E} - \nabla(\nabla \cdot \mathbf{E}) + \frac{\omega^2}{c^2} \eta^2(x)\mathbf{E} = 0, \tag{118}$$

where ϵ_{eff} has been written as η^2 (x). Taking $\nabla.$ of this equation, we get $\nabla \cdot (\eta^2 \mathbf{E}) = 0$, or

$$\nabla \cdot \mathbf{E} = -\frac{E_x}{\eta^2}\frac{d\eta^2}{dx}.$$

We may restrict (without any loss of generality) wave propagation to x-z plane (i.e. $\partial / \partial y = 0$) and eq. (118), in component form, can be written as

$$\nabla^2 E_x - \frac{\partial}{\partial x}\left(-\frac{E_x}{\eta^2}\frac{\partial \eta^2}{\partial x}\right) + \frac{\omega^2}{c^2}\eta^2 E_x = 0,$$

$$\nabla^2 E_y + \frac{\omega^2}{c^2}\eta^2 E_y = 0,$$

$$\nabla^2 E_z - \frac{\partial}{\partial z}\left(-\frac{E_x}{\eta^2}\frac{\partial \eta^2}{\partial x}\right) + \frac{\omega^2}{c^2}\eta^2 E_z = 0. \tag{119}$$

One may note that the wave equation for E_y is independent of E_x or E_z and has $\nabla (\nabla \cdot \mathbf{E}) = 0$. On the other hand, the E_z equation depends on E_x, i.e., if E_x is finite E_z is also finite. Thus, we can treat the two cases separately

(i) **E** $\|\hat{y}$, called S-polarisation,
(ii) **E** $\perp\hat{y}$, called P-polarisation.

In the second case, when electric field is in the plane of wave propagation, $\nabla(\nabla.\mathbf{E})$ term in the wave equation is finite and could be large near $\eta^2 \to 0$. If the refractive index is away from zero and varies gradually over a wavelength, $(c/\omega\eta)\, d\eta/dx << 1$, we may ignore this term in the wave equation and the propagation of S and P polarisations becomes same. Let us rewrite the E_y equation,

$$\frac{\partial^2 E_y}{\partial x^2} + \frac{\partial^2 E_y}{\partial z^2} + \frac{\omega^2}{c^2}\eta^2 E_y = 0. \tag{120}$$

Let $E_y = f_1(x)\, f_2(z)\exp(-i\omega t)$. Substituting it in eq. (120), dividing the resulting equation by $f_1 f_2$ and rearranging terms, we get

$$\frac{1}{f_1}\frac{d^2 f_1}{dx^2} + \frac{\omega^2}{c^2}\eta^2(x) = \frac{-1}{f_2}\frac{d^2 f_2}{dz^2}. \tag{121}$$

The left-hand side is independent of z while the right-hand side is independent of x. Hence both must be independent of x and z, i.e. a constant. Let the constant be $-k_z^2$. Then, from eq. (121), we have

$$\frac{d^2 f_2}{dz^2} + k_z^2 f_2 = 0, \tag{122}$$

$$\frac{d^2 f_1}{dx^2} + k_x^2 f_1 = 0, \tag{123}$$

where $k_x^2 = (\omega^2/c^2)\,\eta^2(x) - k_z^2$ is a function of x. The solution of eq. (122) is

$$f_2 = A_2 e^{ik_z z}.$$

Similarly, if k_x^2 were a constant, f_1 would be $f_1 A_1 \exp(i\, k_x\, x)$ with A_1 a constant. Now since k_x^2 is a slowly varying function of x, we express f_1 as

$$f_1 = A_1(x)e^{i\psi(x)},$$

where A_1 and ψ are both real and A_1 is considered to be a slowly varying function of x. Substituting for f_1 in eq. (123), separating the real and imaginary parts and neglecting $d^2 A_1/ dx^2$, we have

$$\left(\frac{d\psi}{dx}\right)^2 = k_x^2, \tag{124}$$

$$2\frac{dA_1}{dx}\frac{d\psi}{dx} + \frac{d^2\psi}{dx^2}A_1 = 0. \tag{125}$$

The first equation gives

$$\psi = \int k_x dx.$$

Multiplying eq. (125) by A_1 we can rewrite it as

$$\frac{d}{dx}\left(A_1^2\frac{d\psi}{dx}\right) = 0,$$

giving $A_1^2 k_x = \text{constant} = C_1^2$. Hence E_y can be written as

$$E_y = \frac{C_1 A_2}{k_x^{1/2}} e^{i\int k_x dx} e^{-i(\omega t - k_z z)}.$$

If initially at $x = 0$, $E_y = A_0 \exp[-i(\omega t - k_z z)]$, then at any arbitrary x

$$E_y = \frac{A_0}{(k_x/k_{xo})^{1/2}} \exp\left[-i\left(\omega t - k_z z - \int_0^x k_x dx\right)\right], \tag{126}$$

where $k_{x0} = k_x\ (x = 0)$. This is known as the WKB solution. It is valid when $\frac{dk_x^2}{dx} << k_x^3$.

One may still treat k_x as the x component of propagation constant locally. As the wave propagates, its k_z (the z-component of wave vector) remains constant and only k_x changes. If η decreases with x, then k_x also decreases and wave moves away from x axis as it penetrates deeper into the rarer medium. The magnetic field of the wave is given by

$$\mathbf{B} = \frac{1}{\mu_0\omega}\left(-k_z A\hat{x} + \left(k_x A - i\frac{\partial A}{\partial x}\right)\hat{z}\right)\exp\left[-i\left(\omega t - k_z z - \int_0^x k_x dx\right)\right], \tag{127}$$

where $A = A_0\, k_{xo}/\, k_x$. The time average power flow per unit area is

$$\mathbf{S}_{av} = \frac{1}{2}\mathrm{Re}[\mathbf{E}^* \,x\mathbf{H}] = \frac{A_0^2 k_{x0}}{2Z_0 k_x}\ \mathbf{ck}/\omega. \tag{128}$$

The Poynting vector bends as **k** changes. Also, its magnitude changes. However, the total power carried by the wshole wavefront of the wave remains constant, as width of the wavefront decreases.

The trajectory of a ray can be deduced from $dx/\,dt = v_{gx}$, $dz/\,dt = v_{gz}$, i.e.,

$$\frac{dx}{dz} = \frac{v_{gx}}{v_{gz}} = \frac{k_x}{k_z} = \left(\frac{\omega^2\eta^2}{c^2k_z^2} - 1\right)^{1/2}. \tag{129}$$

$$\text{or}\quad z = \int \left(\frac{\omega^2\eta^2}{c^2k_z^2} - 1\right)^{-1/2} dx + c_2, \tag{130}$$

where c_2 is a constant, determined from the initial conditions.

WKB solution provides an insight into wave propagation in inhomogeneous media. In contrast to sharp discontinuity in refractive index where strong reflection occurs, diffuse boundary or gradual change in η can diminish reflection significantly.

PROBLEM 2.8. Half space $z > 0$ contains an inhomogeneous plasma with $\omega_p^2 = \omega_{p0}^2 z/L$. (Fig 2.21). A wave of frequency $\omega = \omega_{p0}$ is launched from free space on to $z = 0$ interface at an angle of incidence θ_i. Obtain the depth z at which ray turns back. Deduce the equation of the ray that enters at x=0, z=0.

Solution: As density does not vary with x, k_x is constant throughout the ray path, and is given as $k_x = (\omega/\,c)\sin\theta_i$. Then

$$k_z^2 = k^2 - k_x^2 = \frac{\omega^2}{c^2}\left[1 - \frac{\omega_{p0}^2}{\omega^2}\frac{z}{L} - \sin^2\theta_i\right] = \frac{\omega^2}{c^2}\left(\cos^2\theta_i - \frac{z}{L}\right).$$

Ray bends at the layer where $k_z = 0$ i.e., $z = z_m \equiv L\cos\theta_i$.

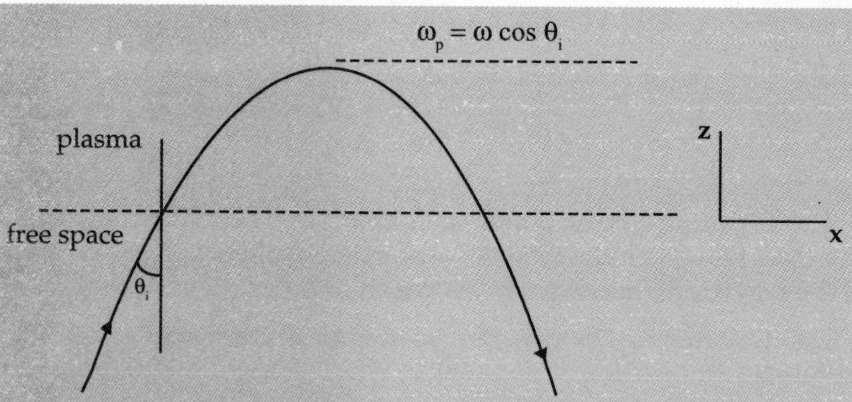

Figure 2.21: Trajectory of a ray incident on plasma half space. Plasma density increases with z.

The equation of the ray is

$$\frac{dz}{dx} = \frac{k_z}{k_x} = \pm\left(\cos^2\theta_i - \frac{z}{L}\right)^{1/2} / \sin\theta_i.$$

Upto the turning point, dz/ dx is positive and the above equation integrates to give

$$x = x_m\left[1-\left(1-\frac{z}{z_m}\right)^{1/2}\right]; \quad x_m = L\sin 2\theta_i.$$

Beyond the turning point ($x > x_m$), dx/ dz is negative and the ray trajectory is given by

$$x = x_m\left[1+\left(1-\frac{z}{z_m}\right)^{1/2}\right].$$

PROBLEM 2.9. An electromagnetic pulse of frequency ω is sent vertically upwards from the ground into the ionosphere (fig. 2.22), where plasma frequency varies with height z as

$$\begin{aligned} \omega_p^2 = & \quad 0 & \text{for } z < h_0. \\ = & \ \omega_{p0}^2 \exp\left[(z-h_0)/L\right] & \text{for } z > h_0, \end{aligned}$$

and L is density scale length. The pulse returns back after a time $t_g = 2\int_0^{h_c} dx/v_g$, where $h_c = h_0 + 2L\ln(\omega/\omega_{po})$ is the height at which $\omega_p = \omega$, i.e., the height at which reflection occurs and v_g is the group velocity. Evaluate t_g. Neglect collisions.

Solution: $v_g = \begin{cases} c, & z < h_0 \\ c\left[1 - \beta\exp\left\{\frac{(z-h_0)}{L}\right\}\right]^{\frac{1}{2}}, & z > h_0 \end{cases}$ where $\beta = \omega_{p0}^2/\omega^2$.
Hence,

$$t_g = \frac{2h_0}{c} + \frac{2}{c}\int_{h_0}^{h_c}\left(1 - \beta\exp\left[(z-h_0)/L\right]\right)^{\frac{1}{2}} dz$$
$$= \frac{2h_0}{c} + \frac{2L}{c}\ln\frac{1+\alpha}{1-\alpha}$$

where $\alpha = \left(1 - \omega_{p0}^2/\omega^2\right)^{\frac{1}{2}}$.

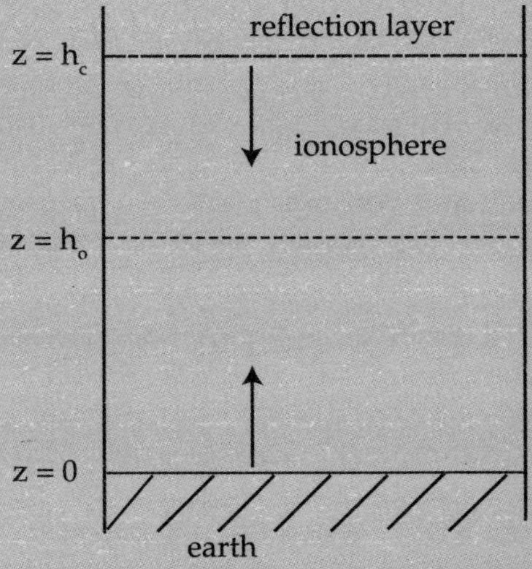

Figure 2.22: An electromagnetic pulse, launched vertically upwards from a ground-based transmitter, suffers reflection after penetrating a distance into the ionosphere.

One may notice that higher frequencies incur greater group delays t_g, as they are reflected back from higher heights. For $\omega^2 >> \omega_{p0}^2$

$$t_g \approx \frac{2\,(h_c + L\ln 4)}{c} \sim \frac{2h_c}{c}.$$

This relation between t_g and h_c is fairly independent of the nature of electron density profile. In fact, by measuring t_g experimentally, one can deduce the height h_c at which plasma frequency $\omega_p = \omega$. Since ω is known, one obtains the variation of ω_p (hence electron density) with height. This is the technique employed in ionosonde to determine the electron density profile in the ionosphere upto a height of 300 km, at which electron density has the highest value $\sim 10^{12}\,m^{-3}$ ($\omega_{pmax}/2\pi \approx 10$ MHz).

2.12 Conclusions

We have studied the propagation of plane electromagnetic waves in free space, dielectrics, conductors, semiconductors and plasmas. A homogeneous plane wave has plane of constant amplitude, coincident with the wavefront. However, for an inhomogeneous wave, the two are different. In a nonlossy medium, the planes of constant amplitude and constant phase are perpendicular to each other. Inhomogeneous waves occur in situations of critical reflection.

Dispersion (i.e. the frequency dependence of refractive index) is a common characteristic of all material media. In dielectrics dispersion is weak at infrared and longer wavelengths. It is strong in conducting media. In fact, electromagnetic waves of frequency below the plasma edge, $\omega < \omega_p/\sqrt{\epsilon_r}$, do not propagate in conductors and plasmas. Above the plasma edge, they propagate with frequency-dependent phase and group velocities. As frequency increases beyond $\omega_p/\sqrt{\epsilon_r}$, v_{ph} and v_g tend to c. Dispersion characteristics strongly manifest themselves in reflectivity. Below the plasma edge, reflectivity is nearly unity whereas above the edge, it falls off rapidly with frequency.

The propagation characteristics of waves are strongly influenced by the application of a dc magnetic field. Right circularly polarised waves of frequency less than the electron cyclotron frequency can propagate in overdense plasmas with very little attenuation.

In weakly inhomogeneous media, WKB approximation provides a tractable solution of wave equation. The component of propagation vector perpendicular to the direction of inhomogeneity is constant, whereas the one along it varies with distance. Subsequently, ray trajectories are curved.

UNSOLVED PROBLEMS

1. A 1 GW laser falls normally on a dielectric slab, from which 60% power is reflected back and 40% is transmitted. Estimate the average force experienced by the slab.

[**HINT:** *In free space, momentum carried by the laser per second equals* $\left(\frac{1}{c}\right)$ *times the power carried by it. Then apply the conservation of momentum.*]

[**ANS.** 4N.]

2. Estimate the frequency ω at which conduction current density, in a collisionless plasma of electron density n_0, has the same magnitude as the displacement current density.

[**ANS.** $\omega = \omega_p = (n_0 e^2/m \epsilon_0)^{1/2}$.]

3. In a plasma, the electric field of an electromagnetic wave is $\mathbf{E} = \mathbf{A} \exp[-i(\omega t - \omega z/4c)]$. Obtain the electron density of the plasma in terms of ω.

[**ANS.** $n_0 = (15/16\, e^2)\, m\, \epsilon_0\, \omega^2$.]

4. A metal is characterised by $\epsilon_r = 10$, $\omega_p = 8 \times 10^{15}$ rad/s and $\nu = 10^{12}\, s^{-1}$. Estimate the skin depth $\delta\ \left(= k_i^{-1}\right)$ at

(i) laser frequency $\omega = 2 \times 10^{15}$ rad/s and (ii) microwave frequency $\omega = 6 \times 10^{10}$ rad/s.

[**Ans.** (i) 0.06 μ m, (ii) 0.2 μ m.]

5. In a dielectric, the field of an electromagnetic wave is $\mathbf{E} = \hat{y}\, A_0 \exp(-\omega x/c) \exp[-i(\omega t - 2\omega z/c)]$.
Deduce $\mathbf{H}, \mathbf{S}_{av}$ and ϵ_r.

[**Ans.** $\mathbf{H} = (-2\,\hat{x} + i\hat{z}) E_y/\mu_0 c, \mathbf{S}_{av} = \hat{z}\,(A_0^2/\mu_0 c) \exp(-2\omega x/c)$,
$\epsilon_r = 3.$]

6. A 10 GHz microwave is normally incident on a lossy medium of relative permittivity ϵ_r and conductivity σ (both real). The amplitude reflection coefficient $R = -(0.33 + 0.055i)$. Estimate $\epsilon_r, \sigma/\omega\, \epsilon_0$.

[**Ans.** $\epsilon_r = 4$, $\sigma/\omega\ \ \epsilon_0 = 1$.]

7. A laser is obliquely incident on a free space-plasma boundary (z = 0) (Fig. 2.23). Inside the plasma, $\mathbf{E} = \hat{y}\ A \exp(-2\omega z/c) \exp\left[-i\left(\omega t - \omega x/c\sqrt{2}\right)\right]$. Obtain (i) ω_p/ω, (ii) $\mathbf{S}_{av}$ inside the plasma, (iii) angle of incidence θ_i and (iv) intensity reflection coefficient. Ignore collisions.

[**Hint:** *Write* $\mathbf{k}, k^2$, *hence deduce* ω_p^2/ω^2. *Use continuity of* k_x *to obtain* θ_i.]

[**Ans.** (i) $3/\sqrt{2}$,
(ii) $\hat{x}\left(A^2/2\sqrt{2}\mu_0 c\right) \exp(-4\omega z/c)$,
(iii) $\pi/4$,
(iv) 1.]

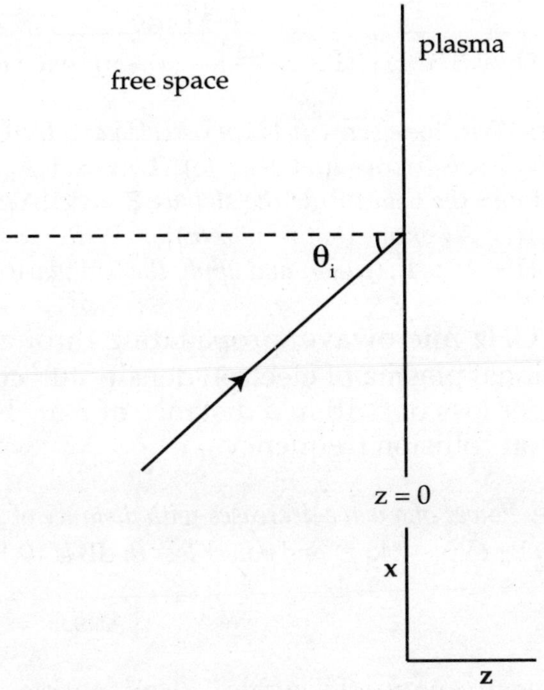

Figure 2.23: An electromagnetic wave obliquely incident on a free space-plasma interface.

8. The electric field amplitude of a laser, propagating along z-axis, has Gaussian distribution in radial coordinate r, $A = A_0 \exp(-r^2/2r_0^2)$. Presuming that r_0 is much bigger than wavelength, obtain the power carried by the laser in vacuum.

[HINT: *Take* $\mathbf{S}_{av} = \hat{z}\,(A^2/2\mu_0 c)$ *and the power carried through a ring of inner and outer radii r and r+dr as* $S_{av} 2\pi r dr$.]

[ANS. $\pi r_0^2 A_0^2/2\mu_0 c$.]

9. An electromagnetic wave $\mathbf{E} = \hat{x} A_0 \exp[-i(\omega t - \omega z/c)]$ is normally incident on a dielectric slab of refractive index η and thickness d. Show that the amplitude transmission coefficient is

$$T = \frac{4\eta}{(1+\eta)^2 \exp\left[i(1-\eta)\frac{\omega d}{c}\right] - (1-\eta)^2 \exp\left[i(1+\eta)\frac{\omega d}{c}\right]}.$$

[**Hint:** *Write the electric fields of reflected and transmitted waves as* $\mathbf{E}_R = \hat{x}\, R\, A_0 \exp[-i(\omega t + \omega z/c)]$, $\mathbf{E}_T = \hat{x}\, T\, A_0 \exp[-i(\omega t - \omega z/c)]$ *and the field inside the slab as* $\mathbf{E} = \hat{x}\, \{A_1 \exp[-i(\omega t - \eta\omega z/c)] + A_2 \exp[-i(\omega t + \eta\omega z/c)]\}$. *Write corresponding* $\mathbf{H}$ *using* $\mathbf{H} = \nabla \times \mathbf{E}/(i\mu_0\omega)$ *and apply the boundary conditions.*]

10. A 15 GHz microwave, propagating through a weakly collisional plasma of electron density $10^{12}\ \text{cm}^{-3}$, suffers a power loss of 5 dB in a distance of 2 m. Estimate the electron collision frequency.

[**Hint:** *Power of a wave decreases with distance of propagation* z *as* $P = P_0 \exp(-2 k_i z)$ *and power loss in* dB *is* $10 \log_{10}(P_0/P)$.]

[**Ans.** $\nu = 5 \times 10^8\ \text{s}^{-1}$.]

11. An electromagnetic wave, propagating through a dielectric, has $\mathbf{E} = \hat{x}\, A \exp[-i(\omega t - (2\omega/c)(y+2z))]$. Obtain (i) ε_r, (ii) $\mathbf{H}$ and (iii) the equation of the wavefront.

[**Ans.** (i) $\varepsilon_r = 20$, (ii) $\mathbf{H} = (2/\mu_0 c)(2\hat{y} - \hat{z})E_x$, (iii) $y + 2z = \text{constant}$.]

12. Two electromagnetic waves in a medium have electric fields

$$\mathbf{E_1} = \mathbf{A_1} \exp[-i(\omega_1 t - (\omega_1/c)(x + 2y))],$$
$$\mathbf{E_2} = \mathbf{A_2} \exp[-i(\omega_2 t - (\omega_2/c)(2x + y))].$$

What is the angle between their respective Poynting vectors?

[**Ans.** $\theta = \cos^{-1}(\mathbf{k_1}.\mathbf{k_2}/k_1 k_2) = \cos^{-1}(4/5)$.]

13. Obtain the magnitude of electron oscillatory velocity due to a linearly polarised laser of frequency

$\omega = 2\times10^{15}$ rad/s and intensity 10^{20} W/m^2, in a plasma of $\omega_P/\omega = 0.8$. Ignore collisions.

[**Ans.** 2.7×10^7 m/s.]

14. A whistler wave of frequency ω propagates along the dc magnetic field in a collisionless plasma with $\omega_p = 10\,\omega$, $\omega_c = 2\,\omega$. Estimate the phase and group velocities of the whistler.

[**Ans.** $v_{ph} = v_g = c/(101)^{1/2}$.]

15. A Langmuir wave of frequency $\omega = 1.1\,\omega_p$ propagates in a hot plasma with electron thermal velocity $v_{th} = 10^9$ cm/s. Estimate the phase and group velocities of the wave. If the wave causes density fluctuation $|n_1|/n_0 = 1\%$, what would be the magnitude of electron oscillatory velocity due to this wave?

[**Ans.** $v_{ph} = 2.3 \times 10^9$ cm/s, $v_g = 4.3 \times 10^8$ cm/s, $|v_1| = 2.3 \times 10^7$ cm/s.

16. A circularly polarised wave in a dielectric has

$$\mathbf{E} = \mathbf{A}(\hat{x} + i\,\hat{y})\,\exp[-i\,(\omega t - 2\omega z/c)].$$

Obtain $\mathbf{S}_{av}$.

[**Ans.** $(2A^2/\mu_0 c)\,\hat{z}$.]

17. An electromagnetic wave is obliquely incident from free space onto a free space-dielectric interface (z = 0). The electric field of the incident wave is

$$\mathbf{E} = A\,(\hat{x} + 4\hat{y} - \hat{z})\,\exp[-i\,(\omega t - (\omega/c\sqrt{2})\,(x + z))].$$

The refractive index of the dielectric is 4. Obtain the field of reflected wave.

[**HINT:** *Decompose the incident wave into* S *and* P *polarisations. Write the reflected fields of the two polarisations and add.*]

[**ANS.** $\mathbf{E_R} = -A(0.47\,\hat{x} + 2.8\,\hat{y} + 0.47\,\hat{z})\,e^{-i(\omega t - (\omega/c\sqrt{2})(x-z))}$].

18. A left circularly polarised pulse of carrier frequency $\omega/2\pi = 10^{10}$ Hz propagates through a vacuum chamber in 1 ns. When the chamber is filled with a collisionless plasma, it takes 2 ns to traverse through the chamber. Obtain ω_p^2/ω^2. If one applies an axial magnetic field with $\omega_c = \omega/2$, what would be the propagation time for the pulse?

[**HINT:** *A pulse propagates with group velocity.*]

[**ANS.** $\omega_p^2/\omega^2 = 3/4, 1.3\,\text{ns}$]

19. In a plasma $\mathbf{E} = \mathbf{A}\,\hat{x}\,\exp[-i(\omega t - kx)]$. Write $\mathbf{B}, \mathbf{J}, \rho$.

[**ANS.** $\mathbf{B} = 0,\ \mathbf{J} = \mathbf{i\omega\varepsilon_0 E},\quad \rho = kJ_x/\omega.$]

20. An electromagnetic wave of frequency ω, normally incident on a collisionless plasma, suffers a phase change of $\pi/2$ on reflection. Estimate ω_p/ω.

[**ANS.** $\omega_p/\omega = \sqrt{2}.$]

21. In a plasma, the electric field of an electromagnetic wave is $\mathbf{E} = \mathbf{A}\exp[-i(\omega t - \omega z/4c)]$. Obtain the electron density of the plasma in terms of ω. Take $\varepsilon_r = 1$.

[**ANS.** $15\omega^2 m\varepsilon_0/(16e^2)$.]

REFERENCES AND SUGGESTED READING

[1] Stix T.H., *Plasma Waves*, New York, Mc Graw Hill, 1962.

[2] Liu C.S. and Tripathi V.K. *Interaction of electromagnetic waves with electron beams and plasmas*, Singapore, World Scientific, 1994.

[3] Krall N. A. and Trivelpice A. W., *Principles of plasma physics*, New York, Mc Graw Hill, 1973.

[4] Ginzburg V. L. *The propagation of electromagnetic waves in plasmas*, Pergamon Press 1970.

[5] Jordan E. C. and Balmain K. G., *Electromagnetic waves and radiating systems*, N. J., USA, Prentice-Hall, 1970.

Unit 3

Guided Waves

Introduction

Point-to-point communication and numerous other applications require electromagnetic waves to remain localised in the transverse direction, as they propagate. However, waves of finite transverse size suffer diffraction divergence. They also suffer scattering losses due to particles, refractive index irregularities and other obstacles in their path. Such losses can be overcome by guiding the waves through coaxial cables, metallic pipes, glass fibres and planar guiding systems. The guiding systems have other advantages too. For instance, wave guides loaded with slow wave structures can reduce the phase velocity of microwaves much below the velocity of light in vacuum, facilitating the amplification of these waves by electron beams (via *Cerenkov interaction*) in travelling wave tubes. Guided slow waves, under suitable conditions, can be used to accelerate electrons to ultra-relativistic energies as in a linear accelerator (LINAC). Microwave and optical passive devices like directional couplers, circulators, attenuators, etc. too involve guided propagation.

In this unit, we study guided propagation of electromagnetic waves through various kinds of guiding systems. We begin with the surface plasma wave (SPW), that is guided along the interface between a conductor and a dielectric. Then, we study the propagation of waves through parallel plane guiding system, rectangular waveguide, cylindrical waveguide, sheath helix, rippled wall guiding systems and coaxial cable. The propagation through optical fibres is taken up separately in the chapter on optical communication.

A common feature of all the guiding systems, except the rippled wall configuration, is that in one direction (chosen as z axis in this chapter) electrical and dielectric properties of the medium or guiding structure do not change. Hence, for a monochromatic wave of frequency ω propagating along the z direction, the t, z variations of fields can be taken as

$$\mathbf{E}, \mathbf{H} \sim e^{-i(\omega t - k_z z)}. \tag{1}$$

Then, the only problem is to solve the wave equation in one or two space variables, with proper boundary conditions, to deduce the dispersion relation (k_z as a function of ω) and the transverse variation of field amplitudes (called mode structure).

3.1 Surface Plasma Waves

A surface plasma wave (also known as a surface wave) is a guided electromagnetic wave that propagates along the boundary between a conductor and a dielectric. Its amplitude is maximum on the boundary and falls off, away from it, in either medium.

Consider a conductor-dielectric interface at $x = 0$ (Fig. 3.1). The conductor, characterised by effective relative permittivity ϵ_{eff}, occupies the half space $x < 0$, while the dielectric of relative permittivity ϵ_r lies in $x > 0$. ϵ_{eff} is related to dielectric constant ϵ_L, electrical conductivity σ of the conductor and the frequency ω of the wave as

$$\epsilon_{eff} = \epsilon_L + \frac{i\sigma}{\omega \epsilon_0}.$$

We presume t, z variations of fields as given in eq. (1), take y-variation to be zero and look for solutions of wave equation that fall off with $|x|$, as one moves away from the interface. If we choose $\mathbf{E} \parallel \hat{y}$, we do not get such a solution hence we discard this case. We consider $\mathbf{E} = E_x \hat{x} + E_z \hat{z}$. The procedure is as follows. We solve

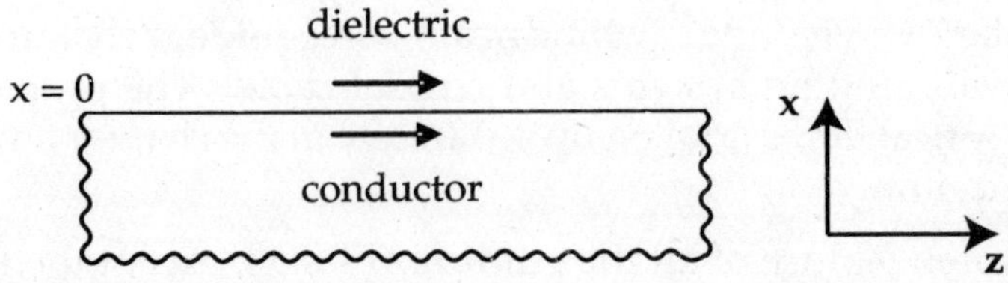

Figure 3.1: Schematic of surface plasma wave propagation over a conductor - dielectric interface

the wave equation for E_z in the two media, deduce E_x, H_y from E_z using the relevant Maxwell's equations and apply the continuity of E_z and H_y across $x = 0$ to obtain the dispersion relation.

The third and fourth Maxwell's equations, on replacing $\partial/\partial t$ with $-i\omega$, can be written as

$$\nabla \times \mathbf{E} = i\omega\mu_0 \, \mathbf{H}, \tag{2}$$

$$\nabla \times \mathbf{H} = i\omega \, \epsilon_0 \, \epsilon'_{eff} \, \mathbf{E}, \tag{3}$$

where $\epsilon'_{eff} = \epsilon_{eff}$ for $x < 0$ and $\epsilon'_{eff} = \epsilon_r$ for $x > 0$. Taking curl of eq. (2), using eq. (3) and the identity $\nabla \times \nabla \times \mathbf{E} = \nabla(\nabla.\mathbf{E}) - \nabla^2\mathbf{E}$, we get

$$\nabla^2\mathbf{E} - \nabla(\nabla.\mathbf{E}) + \frac{\omega^2}{c^2} \, \epsilon'_{eff} \, \mathbf{E} = 0. \tag{4}$$

$\nabla.$ of this equation, in each of the two media, gives $\epsilon'_{eff} \, \nabla. \, \mathbf{E} = 0$, i.e., $\nabla. \, \mathbf{E} = 0$. Hence eq. (4) becomes

$$\nabla^2\mathbf{E} + \frac{\omega^2}{c^2} \, \epsilon'_{eff} \, \mathbf{E} = 0. \tag{5}$$

Using $\partial/\partial z = ik_z$ to write $\nabla^2 = \partial^2/\partial x^2 - k_z^2$ in eq. (5), we obtain for E_z,

$$\frac{\partial^2 E_z}{\partial x^2} - \alpha^2 E_z = 0, \tag{6}$$

$$\text{where} \quad \alpha^2 = \alpha_1^2 \equiv k_z^2 - \omega^2 \frac{\epsilon_r}{c^2} \qquad \text{for } x > 0,$$

$$\alpha^2 = \alpha_2^2 \equiv k_z^2 - \omega^2 \frac{\epsilon_{eff}}{c^2} \qquad \text{for } x < 0. \tag{7}$$

The well-behaved solution of eq. (6), that vanishes at $x \to \pm\infty$, is

$$E_z = Ae^{-\alpha_1 x} e^{-i(\omega t - k_z z)}, \quad x > 0$$
$$= A' e^{\alpha_2 x} e^{-i(\omega t - k_z z)}, \quad x < 0. \tag{8}$$

Demanding the continuity of E_z at x = 0, we get A′ = A. Using eq. (8) in $\nabla . \mathbf{E} = 0$, we obtain

$$E_x = \left(i\frac{k_z}{\alpha_1}\right) Ae^{-\alpha_1 x} e^{-i(\omega t - k_z z)}, \quad x > 0$$
$$= -\left(i\frac{k_z}{\alpha_2}\right) Ae^{\alpha_2 x} e^{-i(\omega t - k_z z)}, \quad x < 0. \tag{9}$$

Substituting for E_x and E_z in eq. (2), we get the magnetic field of the wave as

$$H_y = i\left(\omega\frac{\epsilon_r}{\mu_0 c^2 \alpha_1}\right) Ae^{-\alpha_1 x} e^{-i(\omega t - k_z z)}, \quad x > 0$$
$$= -i\left(\omega\frac{\epsilon_{eff}}{\mu_0 c^2 \alpha_2}\right) Ae^{\alpha_2 x} e^{-i(\omega t - k_z z)}, \quad x < 0. \tag{10}$$

Applying the continuity of H_y (or $\epsilon'_{eff} E_x$) at x = 0, we get $\epsilon_{eff}/\epsilon_r = -\alpha_2/\alpha_1$ or

$$k_z^2 = \frac{\omega^2}{c^2}\frac{\epsilon_r \epsilon_{eff}}{\epsilon_r + \epsilon_{eff}}. \tag{11}$$

This is the dispersion relation for surface plasma wave. If ϵ_r and ϵ_{eff} are real, k_z^2 is real. The decay rates of wave amplitude in the two media are α_1 and α_2 respectively, given by

$$\alpha_1 = \frac{(\omega\ \epsilon_r\ /c)}{[-(\epsilon_r + \epsilon_{eff})]^{\frac{1}{2}}}$$
$$\alpha_2 = \frac{(\omega/c)}{[-(\epsilon_r + \epsilon_{eff})/\ \epsilon_{eff}^2]^{\frac{1}{2}}}. \tag{12}$$

For α_1, α_2 to be real one must have $\epsilon_r + \epsilon_{eff} < 0$. In a conductor, if collisions are ignored, $\sigma = -n_0 e^2/\ mi\omega$, $\epsilon_{eff} = \epsilon_L - \omega_p^2/\omega^2$. This condition gives

$$\omega < \omega_{sc}; \quad \omega_{sc} = \frac{\omega_p}{(\epsilon_r + \epsilon_L)^{\frac{1}{2}}} \tag{13}$$

where ω_{sc} is the *surface wave cutoff frequency*, $\omega_p = (n_0 e^2 / m \in_0)^{1/2}$ is the electron plasma frequency and n_0, $-e$ and m are the density, charge and effective mass of free electrons respectively. Condition (13) is essential for the existence of surface plasma wave. Under this condition, $\in_{eff}$ is negative and eq. (11) can be written as

$$k_z^2 = \frac{\omega^2 \in_r}{c^2} \frac{\omega_p^2/\omega^2 - \in_L}{\omega_p^2/\omega^2 - (\in_L + \in_r)}. \tag{14}$$

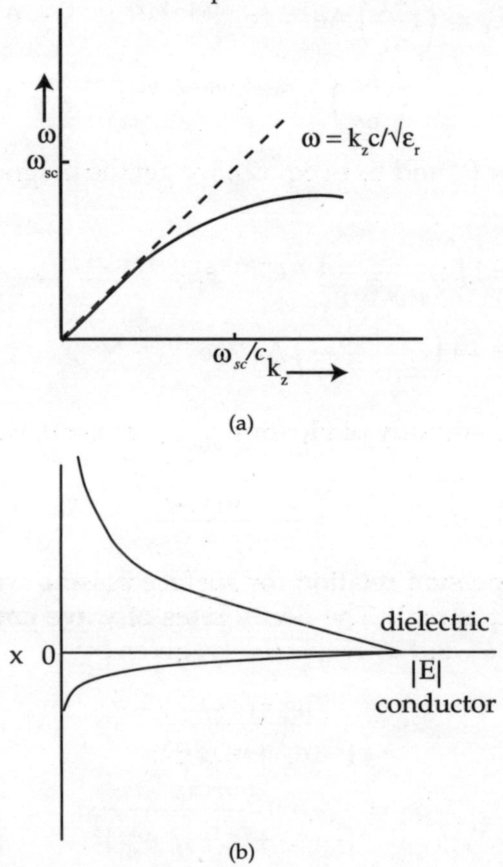

Figure 3.2: (a) Surface plasma wave dispersion relation
(b) amplitude of surface plasme wave as a function of distance from the interface.

We have plotted the variation of ω with k_z in fig. 3.2 . At low frequencies $\omega << \omega_p/(\in_L + \in_r)^{1/2}$, ω varies linearly with k_z and the phase velocity of the wave is $c/\sqrt{\epsilon_r}$. In this case $\alpha_1 = \omega^2 \epsilon_r / c\, \omega_p$, $\alpha_2 = \omega_p \epsilon_r / c$, i.e. $\alpha_1 << \alpha_2$, and the wave is localised in a thin layer of thickness $\sim \alpha_2^{-1}$ in the conductor. However, it extends to a larger width $\sim \alpha_1^{-1}$ in the dielectric. As k_z increases, ω increases less than linearly and asymptotically approaches the value $\omega_{sc} = \omega_p/(\epsilon_L + \epsilon_r)^{1/2}$ as $k_z \rightarrow \infty$. Near this cutoff frequency, the surface plasma wave is strongly localised near the interface.

At optical frequencies, surface plasma waves are usually excited over metals by a laser, using attenuated total reflection configuration (Fig. 3.3). In this scheme, a gold film of thickness $d \leq 0.1\ \mu m$ is deposited over a prism of refractive index η_g. Laser is launched through the other face of the prism onto the glass-metal interface, at angle of incidence θ_i. Usually, 99% of laser light gets reflected back.

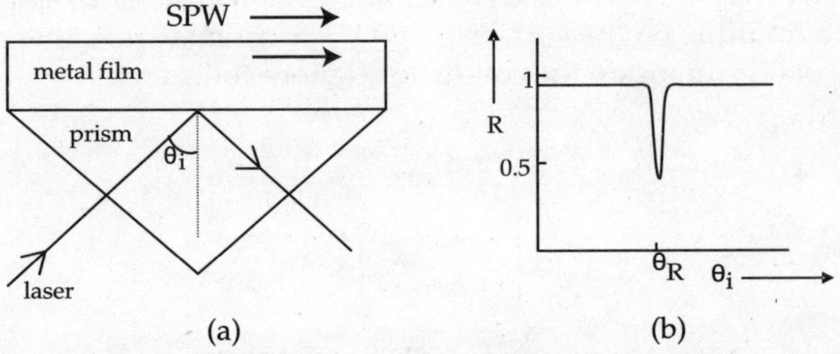

Figure 3.3: (a) Schematic of attenuated total reflection (ATR) geometry for the excitation of a surface plasma wave.
(b) Intensity reflection coefficient as a function of angle of incidence.

However, at a specific angle $\theta_i = \theta_R = \sin^{-1}(c\, k_z / \omega\, \eta_g)$ where k_z is given by eq. (14), the reflectivity drops down drastically below 50%. At this angle, surface wave is excited at the metal-free space boundary, as the component of laser wave vector along the, interface equals the wave vector of the surface plasma wave (SPW) and laser energy tunnels through the metal to excite the SPW. If there is small contamination or deposition of a film over the metal surface, the SPW resonance angle θ_R changes significantly. For this reason, SPW is widely used for sensor applications, specially bio-sensors and nano-particle detection. High amplitude surface plasma waves are effective in rapid ablation of materials. A sub-picosecond intense pulse, with fluence $\geq 0.3\,\text{mJ/cm}^2$, is usually employed for material ablation. When the laser energy is mode converted into surface plasma wave, either by employing ATR configuration or by creating short wavelength ripples over the target metal surface, ten-fold enhancement in ablation yield has been observed.

A major traditional application of surface plasma waves is in medium frequency radio communication, along the surface of the Earth. We discuss this problem in detail, in the unit on radio communication.

PROBLEM 3.1. A surface plasma wave (SPW), of $\omega = 2 \times 10^{15}$ rad/s, propagates over a conductor-free space interface. The conductor has $\epsilon_r = 10$, $\omega_p = 8 \times 10^{15}$ rad/s. Estimate the wavelength of SPW. If the amplitude on the surface is 100 V/m, estimate its value at a depth of 0.1 μm inside the conductor. Ignore collisions.

Solution :

$$\epsilon_{eff} = \epsilon_r - \frac{\omega_p^2}{\omega^2} = -6,$$

$$k_z = \left(\frac{\omega}{c}\right)\left[\frac{\epsilon_{eff}}{(1+\epsilon_{eff})}\right]^{\frac{1}{2}} = \sqrt{\frac{6}{5}}\frac{\omega}{c},$$

$$\lambda \approx \frac{2\pi}{k_z} = 0.85\mu\text{m}.$$

The field amplitude at depth d = 0.1 μm inside the conductor is $A_d = A \exp(-\alpha_2 d)$, where

$$\alpha_2 = \left(\frac{\omega}{c}\right)\left[-\frac{\epsilon_{eff}^2}{(1+\epsilon_{eff})}\right]^{\frac{1}{2}}, \alpha_2 d = 1.8.$$

Hence $A_d \approx 15$ V / m.

3.2 Parallel Plane Guiding System

A parallel plane guiding system comprises two parallel metallic planes, say at x = 0 and x = a (Fig. 3.4). The electromagnetic waves

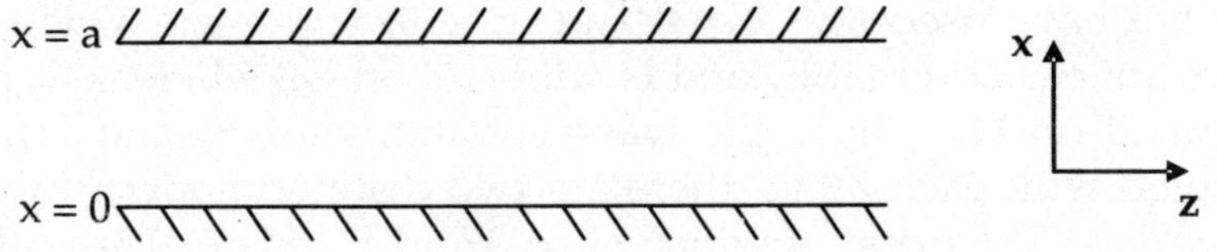

Figure 3.4: A parallel plane guiding system

propagate through the space between them. We take the direction of propagation as z-axis and write $\partial/\partial y = 0$. x-variation of fields, however, cannot be ignored as x-independent fields do not satisfy the metallic boundary conditions at $x = 0$ and $x = a$, except in a special case when the electric field of the wave is polarised along $\hat{x}$. We presume t, z variations of fields as $\exp[-i(\omega t - k_z z)]$ and replace $\partial/\partial t$ by $-i\omega$ and $\partial/\partial z$ by ik_z. Then the third and fourth Maxwell's equations, in the free space between the guiding planes, take the form of eqs. (2) and (3) with $\epsilon'_{eff} = 1$. The x and y components of these equations facilitate the writing of transverse components of **E** and **H**, in terms of their longitudinal components E_z, H_z as

$$E_x = \frac{ik_z}{k_\perp^2}\frac{\partial E_z}{\partial x}, \quad H_y = \frac{i\omega \epsilon_0}{k_\perp^2}\frac{\partial E_z}{\partial x},$$
$$E_y = -\frac{i\omega\mu_0}{k_\perp^2}\frac{\partial H_z}{\partial x}, \quad H_x = \frac{ik_z}{k_\perp^2}\frac{\partial H_z}{\partial x}, \tag{15}$$

where $k_\perp^2 = (\omega^2/c^2) - k_z^2$. The wave equations governing **E** and **H** are

$$\nabla^2\mathbf{E} + \frac{\omega^2}{c^2}\mathbf{E} = 0, \tag{16}$$
$$\nabla^2\mathbf{H} + \frac{\omega^2}{c^2}\mathbf{H} = 0. \tag{17}$$

For E_z and H_z, these equations can be written as

$$\frac{\partial^2 E_z}{\partial x^2} + k_\perp^2 E_z = 0, \tag{18}$$
$$\frac{\partial^2 H_z}{\partial x^2} + k_\perp^2 H_z = 0, \tag{19}$$

where we have used $\partial/\partial y = 0$ and $\partial^2 / \partial z^2 = -k_z^2$. One may notice from eq. (15) that E_x and H_y depend on E_z, whereas E_y and H_x depend on H_z. Since the wave equations for E_z and H_z are uncoupled with each other, there are two distinct modes of wave propagation. The ones for which $H_z = 0$ and $E_x, H_y \neq 0$ are called *transverse magnetic* (TM) modes, whereas the ones with $E_z = 0$, H_x, $E_y \neq 0$ are called *transverse electric* (TE) modes.

3.2.1 TM Modes

For TM modes, eq. (18) gives

$$E_z = \left(A_1 e^{ik_\perp x} + A_2 e^{-ik_\perp x}\right) e^{-i(\omega t - k_z z)}.$$

Applying the boundary condition, $E_z = 0$ at $x = 0$, one has $A_1 = -A_2$ and E_z can be written as

$$E_z = A \sin(k_\perp x)\, e^{-i(\omega t - k_z z)}, \tag{20}$$

where $A = 2iA_1$ is an arbitrary constant. The boundary condition, $E_z = 0$ at $x = a$ demands $k_\perp a = n\pi$ or

$$\omega^2 = k_z^2 c^2 + \frac{n^2 \pi^2 c^2}{a^2}, \tag{21}$$

where n is an integer. A TM mode with a particular value of n is called TM_{n0} mode, where second subscription 0 implies that there is no y-variation. Its minimum frequency, called the *cutoff*, is

$$\omega_{n0} = \frac{n\pi c}{a}. \tag{22}$$

As k_z increases from $k_z = 0$ to large values, ω increases from ω_{n0} to $k_z c$. The cutoff frequencies for higher order modes are higher (Fig. 3.5). For odd modes ($n = 1, 3, 5$ ...), E_z is maximum at the centre of the guiding system whereas for even modes, $E_z = 0$ at $x = a/2$.

E_x and H_y for TM modes are

$$E_x = iA \frac{k_z}{k_\perp} \cos(k_\perp x)\, e^{-i(\omega t - k_z z - \pi/2)},$$

$$H_y = iA \frac{\omega \in_0}{k_\perp} \cos(k_\perp x)\, e^{-i(\omega t - k_z z - \pi/2)}. \tag{23}$$

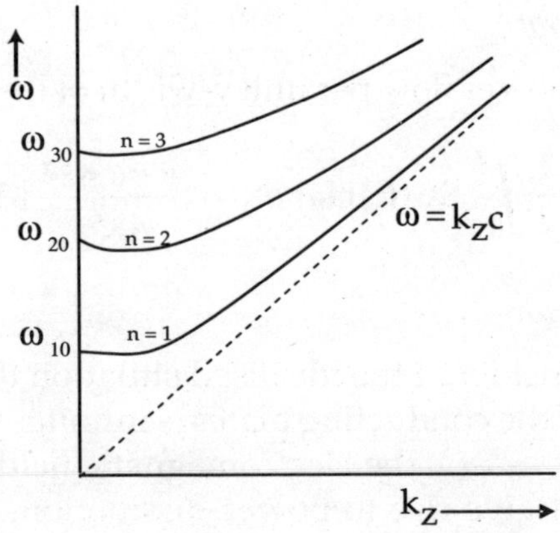

Figure 3.5: Dispersion relation for TM_{10}, TM_{20} and TM_{30} modes, corresponding to n = 1, 2, 3 respectively, in a parallel plane guiding system

These lag behind E_z in phase by $\pi/2$. Further, at values of x where E_z is maximum, E_x, H_y have minima and vice versa. In fig. 3.6, we have plotted $|E_z|$ and $|E_x|$ as a function of x for TM_{10} and TM_{20} modes.

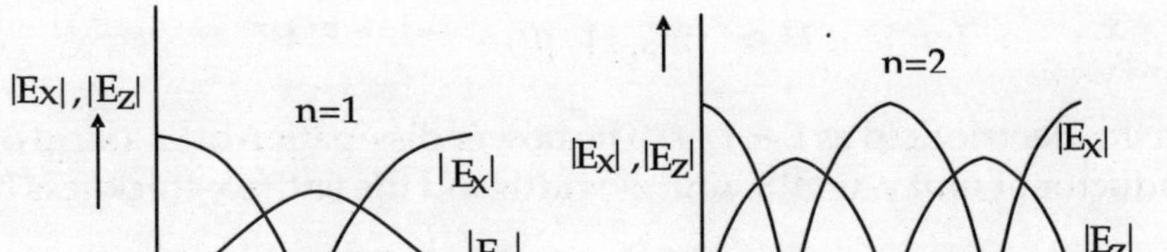

Figure 3.6: Electric field profiles for TM_{10} and TM_{20} modes.

The case of n = 0 i.e. $k_\perp = 0$ is special. For this mode, $E_z = 0$, $E_x = A' e^{-i(\omega t - \omega z/c)}$, $H_y = (A' / \mu_0 c)\, e^{-i(\omega t - \omega z/c)}$. This is like a plane wave in an unbounded medium, with purely transverse electric and transverse magnetic fields and is known as *TEM mode*.

3.2.1.1 *Power flow*

The time average power flow per unit y-width of the guiding system is

$$P_z = \frac{1}{2}\int_0^a \mathrm{Re}\left(E_x^* H_y\right) dx = \frac{\omega \in_0 k_z a}{4k_\perp^2}\,|A|^2 . \qquad (24)$$

3.2.1.2 *Attenuation*

So far, we have considered the idealised situation that the electrical conductivity, σ, of the conducting plates is infinite. When σ is finite, yet large ($\sigma >> \omega \in_0 \in_L$), the electromagnetic fields penetrate into the conductor and give rise to power dissipation. An estimate of attenuation of waves can be obtained as follows.

Consider the x=0 interface. Just above x=0, $H_y = H_y(0)\, e^{-i(\omega t - k_z z)}$ for a TM mode, where $H_y(0) = \frac{iA\omega\in_0}{k_\perp}$. By the continuity of H_y across the interface, H_y inside the conductor (x < 0) can be written as

$$H_y = H_y(0) e^{-i(\omega t - k_z z + k_c x)}, \qquad (25)$$

where $k_c = \left(\omega^2 \varepsilon_{eff}/c^2 - k_z^2\right)^{1/2} \approx \alpha\,(1+i)$, $\alpha = (\omega/c)\,(\sigma/2\omega\varepsilon_0)^{1/2}$. Using generalised Ampere's law, the current density can be written (on ignoring the displacement current density) as

$$\mathbf{J} \approx \nabla \times \mathbf{H} \approx -\hat{z} i k_c H_y(0)\, e^{-i(\omega t - k_z z + k_c x)}$$

and the electric field as $\mathbf{E} \approx \mathbf{J}/\sigma$. The power dissipation over a strip of conductor of unit y-width, unit z- width and the entire x-thickness is

$$P_A = \frac{1}{2\sigma}\int_0^{-\infty} \mathbf{J}.\mathbf{J}^* dx = \frac{\alpha}{2\sigma}\left|H_y(0)\right|^2 . \qquad (26)$$

In the case of a parallel plane guiding system, the power is equally dissipated in the lower plane as well as the upper plane. Let the time average power flow through the guiding system (per unit y-width) at z be P_z and at z + dz, be $P_z + d\,P_z$. In the conductor strips (x < 0 and x > a) of y-width unity and z-width dz, by energy conservation, $dP_z = -2\,P_A\,dz$ or

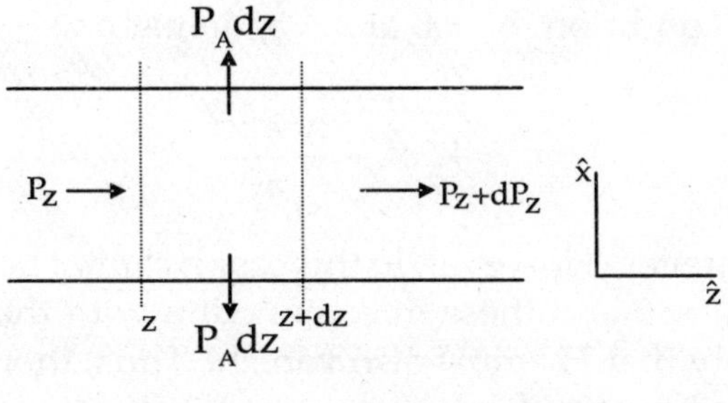

Figure 3.7: Power flow in a parallel plane guiding system. P_z is the power crossing a cross-sectional plane of height a and width 1 unit.

$$\frac{dP_z}{dz} = -\frac{\alpha}{\sigma}\left|H_y(0)\right|^2 = -2k_iP_z \tag{27}$$

where $k_i = (2\alpha\omega\,\epsilon_0)/(\sigma k_z a) = (\omega/ck_z a)(2\omega\,\epsilon_0/\sigma)^{1/2}$. The solution of above equation is $P_z = P_0\,e^{-2k_i z}$. k_i is called the amplitude attenuation constant. It goes as $\sigma^{-1/2}$ with the conductivity of the guiding planes. Its dependence on wave frequency is complex. Near the cutoff frequency ($\omega \sim \omega_{n0}$) it is large, then it falls off with ω and at $\omega >> \omega_{n0}$, it scales as $\omega^{1/2}$. Thus, the higher frequency waves suffer stronger attenuation.

3.2.2 TE Modes

For TE modes, eq. (17) gives

$$H_z = \left(A_1 e^{ik_\perp x} + A_2 e^{-ik_\perp x}\right) e^{-i(\omega t - k_z z)}. \tag{28}$$

Using eq. (15), E_y can be written as

$$E_y = \frac{\omega\,\mu_0}{k_\perp}\left(A_1 e^{ik_\perp x} - A_2 e^{-ik_\perp x}\right) e^{-i(\omega t - k_z z)}.$$

Applying the boundary condition, $E_y = 0$ at $x = 0$, we get $A_1 = A_2$ and E_y can be written as

$$E_y = A\sin(k_\perp x)\, e^{-i(\omega t - k_z z)}. \tag{29}$$

The boundary condition, $E_y = 0$ at $x = a$, demands $k_\perp a = n\pi$ or

$$\omega^2 = k_z^2 c^2 + \frac{n^2\pi^2 c^2}{a^2}, \tag{30}$$

where n is an integer. However, in this case n cannot be zero because then $E_y = 0$, $H_x = 0$ and these in conjunction with the Gauss's law $\nabla \cdot \mathbf{H} = 0$, imply that H_z must also vanish. Thus, there is no wave. The TE_{n0} mode has cutoff frequency

$$\omega_{n0} = \frac{n\pi c}{a}, \tag{31}$$

that increases linearly with the mode number. For odd order modes, E_y has a maximum on the axis whereas for even order modes, E_y is minimum at $x = a/2$. The plot of ω vs. k_z for TE_{n0} mode is the same as that for TM_{n0} mode.

The time average power flow per unit y-width is

$$P_z = \frac{1}{2}\int_0^a \mathrm{Re}\left(-H_x E_y^*\right) dx = \frac{k_z a}{4\omega\mu_0}|A|^2, \tag{32}$$

where we have employed (from eqs. (15)), $H_x = -(k_z/\omega\mu_0)E_y$. For metallic planes of finite conductivity, the power dissipation in the strips of unit y-width, unit z-width and full depth in x

$$P_A = \frac{\alpha}{2\sigma}|H_z(0)|^2 = \frac{\alpha}{2\sigma}\frac{k_\perp^2}{\omega^2\mu_0^2}|A|^2 = k_i P_z,$$

where $k_i = \left(2\alpha \in_0 k_\perp^2 c^2\right)/(\sigma k_z a\omega)$. Following the treatment of the TM mode, the equation for dP_z/dz can be obtained as

$$\frac{dP_z}{dz} = -2P_A = -2k_i P_z. \tag{33}$$

The power flow diminishes with z as $P_z = P_0\, e^{-2k_i z}$.

A parallel plane guiding system localises the waves in one dimension and is useful when power from a source of radiation is to be delivered to a larger number of receivers on the y-z plane. In fact, there exists a natural waveguide on the Earth, with Earth as one conducting plane and ionosphere (the ionised region of atmosphere at a height of >90 km) as the second. Extremely low frequency waves (≤ 100 Hz) propagate as TEM mode through this waveguide around the globe. These waves penetrate into sea water and are used for submarine communication.

3.3 Rectangular Waveguide

A rectangular waveguide is a hollow metallic pipe with rectangular cross-section. It is widely used for microwave propagation, microwave characterisation of materials and device fabrication. We consider a waveguide, with its axis along $\hat{z}$ and the interior walls at x = 0, a and y = 0, b as shown in fig. 3.8. An electromagnetic wave propagates through it with t, z dependence as $\exp[-i(\omega t - k_z z)]$. We express transverse components of **E** and **H** in terms of E_z, H_z using the third and fourth Maxwell's equations.

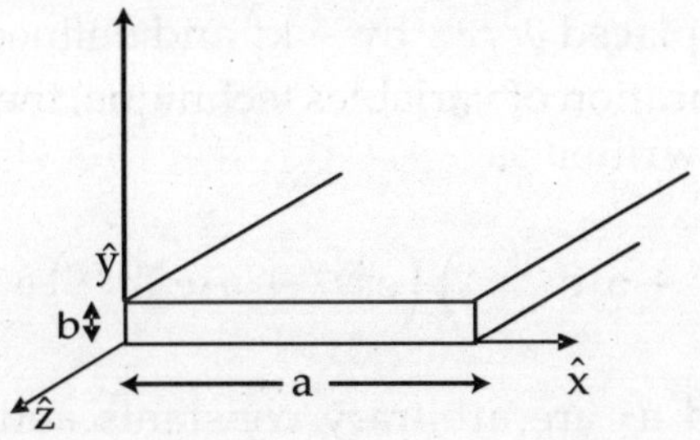

Figure 3.8: A rectangular waveguide.

Eqs. (2) and (3) yield

$$E_x = \frac{i}{k_\perp^2}\left(\omega\,\mu_0 \frac{\partial H_z}{\partial y} + k_z \frac{\partial}{\partial x} E_z\right),$$

$$H_y = \frac{i}{k_\perp^2}\left(k_z \frac{\partial H_z}{\partial y} + \omega\,\epsilon_0\,\frac{\partial}{\partial x} E_z\right),$$

$$E_y = -\frac{i}{k_\perp^2}\left(\omega\,\mu_0\frac{\partial H_z}{\partial x} - k_z\frac{\partial}{\partial y}E_z\right),$$

$$H_x = \frac{i}{k_\perp^2}\left(k_z\frac{\partial H_z}{\partial x} - \omega\,\epsilon_0\,\frac{\partial}{\partial y}E_z\right), \tag{34}$$

where $k_\perp^2 = (\omega^2/c^2) - k_z^2$. Since the wave equations governing E_z and H_z are independent of each other and there is no coupling between these components through the boundary conditions, we study the propagation of waves with (i) $H_z = 0$, $E_z \neq 0$ (TM mode) and (ii) $H_z \neq 0$, $E_z = 0$ (TE mode) separately. Any general wave would be a linear superposition of the two.

3.3.1 TM Mode

Equation (16) for E_z can be written as

$$\frac{\partial^2 E_z}{\partial x^2} + \frac{\partial^2 E_z}{\partial y^2} + k_\perp^2 E_z = 0, \tag{35}$$

where we have replaced $\partial^2/\partial z^2$by $-\,k_z^2$ and defined $k_\perp^2 = (\omega^2/c^2) - k_z^2$. Using the separation of variables technique, the general solution of eq. (35) can be written as

$$E_z = A_1\left(e^{ik_x x} + a_1 e^{-ik_x x}\right)\left(e^{ik_y y} + a_2 e^{-ik_y y}\right)e^{-i(\omega t - k_z z)}, \tag{36}$$

where A_1, a_1 and a_2 are arbitrary constants and $k_x^2 + k_y^2 = k_\perp^2$. Applying the boundary conditions, $E_z = 0$ at $x = 0$, a, we obtain $a_1 = -1$,

$$k_x a = m\pi, \tag{37}$$

where m is an integer. Similarly, the boundary conditions, $E_z = 0$ at $y = 0$, b give $a_2 = -1$,

$$k_y b = n\pi,, \tag{38}$$

where n is an integer. Eq. (36) can now be written as

$$E_z = A\sin\left(\frac{m\pi}{a}x\right)\sin\left(\frac{n\pi}{b}y\right)e^{-i(\omega t - k_z z)}, \tag{39}$$

whereas using the values of k_x and k_y in the expression for $k_\perp^2$, we obtain the dispersion relation

$$\omega^2 = k_z^2 c^2 + \omega_{mn}^2, \tag{40}$$

where $\omega_{mn}^2 = \pi^2 c^2 \left(\frac{m^2}{a^2} + \frac{n^2}{b^2}\right)$.

A TM mode with specific values of m and n is denoted as TM_{mn} mode. The minimum value of frequency for a propagating TM_{mn} mode $(k_z^2 \geq 0)$ is ω_{mn}. It is also called the cutoff frequency. ω_{mn} is higher for higher modes. At the cutoff frequency, the phase velocity of a particular mode is infinite. As ω increases, phase velocity decreases and approaches c asymptotically. One may note from eq. (39), that for a non-trivial solution $m \geq 1$ and $n \geq 1$. For the lowest order mode (TM_{11} mode), E_z peaks at $x = a/2$, $y = b/2$ and the dispersion relation simplifies to

$$\omega^2 = k_z^2 c^2 + \pi^2 c^2 \left(\frac{1}{a^2} + \frac{1}{b^2}\right).$$

One may explicitly express the transverse field components of a TM_{mn} mode, using eqs. (34), as:

$$E_x = \left(\frac{i k_z k_x}{k_\perp^2}\right) A \cos(k_x x) \sin(k_y y),$$

$$H_y = \left(\frac{i\omega \epsilon_0 k_x}{k_\perp^2}\right) A \cos(k_x x) \sin(k_y y),$$

$$E_y = \left(\frac{i k_z k_y}{k_\perp^2}\right) A \sin(k_x x) \cos(k_y y),$$

$$H_x = -\left(\frac{i\omega \epsilon_0 k_y}{k_\perp^2}\right) A \sin(k_x x) \cos(k_y y), \tag{41}$$

where $\exp[-i(\omega t - k_z z)]$ factor is implicit. The dispersion relation may also be expressed in terms of guide wavelength $\lambda_g = 2\pi k_z^{-1}$, free space wavelength $\lambda_0 = 2\pi c / \omega$ and cutoff wavelength $\lambda_{mn} = 2\pi c/\omega_{mn}$ as

$$\lambda_g^2 = \lambda_0^2 / \left(1 - \frac{\lambda_0^2}{\lambda_{mn}^2}\right). \tag{42}$$

If the waveguide is filled with a dielectric of relative permittivity ϵ_r, then eqs. (39) - (41), are valid with ϵ_0 replaced by $\epsilon_0\epsilon_r$, c^2 by c^2 / ϵ_r and $k_\perp^2 = (\omega^2/c^2)\ \epsilon_r - k_z^2$. The guide wavelength is given by

$$\lambda_g^2 = \left(\frac{\lambda_0^2}{\epsilon_r}\right) \Big/ \left(1 - \frac{\lambda_0^2}{\lambda_{mn}^2}\right),$$

where $\lambda_{mn} = 2\sqrt{\epsilon_r}/\left(\frac{m^2}{a^2} + \frac{n^2}{b^2}\right)^{1/2}$.

3.3.1.1 *Power Flow*

The time average power flow through the waveguide is

$$P_z = \frac{1}{2}\int_0^a\int_0^b \mathrm{Re}\left(E_x^* H_y - E_y^* H_x\right) dx\, dy = \frac{k_z\omega\ \epsilon_0}{8k_\perp^2} ab\ A^2. \qquad (43)$$

3.3.1.2 *Attenuation*

When the conductivity, σ, of the waveguide walls is finite, then there is power dissipation in the walls. Following eq. (26), the power dissipation per unit axial length in the waveguide is

$$\begin{aligned} P_A = \frac{1}{2c}\left(\frac{\omega}{2\sigma\ \epsilon_0}\right)^{\frac{1}{2}} \Bigg[\int_0^b \left(|H_y(x=0)|^2 + |H_y\,(x=a)|^2\right) dy \\ + \int_0^a \left(|H_x\,(y=0)|^2 + |H_x\,(y=b)|^2\right) dx \Bigg] \\ = \frac{1}{2c}\left(\frac{\omega}{2\sigma\ \epsilon_0}\right)^{\frac{1}{2}} \frac{\omega^2\ \epsilon_0^2}{k_\perp^4} A^2 \left(bk_x^2 + ak_y^2\right). \qquad (44) \end{aligned}$$

Consider two cross-sectional planes at z and z + dz. The power entering through the former is P_z and that leaving through the latter is $P_z + dP_z$, while the power dissipation in the walls in the region is P_A dz. Hence, by energy conservation,

$$\frac{dP_z}{dz} = -P_A = -2k_iP_z \qquad (45)$$

where $k_i = \frac{\omega}{c}\left(\frac{2\omega\ \epsilon_0}{\sigma}\right)^{1/2} \frac{\pi^2}{k_\perp^2 k_z}\left(\frac{m^2}{a^3} + \frac{n^2}{b^3}\right)$.

The solution of eq. (45) is

$$P_z = P_0\, e^{-2k_i z}. \tag{46}$$

The power diminishes exponentially with the distance of propagation.

3.3.1.3 *Characteristic Impedance*

The ratio of total transverse electric field strength to magnetic intensityof the TM_{mn} mode is called *characteristic impedance of the mode.*

$$Z_c = \frac{(E_x^2 + E_y^2)^{\frac{1}{2}}}{(H_x^2 + H_y^2)^{\frac{1}{2}}} = \frac{k_z}{\omega\varepsilon_0} \tag{47}$$

Using the values of k_z for TM_{mn} mode we obtain,

$$Z_c\,(TM) = Z_0 \left(1 - \frac{\omega_{mn}^2}{\omega^2}\right)^{\frac{1}{2}} \tag{48}$$

where Z_0 is the free space impedance.

3.3.2 TE Mode

Equation (17) for H_z can be written as

$$\frac{\partial^2 H_z}{\partial x^2} + \frac{\partial^2 H_z}{\partial y^2} + k_\perp^2 H_z = 0, \tag{49}$$

where $k_\perp^2 = \omega^2/c^2 - k_z^2$. Its solution is

$$H_z = A_1 \left(e^{ik_x x} + a_1 e^{-ik_x x}\right)\left(e^{ik_y y} + a_2 e^{-ik_y y}\right).e^{-i(\omega t - k_z z)}, \tag{50}$$

where $k_x^2 + k_y^2 = k_\perp^2$. The x and y components of **E**, from eqs. (41), are

$$E_x = -\frac{\omega\,\mu_0\,k_y}{k_\perp^2} A_1 \left(e^{ik_x x} + a_1 e^{-ik_x x}\right)\left(e^{ik_y y} - a_2\, e^{-ik_y y}\right),$$

$$E_y = \frac{\omega\,\mu_0\,k_x}{k_\perp^2} A_1 \left(e^{ik_x x} - a_1 e^{-ik_x x}\right)\left(e^{ik_y y} + a_2\, e^{-ik_y y}\right). \tag{51}$$

Applying the boundary conditions, $E_x = 0$ at $y = 0, b$ and $E_y = 0$ at $x = 0, a$, we obtain $a_2 = 1$, $a_1 = 1$,

$$k_x = \frac{m\pi}{a}, k_y = \frac{n\pi}{b}, \tag{52}$$

and the field components can be written as

$$\begin{aligned}
H_z &= A \cos(k_x x)\cos(k_y y),\\
E_x &= -\frac{i\omega\,\mu_0\,k_y}{k_\perp^2} A\cos(k_x x)\sin(k_y y),\\
E_y &= -\frac{i\,\omega\,\mu_0\,k_x}{k_\perp^2} A\sin(k_x x)\cos(k_y y),\\
H_x &= -\frac{ik_z\,k_x}{k_\perp^2} A\sin(k_x x)\cos(k_y y),\\
H_y &= -\frac{ik_z\,k_y}{k_\perp^2} A\cos(k_x x)\sin(k_y y),
\end{aligned} \tag{53}$$

where $\exp[-i(\omega t - k_z z)]$ factor is implicit in all the field components. Using the values of k_x and k_y in $k_\perp^2$, we obtain the dispersion relation for the TE_{mn} mode as

$$\omega^2 = k_z^2\,c^2 + \omega_{mn}^2 \tag{54}$$

where ω_{mn} is the same as that for TM_{mn} mode. However, in this case, either m or n can take a zero value as only $m = 0, n = 0$ is a trivial solution. In most of the experiments, one often excites TE_{10} mode with

$$E_x = 0, \;\; E_y = A' \sin\left(\frac{\pi x}{a}\right) e^{i(\omega t - k_z z)}, \tag{55}$$

$$\omega^2 = k_z^2 c^2 + \pi^2 \frac{c^2}{a^2}.$$

Fig. 3.9 shows the mode structure of the mode.

Every TE mode has a cutoff frequency, ω_{mn}, below which wave does not propagate. At $\omega = \omega_{mn}$, $k_z = 0$ and phase velocity $v_{ph} = \omega / k_z = \infty$. As k_z increases, ω increases and approaches the light line $\omega = k_z c$ asymptotically. If one obtains the group velocity $v_g = \partial\omega / \partial k_z$ of TE_{mn} mode, one finds that

$$v_g v_{ph} = c^2.$$

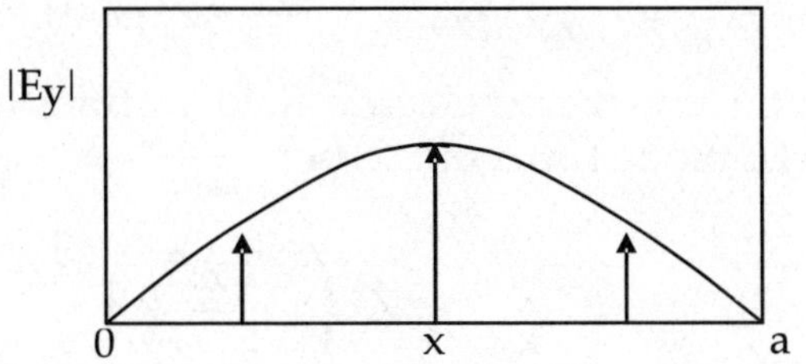

Figure 3.9: Electric field distribution of TE_{10} mode in a rectangular waveguide.

3.3.2.1 *Power Flow*

The time average power flow in a TE_{mn} mode is

$$P_z = \frac{1}{2}\int_0^b\int_0^a \mathrm{Re}\left(E_x^* H_y - E_y^* H_x\right) dx\, dy = \frac{\omega\, \mu_0\, k_z}{8k_\perp^2} ab\, A^2. \quad (56)$$

3.3.2.2 *Attenuation*

When the conductivity σ of the metallic walls is finite then, following the procedure of TM mode, the power dissipation per unit axial length turns out to be

$$P_A = \frac{1}{2c}\left(\frac{\omega}{2\sigma\, \epsilon_0}\right)^{\frac{1}{2}} A^2 \left(a + b + \frac{k_z^2}{k_\perp^4}\left(bk_y^2 + ak_x^2\right)\right). \quad (57)$$

The equation governing P_z is the same as eq. (45), giving $P_z = P_0$ $e^{-2k_i z}$ with

$$k_i = \frac{1}{c}\left(\frac{2\omega}{\sigma\, \epsilon_0}\right)^{\frac{1}{2}} \frac{k_\perp^2}{\omega\, \mu_0\, k_z\, ab}\left(a + b + \frac{k_z^2}{k_\perp^4}\left(bk_y^2 + ak_x^2\right)\right). \quad (58)$$

The power diminishes with z at a characteristic damping length $\frac{1}{2k_i}$.

3.3.2.3 *Characteristic Impedance*

The ratio of total transverse electric field to transverse magnetic intensity for a TE mode turns out to be

$$Z_c\,(\mathrm{TE}) = \frac{\omega\,\mu_0}{k_z} = Z_0\left(1 - \frac{\omega_{mn}^2}{\omega^2}\right)^{-\frac{1}{2}}. \tag{59}$$

One may notice that

$$Z_c\,(\mathrm{TM})\,Z_c\,(\mathrm{TE}) = Z_0^2.$$

PROBLEM 3.2. Estimate the phase velocity of TE_{10} mode in a rectangular waveguide at twice the cutoff frequency of this mode. What will be the phase velocity of the TE_{30} mode at this frequency?

Solution : For TE_{10} mode, $\omega^2 = \omega_{10}^2 + k_z^2\,c^2$.
At $\omega = 2\,\omega_{10}$,

$$k_z = \frac{\sqrt{3}\omega_{10}}{c}, v_{ph} = \frac{2c}{\sqrt{3}}.$$

For TE_{30} mode,

$$\omega^2 = \omega_{30}^2 + k_z^2 c^2.$$

Since $\omega_{30} = 3\omega_{10}$ and $\omega = 2\omega_{10}$, k_z^2 is negative. Hence TE_{30} mode does not exist.

PROBLEM 3.3. A rectangular waveguide of width a, height b and length l, closed at z = 0 and z = l by conducting planes, becomes a resonator. The modes with $H_z = 0$ have

$$E_z = A_1 \sin\left(\frac{m\pi x}{a}\right) \sin\left(\frac{n\pi y}{b}\right) [\exp(ik_z z) + A_2 \exp(-ik_z z)] e^{-i\omega t},$$

$$k_z{}^2 = \left(\omega^2/c^2\right) - m^2\pi^2/a^2 - n^2\pi^2/b^2.$$

Deduce E_x. Applying the boundary conditions on E_x at z = 0, z = l, deduce the eigen frequencies of the resonator.

Solution : From eqs. (41), one may note that if E_z has only one term with exp (i k_z z), E_x would be

$$E_x = ik_z\left(\frac{m\pi}{ak_\perp^2}\right) A_1 \cos\left(\frac{m\pi x}{a}\right) \sin\left(\frac{n\pi y}{b}\right) e^{i(\omega t - k_z\,z)}.$$

where $k_\perp^2 = m^2\pi^2/a^2 + n^2\pi^2/b^2$. When E_z has the other term, E_x would have an additional term that can be obtained from this expression by replacing k_z by - k_z and A_1 by A_1 A_2. Hence

$$E_x = ik_z\left(\frac{m\pi}{ak_\perp^2}\right)A_1\cos\left(\frac{m\pi x}{a}\right)\sin\left(\frac{n\pi y}{b}\right)(e^{ik_z z} - A_2\, e^{-ik_z z})\exp(-i\,\omega\, t).$$

At $z = 0$, $E_x = 0$ gives $A_2 = 1$. At $z = l$, $E_x = 0$ gives $k_z l = p\pi$ where p is an integer. Using this in the expression for $k_\perp^2$, we obtain the eigen frequencies $\omega = \omega_{mnp}$, where

$$\omega_{mnp}^2 = \left(m^2\frac{\pi^2}{a^2} + n^2\frac{\pi^2}{b^2} + p^2\frac{\pi^2}{l^2}\right)c^2.$$

3.4 Cylindrical Waveguide

Waveguides with circular cross-section (Fig. 3.10) are widely used in active and passive microwave devices. We choose the axis of the waveguide as z-axis and employ cylindrical polar coordinates (r, ϕ, z, *where r is the perpendicular distance of a point from the z-axis and ϕ is the angle this perpendicular makes with the x-axis*) to study the propagation of waves through the waveguide. The radius of the waveguide is a. We take the t, z variation as exp [-i (ωt - k_z z)]. Then the wave equations (16), (17) for E_z and H_z can be cast into the form,

$$\frac{\partial^2 F}{\partial r^2} + \frac{1}{r}\frac{\partial F}{\partial r} + \frac{1}{r^2}\frac{\partial^2 F}{\partial \phi^2} + k_\perp^2 F = 0, \qquad (60)$$

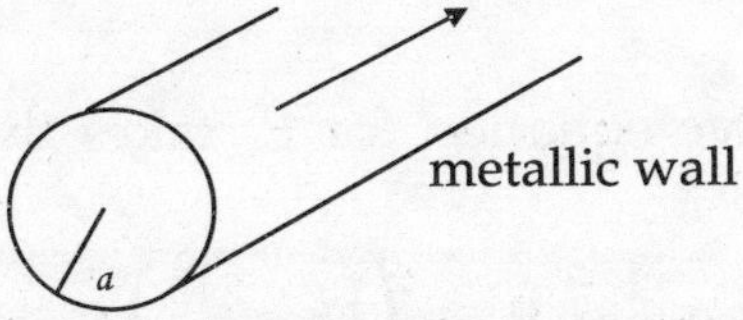

Figure 3.10: A cylindrical waveguide.

where $F = E_z$ or H_z and $k_\perp^2 = (\omega^2/c^2) - k_z^2$. We solve this equation by using the separation of variables technique. Writing $F(r, \theta) = F_1(r)\, F_2(\theta)$ in eq. (60) and dividing the resulting equation by $F_1 F_2$, we get

$$\frac{1}{F_1}\left(\frac{\partial^2 F_1}{\partial r^2} + \frac{1}{r}\frac{\partial F_1}{\partial r}\right) + k_\perp^2 = -\frac{1}{r^2}\frac{1}{F^2}\frac{\partial^2 F_2}{\partial \phi^2}. \qquad (61)$$

Since the left-hand side is independent of ϕ, the right-hand side must also be independent of ϕ, i.e.,

$$\frac{1}{F_2}\frac{\partial^2 F_2}{\partial \phi^2} = -l^2. \qquad (62)$$

The general solution of this equation is of the form $e^{\pm il\phi}$. Since ϕ and $(\phi + 2\pi)$ represent the same point, hence $F_2(\phi) = F_2(\phi + 2\pi)$. This is possible only when l is an integer. Thus, the general t, z, ϕ dependence of fields can be taken as $\exp[-i(\omega t - k_z z - l\phi)]$. Now let us express the transverse components of **E** and **H** in terms of E_z and H_z by using the third and fourth Maxwell's equations,

$$E_r = \frac{1}{k_\perp^2}\left(ik_z \frac{\partial}{\partial r}E_z - \frac{l\omega\,\mu_0}{r}H_z\right).$$

$$H_\phi = \frac{1}{k_\perp^2}\left(i\omega\,\epsilon_0 \frac{\partial}{\partial r}E_z - \frac{l\,k_z}{r}H_z\right),$$

$$E_\phi = -\frac{1}{k_\perp^2}\left(\frac{k_z l}{r}E_z + i\omega\,\mu_0 \frac{\partial}{\partial r}H_z\right),$$

$$H_r = \frac{1}{k_\perp^2}\left(\frac{l\omega\,\epsilon_0}{r}E_z + ik_z \frac{\partial}{\partial r}H_z\right). \qquad (63)$$

Again there are two distinct modes of wave propagation:

(i) $H_z = 0$, $E_z \neq 0$ (TM mode)

(ii) $H_z \neq 0$, $E_z = 0$ (TE mode).

3.4.1 TM Mode

In this case, the wave equation for E_z takes the form of a Bessel equation,

$$\frac{\partial^2 E_z}{\partial r^2} + \frac{1}{r}\frac{\partial}{\partial r}E_z + \left(k_\perp^2 - \frac{l^2}{r^2}\right)E_z = 0. \qquad (64)$$

Its general solution is

$$E_z = AJ_l(k_\perp r) + A'Y_l(k_\perp r),$$

where J_l ($k_\perp$ r) and Y_l ($k_\perp$ r) are Bessel functions of first and second kind, of order l, respectively. However, Y_l goes to ∞ as $r \to 0$, hence we discard this solution ($A' = 0$). Thus

$$E_z = AJ_l(k_\perp r)e^{-i(\omega t - k_z z - l\phi)}. \tag{65}$$

The boundary condition, $E_z = 0$ at r = a, demands

$$J_l\,(k_\perp a) = 0.$$

Since J_l is an oscillatory function of its argument, this equation gives many roots

$$k_\perp a = \chi_{lm}, \tag{66}$$

where χ_{lm} are the values of χ at which $J_l(\chi) = 0$. As an example, the zeros of $J_0(\chi)$ occur at $\chi = \chi_{01} = 2.4$, $\chi = \chi_{02} = 5.5$, $\chi = \chi_{03} = 8.6$ whereas the zeros of $J_1(\chi)$ occur at $\chi = \chi_{11}, \chi_{12}, \chi_{13} = 3.8, 7.0, 10.2...$ Using the value of $k_\perp$ from eq. (66) in $k_\perp^2 = (\omega^2/c^2) - k_z^2$, one obtains the dispersion relation for TM_{lm} mode as

$$\omega^2 = k_z^2 c^2 + \chi_{lm}^2 \frac{c^2}{a^2}. \tag{67}$$

Fig. 3.11 shows the variation of ω with k_z. There is no propagation for $\omega < \omega_{lm} = \chi_{lm}\, c/a$. As ω increases beyond the cutoff frequency ω_{lm}, the dispersion curve asymptotically approaches the light line $\omega = k_z\, c$.

The r, ϕ components of **E** and **H** for a TM mode are,

$$E_r = \frac{ik_z}{k_\perp} A\, J_l'(k_\perp r), \quad H_\phi = \frac{i\omega\, \epsilon_0}{k_\perp} A\, J_l'(k_\perp r),$$
$$E_\phi = -\frac{k_z l}{r k_\perp^2} A\, J_l(k_\perp r), \quad H_r = \frac{l\omega\, \epsilon_0}{r k_\perp^2} A\, J_l(k_\perp r), \tag{68}$$

where prime over a Bessel function denotes its differentiation with respect to argument.

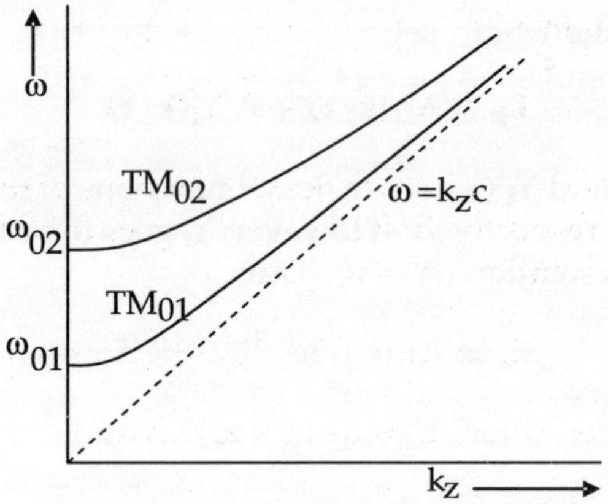

Figure 3.11: Frequency versus wave number for TM_{01} and TM_{02} modes in a cylindrical waveguide.

3.4.1.1 *Power Flow*

The time average power flow in a TM_{lm} mode is

$$P_z = \frac{1}{2}\text{Re}\int_0^a \left(E_r^* H_\phi - E_\phi^* H_r\right) 2\pi r\, dr = \pi \frac{k_z \omega \in_0}{k_\perp^4} A^2\, I, \tag{69}$$

where

$$I = \int_0^{k_\perp a} \left[x J_l^2(x) + l^2 \frac{J_l^2(x)}{x} \right] dx = \frac{1}{4}\left[J_l'(k_\perp a)\right]^2 \chi_{lm}^2.$$

3.4.1.2 *Attenuation*

When the conductivity, σ, of the wall is finite, there is power dissipation in the wall. Following eq. (26), the time average power dissipation in the wall (per unit axial length) is

$$P_A = \frac{1}{2c}\sqrt{\left(\frac{\omega}{\sigma \in_0}\right)} \left|H_\phi (r = a)\right|^2 2\pi a. \tag{70}$$

If we consider a length segment, dz, of the waveguide, then the power *in* is P_z, power *out* is $P_z + dP_z$ and the power dissipated is P_A dz. Power balance gives

$$\frac{dP_z}{dz} = -2k_i\, P_z, P_z = P_0\, e^{-2k_i z}, \tag{71}$$

where

$$k_i = \frac{P_A}{2P_z} = \frac{1}{c}\left(\frac{\omega}{\sigma \in_0}\right)^{\frac{1}{2}} \frac{k_\perp^2 \omega \in_0 2a}{k_z \chi_{lm}^2}.$$

3.4.1.3 Characteristic Impedance

The characteristic impedance of the waveguide is

$$Z_c\,(TE) = \frac{E_r}{H_\phi} = -\frac{E_\phi}{H_r} = \sqrt{\frac{\left(E_r^2 + E_\phi^2\right)}{\left(H_\phi^2 + H_r^2\right)}} = \frac{k_z}{\omega \in_0} = Z_0\sqrt{\left(1 - \frac{\omega_{lm}^2}{\omega^2}\right)}. \tag{72}$$

This expression is the same as that for a rectangular waveguide, with the only difference in the value of ω_{lm}– the cutoff frequency for the TM_{lm} mode.

3.4.2 TE Mode

For the TE mode ($E_z = 0$, $H_z \neq 0$), the wave equation for H_z gives the solution

$$H_z = AJ_l\,(k_\perp r)\, e^{-i(\omega t - k_z z - l\phi)} \tag{73}$$

and the transverse field components are

$$E_r = -\frac{l\omega\,\mu_0}{rk_\perp^2} A\, J_l\,(k_\perp r) \quad H_\phi = -\frac{lk_z}{rk_\perp^2} A\, J_l\,(k_\perp r)\,,$$

$$E_\phi = -\frac{i\omega\,\mu_0}{k_\perp} A\, J_l'\,(k_\perp r) \quad H_r = \frac{ik_z}{k_\perp} A\, J_l'\,(k_\perp r)\,, \tag{74}$$

where $k_\perp^2 = \omega^2/c^2 - k_z^2$, prime over J_l denotes differentiation with respect to the argument, and phase factor exp $[-i\,(\omega t - k_z\, z - l\phi)]$

is implicit. The boundary condition, $E_\phi = 0$ at $r = a$, gives the eigen value condition $J_1'(k_\perp a) = 0$ or

$$k_\perp a = \chi'_{lm} \tag{75}$$

where χ'_{lm} are the zeros of J_1'. Using this value of $k_\perp$ in $k_\perp^2 = \omega^2/c^2 - k_z^2$, we obtain the dispersion relation

$$\omega^2 = k_z^2 c^2 + \omega_{lm}^2; \quad \omega_{lm}^2 = \chi_{lm}'^2 \frac{c^2}{a^2}. \tag{76}$$

3.4.2.1 *Power Flow*

The time average power flow of a TE_{lm} mode is

$$P_z = \frac{1}{2} \int_0^a \mathrm{Re}\left(E_r^* H_\phi - E_\phi^* H_r\right) 2\pi\, r\, dr = \frac{\pi \omega \mu_0 k_z}{k_\perp^4} A^2 I' \tag{77}$$

where

$$I' = \int_0^{k_\perp a} \left[x J_1'^2(x) + \frac{1^2}{x} J_1^2(x) \right] dx = \frac{1}{4}\left(\chi_{lm}'^2 - 1^2\right) J_1^2(\chi'_{lm}).$$

3.4.2.2 *Attenuation*

In a waveguide of finite wall conductivity, σ, the power dissipation in the wall (per unit axial length) is (refer to eq. 26),

$$\begin{aligned} P_A &= \frac{1}{2c}\sqrt{\left(\frac{\omega}{\sigma \in_0}\right)} \left[|H_\phi(r=a)|^2 + |H_z(r=a)|^2 \right] 2\pi a \\ &= \frac{1}{2c}\sqrt{\left(\frac{\omega}{\sigma \in_0}\right)} \left(1 + \frac{1^2 k_z^2}{a^2 k_\perp^4}\right) A^2 J_1^2(k_\perp a)\, 2\pi a. \end{aligned} \tag{78}$$

Subsequently, the power flow P_z decreases with the distance of propagation as

$$P = P_0 e^{-2k_i z}, \tag{79}$$

where

$$k_i = \frac{P_A}{2P_z} = \frac{2}{c}\sqrt{\left(\frac{\omega}{\sigma \in_0}\right)} \left(1 + \frac{1^2 k_z^2}{a^2 k_\perp^4}\right) \frac{a k_\perp^4}{\omega \mu_0 k_z} \frac{1}{(\chi_{lm}'^2 - 1^2)}.$$

3.5 Sheath Helix

Sheath helix is a slow wave structure, commonly used in travelling wave tubes. It comprises a helical winding of a wire of high conductivity, to form a helix of radius a and pitch ψ; ψ is the angle between the wire and the cross-sectional plane of the helix. Also, $\tan\psi = L/2\pi\, a$, where L is the separation between two turns.

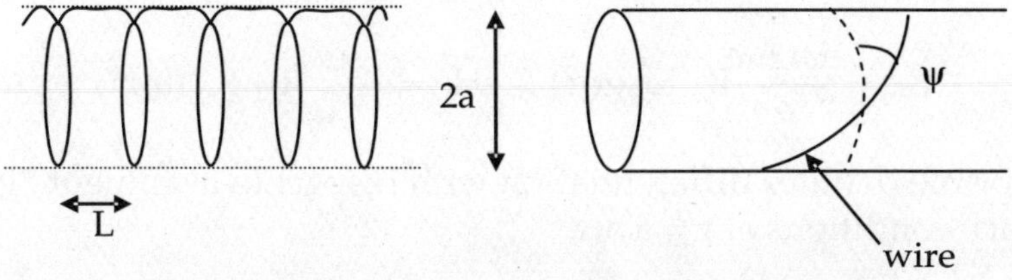

Figure 3.12: Schematic of sheath helix winding.

For waves of long wavelength ($k_z L << 1$), we may treat the helix as a continuous anisotropic conductor with $\sigma = \infty$ along the line of the wire and $\sigma = 0$ perpendicular to it. The TE and TM modes in this case are coupled.

We assume (t, z) variations as $\exp[-i(\omega t - k_z z)]$ and consider only azimuthally symmetric ($\phi-$ independent) modes. Then, the wave equations (16), (17) for E_z, H_z take the form

$$\frac{\partial^2}{\partial r^2} E_z + \frac{1}{r}\frac{\partial E_z}{\partial r} - \alpha^2 E_z = 0,$$
$$\frac{\partial^2 H_z}{\partial r^2} + \frac{1}{r}\frac{\partial H_z}{\partial r} - \alpha^2 H_z = 0 \tag{80}$$

where $\alpha^2 = k_z^2 - \omega^2/c^2$. The well-behaved solutions of these equations in the inner and outer regions are

$$\mathbf{r < a}: E_z = A_1 I_0(\alpha r),\ H_z = B_1 I_0(\alpha r),$$
$$\mathbf{r > a}: E_z = A_2 K_0(\alpha r),\ H_z = B_2 K_0(\alpha r), \tag{81}$$

where I_0, K_0 are modified Bessel functions. The transverse field components are

$$\mathbf{r} < \mathbf{a} : E_r = -\frac{ik_z}{\alpha} A_1 I_0'(\alpha r), \quad H_\phi = -\frac{i\omega \epsilon_0}{\alpha} A_1 I_0'(\alpha r),$$

$$E_\phi = \frac{i\omega \mu_0}{\alpha} B_1 I_0'(\alpha r), \quad H_r = -\frac{ik_z}{\alpha} B_1 I_0'(\alpha r),$$

$$\mathbf{r} > \mathbf{a} : E_r = -\frac{ik_z}{\alpha} A_2 K_0'(\alpha r), \quad H_\phi = -\frac{i\omega \epsilon_0}{\alpha} A_2 K_0'(\alpha r),$$

$$E_\phi = \frac{i\omega \mu_0}{\alpha} B_2 K_0'(\alpha r), \quad H_r = -\frac{ik_z}{\alpha} B_2 K_0'(\alpha r), \tag{82}$$

where prime denotes differentiation with respect to argument. The boundary conditions at r = a are :

a. E_ϕ and E_z are continuous,

$$A_1 I_0(\alpha a) = A_2 K_0(\alpha a), \quad B_1 I_0'(\alpha a) = B_2 K_0'(\alpha a).$$

b. The component of **E** along the conduction line vanishes at r = a,

$$E_\phi \cos\psi + E_z \sin\psi = 0, \frac{i\omega \mu_0}{\alpha} B_1 I_0'(\alpha a) = -\tan\psi A_1 I_0(\alpha a).$$

The component of **H** along the conduction line, $H_\phi \cos\psi - H_z \sin\psi$ is continuous at r = a,

$$\frac{i\omega \epsilon_0}{\alpha} A_1 I_0'(\alpha a) + B_1 I_0(\alpha a) \tan\psi$$
$$= -\frac{i\omega \epsilon_0}{\alpha} A_2 K_0'(\alpha a) + B_2 K_0(\alpha a) \tan\psi.$$

They give the dispersion relation

$$\frac{I_0'(\alpha a) K_0'(\alpha a)}{I_0(\alpha a) K_0(\alpha a)} = -\frac{\alpha^2 c^2}{\omega^2} \tan^2\psi. \tag{83}$$

This equation can be solved graphically to obtain (αa) as a function of ω a/c. Choose a value of ω a/c, plot LHS and RHS as functions of αa respectively and determine the value of αa at which the two curves intersect. This is the value of αa corresponding to a given ω. From this, one can plot ω versus k_z, the dispersion curve. In fig. 3.13, we have plotted $ck_z/\omega = c/v_{ph}$ as a function of $(\omega a/c)$ for cot ψ = 15, 7. The waves of higher frequencies travel with slower phase velocity v_{ph}. As cot ψ increases, v_{ph} becomes smaller. In fact, the dispersion relation in the limit $\alpha a >> 1$ goes over to $\alpha^2 c^2/\omega^2 = \cot^2 \psi$, hence $k_z c/\omega = \text{cosec}\, \psi$. However, one may note that the fields of slow waves (with large αa) are localised very near the helix surface (r = a) and fall off almost exponentially, away from it. This becomes a major constraint in the generation of high frequencies by electron beams as electron beams are required to propagate through the regions where longitudinal field E_z is large.

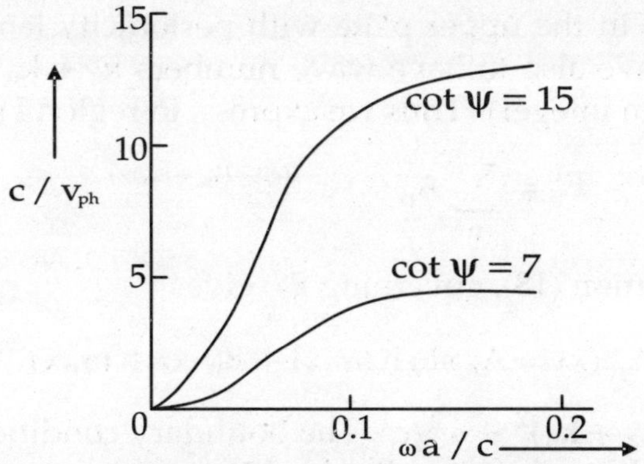

Figure 3.13: c/v_{ph} as a function of a ω/c in a sheath helix, for a slow wave.

3.6 Flat Periodic Waveguide

A flat periodic waveguide also supports slow waves and is employed in active devices producing microwaves. It comprises a flat metallic plate at x = 0 (Fig. 3.14). Another metal plate, with periodic grooves of depth h and width d, runs parallel to it. The periodicity length is λ_w.

We consider the propagation of a TM mode with (t, z) variation as $\exp[-i(\omega t - k_z z)]$ and assume $\partial/\partial y = 0$. Due to the periodicity

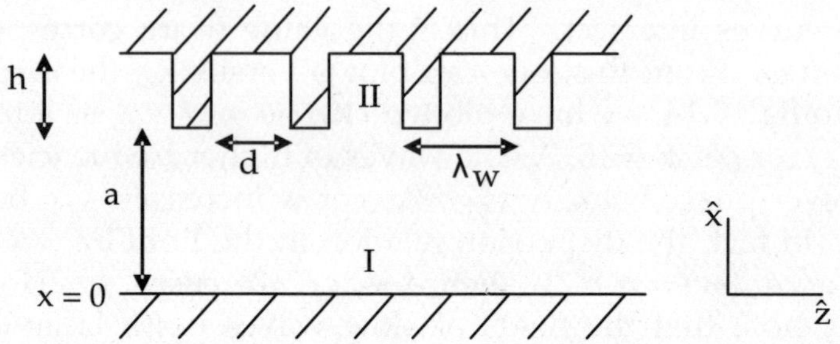

Figure 3.14: A flat periodic wave guide with a flat metallic plate at x= 0 and another metal plate with periodic grooves of depth h and spatial period λ_w.

of the grooves in the upper plate with periodicity length λ_w, one expects the wave also to have wave numbers $k_z + k_n$ where $k_n = 2n\pi/\lambda_w$ (n is an integer). Thus we express, in region I $(0 < x < a)$,

$$E_z^I = \sum_n A_n'(x)\, e^{-i(\omega t-(k_z+k_n)z)}. \tag{84}$$

The wave equation (18), governing E_z, gives

$$A_n'(x) = A_n \sinh(\alpha_n x) + B_n \cosh(\alpha_n x)$$

where $\alpha_n^2 = (k_z + k_n)^2 - \omega^2/c^2$. The boundary condition $E_z^I = 0$ at $x = 0$ for all values of z, gives $B_n = 0$. Hence,

$$A_n'(x) = A_n \sinh(\alpha_n x).$$

The transverse field components in region I are :

$$E_x^I = -\sum_n \frac{i(k_z + k_n)}{\alpha_n} A_n \cosh(\alpha_n x)\, e^{-i[\omega t-(k_z+k_n)z]},$$

$$H_y^I = -\sum_n \frac{i\omega \epsilon_0}{\alpha_n} A_n \cosh(\alpha_n x)\, e^{-i[\omega t-(k_z+k_n)z]}. \tag{85}$$

Inside a slot, we assume the field to be z-independent and propagating forward/backward in the x-direction with **E** perpendicular to the slot walls at $z = \pm\frac{d}{2}$, i.e., $E \,||\, \hat{z}$. Since E_z must vanish at $x = a + h$, we have inside the sth slot (region II),

$$E_z^{II} = A \sin\left(\frac{\omega}{c}(x - a - h)\right) e^{-i(\omega t - \theta_s)},$$

$$H_y^{II} = i\frac{A}{\mu_0 c} \cos\left(\frac{\omega}{c}(x - a - h)\right) e^{-i(\omega t - \theta_s)}. \qquad (86)$$

Since the phase difference over z-distance = λ_w in region I is $k_z \lambda_w$, hence $\theta_s = sk_z \lambda_w$ and the continuity of E_z at x = a requires

$$\sum_n A_n \sinh(\alpha_n a) e^{i(k_z+k_n)z} = F \qquad (87)$$

where $F = - A \sin(\omega h/c) \exp(i\theta_s)$ at the mouth of the slot ($-d/2 < z > d/2$ for the zeroth slot, $s\lambda_w - d/2 < x < s\lambda_w + d/2$ for the sth slot), while F = 0 otherwise (at the conductor interface). To determine A_m, we multiply both sides of eq. (87) by $\exp[-i(k_z + k_m) z]$ and integrate over z from $-\lambda_w/2$ to $\lambda_w/2$. We obtain

$$A_m = -\frac{2A \sin(\omega h/c)}{(k_z + k_m) L \sinh(\alpha_m a)} \sin((k_z + k_m) d/2), \qquad (88)$$

where we have used the property $\int_{-\lambda_w/2}^{\lambda_w/2} \exp[-i(k_n - k_m) z] dz = 0$ for $n \neq m$ and neglected stray fields of the slot. At x = a, we also demand $\langle H_y^I \rangle = H_y^{II}$, where <> denotes average over z from $-d/2$ to $d/2$,

$$\sum_n \frac{2i\omega \epsilon_0 A_n \cosh(\alpha_n a)}{\alpha_n (k_z + k_n)} \sin[(k_z + k_n) d/2] = -\frac{iAd}{\mu_0 c} \cos(\omega h/c). \qquad (89)$$

Using the value of A_n from eq. (88) in eq. (89), we obtain the dispersion relation

$$\sum_n \frac{\omega d}{c\alpha_n \lambda_w} \coth(\alpha_n a) \left[\frac{\sin((k_z + k_n) d/2)}{(k_z + k_n) d/2}\right]^2 = \cot\left(\frac{\omega}{c}h\right). \qquad (90)$$

This equation can be solved numerically to obtain k_z as a function of ω. A mode comprises many wave components, propagating with different phase velocities $v_{ph} = \omega/(k_z + k_n)$. In fig. 3.15, we have plotted $c/v_{ph} = (k_z \lambda_0)/(2\pi) + (n\lambda_0)/(\lambda_w)$ as a function of λ_0/λ_w (where $\lambda_0 = 2\pi c/\omega$ is the free space wavelength) for $h/\lambda_w = 15$, $d/\lambda_w = 0.5$, $a/\lambda_w = 0.3$. We see that the phase velocity of n = 1 component is much smaller than c. However, the amplitudes of the n = 1 field components in region I are strongly localised near the x = a interface.

For $\omega\lambda_w/c << 1$, one may retain only the n = 0 term in the dispersion relation as the terms with higher n are smaller. Then the dispersion relation reduces to

$$\frac{\omega d}{c} \tan\left(\frac{\omega h}{c}\right) = \alpha_0 \lambda_w \tanh(\alpha_0 a) \tag{91}$$

where $\alpha_0 = \sqrt{(k_z^2 - \omega^2/c^2)}$. Further, if $\alpha_0 a << 1$ and $\omega h/c << 1$, one obtains $c/v_{ph} = ck_z/\omega \approx \sqrt{\frac{dh}{\lambda_w a}}$

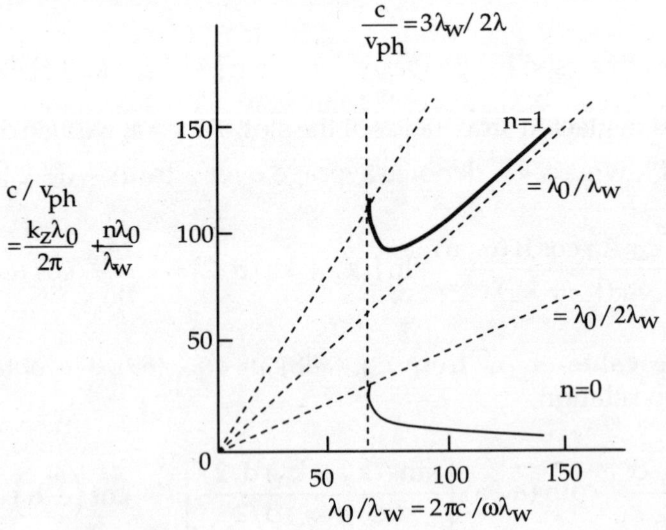

Figure 3.15: Variation of c/v_{ph} of a slow wave with normalised free space wavelength λ_0/λ_w, in a flat periodic waveguide of ripple period λ_w.

PROBLEM 3.4. Consider a cylindrical metal pipe (called corrugated cylindrical waveguide) whose inner radius is R (z) = a + hcos qz, where $\hat{z}$ is along the axis of the waveguide, h is the corrugation amplitude, $q = 2\pi/\lambda_w$ and λ_w is the period of corrugation. Obtain the dispersion relation for an azimuthally symmetric ($\partial/\partial\phi = 0$) TM mode. Such a waveguide is used in a backward wave oscillator to produce high power microwaves by relativistic electron beams.

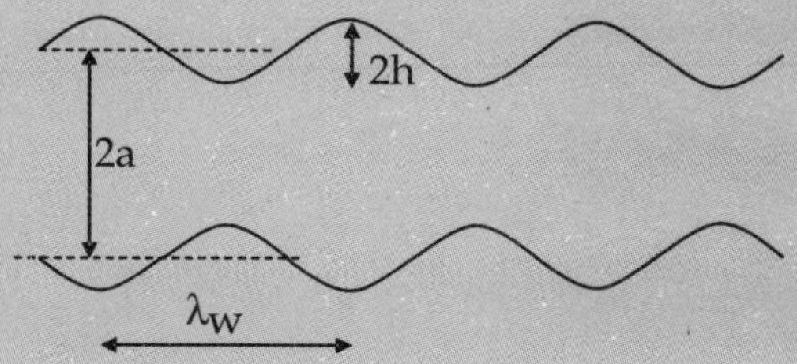

Figure 3.16: A corrugated wall cylindrical wave guide of radius a + h cos (qz) and corrugation wavelength λ_w.

Solution: For the TM mode ($H_z = 0$, $E_z \neq 0$) we presume t, z dependence as exp $[-i(\omega t - k_z z)]$. However, as the waveguide has periodicity with period λ_w. Hence, one expects the fields also to have wave numbers of the form $k_z + nq$, where n is an integer. Thus we express E_z as

$$E_z = \sum_n A_n'(r)\, e^{-i[\omega t - (k_z + nq)z]}. \tag{92}$$

Using eq. (92) in the wave equation, we obtain the equation governing A_n' as

$$\frac{\partial^2 A_n'}{\partial r^2} + \frac{1}{r}\frac{\partial A_n'}{\partial r} + k_{\perp n}^2 A_n' = 0, \tag{93}$$

where $k_{\perp n}^2 = \omega^2/c^2 - (k_z + nq)^2$. This is the *Bessel's equation* and its solution, well-behaved at r = 0, is $A_n' = A_n J_0(k_{\perp n} r)$. Hence,

$$E_z = \sum_{n=-\infty}^{\infty} A_n J_0(k_{\perp n} r)\, e^{-i[\omega t - (k_z + nq)z]}. \tag{94}$$

The non-zero transverse field components can be obtained from eqs. (63),

$$E_r = -\sum_n \frac{i(k_z + nq)}{k_{\perp n}} A_n J_1(k_{\perp n} r)\, e^{-i(\omega t - (k_z + nq)z)},$$

$$H_\phi = -\sum_n \frac{i\omega \epsilon_0}{k_{\perp n}} A_n J_1(k_{\perp n} r)\, e^{-i(\omega t - (k_z + nq)z)}. \qquad (95)$$

The metallic boundary condition at the corrugated wall demands that the tangential component of **E** at r = a + h cos qz must vanish, i.e. $E_z \cos\psi + E_r \sin\psi = 0$ where ψ is the slope of the tangent to the wall at r, z [$\psi = -\tan^{-1}(qh \sin qz)$]. Thus

$$\sum_n A_n \Big[J_0(k_{\perp n}(a + h\cos qz)) + \frac{iqh}{k_{\perp n}}(k_z + nq).(\sin qz) J_1(k_{\perp n}(a + h\cos qz))\, e^{-i[\omega t - (k_z + nq)z]} \Big] = 0 \qquad (96)$$

By expanding the Bessel functions around $k_{\perp n} a$ and equating the coefficients of various exponentials, $\exp[-i(k_z + nq)z]$, to zero we get a set of equations coupling A_ns. For a nontrivial solution of this set, the determinant of the set of coefficients of A_ns must vanish. This gives the dispersion relation for the azimuthally symmetric TM mode.

For $k_{\perp n} h < 1$, we may retain A_0, A_{-1}, A_{+1} terms only. Then the dispersion relation takes the form

$$J_0(k_\perp a) - \frac{h^2}{4} \frac{J_1(k'_{\perp 1} a)(k_{\perp 0} a)}{J_0(k'_{\perp 1} a)} \left(k_{\perp 0} - \frac{qk_z}{k_{\perp 0}}\right) \cdot \left(k'_{\perp 1} - \frac{q(k_z - q)}{k'_{\perp 1}}\right) - \frac{h^2}{4} \frac{J_1(k_{\perp 1} a) J_1(k_{\perp 0} a)}{J_0(k_{\perp 1} a)} \left(k_{\perp 0} - \frac{qk_z}{k_{\perp 0}}\right) \cdot \left(k_{\perp 1} - \frac{q(k_z + q)}{k_{\perp 1}}\right) = 0 \qquad (97)$$

where $k'_{\perp 1} = \left[\frac{\omega^2}{c^2} - (k_z - q)^2\right]^{1/2}$. Fig. 3.17 gives a qualitative plot of dispersion relation.

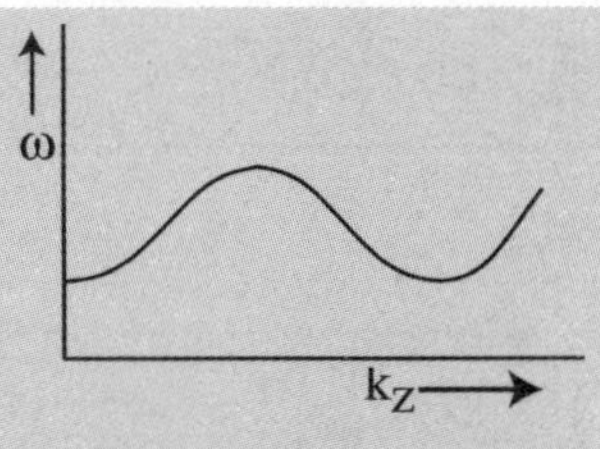

Figure 3.17: A typical dispersion relation for TM mode of a corrugated waveguide.

3.7 Coaxial Cable

The waveguides that we have studied so far support only TE or TM modes or mixed modes and permit the propagation of waves with frequency higher than a cutoff frequency only. The situation changes dramatically when we introduce an inner conductor. A cylindrical waveguide, with a coaxial metallic cylinder of smaller radius inside, is a commonly used guiding structure for the propagation of waves of frequency $\leq$ 10 GHz and is known as a coaxial cable. A coaxial cable permits a transverse electromagnetic mode with $E_z = 0$, $B_z = 0$ (where z refers to the axis of the cable) that has no cutoff frequency. This mode has only E_r and H_ϕ. Here, we obtain the mode structure and dispersion relation of this mode.

Consider a coaxial cable of inner conductor radius a and outer conductor radius b. The space between the two is filled with a dielectric of relative permittivity ϵ_r (Fig. 3.18). Electromagnetic wave propagates through the dielectric region. Let the t-z variation of **E**, **H** be of the form $\exp[-i(\omega t - k_z z)]$. Then the wave equation governing E_x (or E_y) takes the form

$$\frac{\partial^2}{\partial r^2}E_x + \frac{1}{r}\frac{\partial}{\partial r}E_x + \frac{1}{r^2}\frac{\partial}{\partial \phi^2}E_x + \left(\frac{\omega^2 \epsilon_r}{c^2} - k_z^2\right)E_x = 0. \qquad (98)$$

Presuming that $\mathbf{E} = \hat{r}E_r$ and E_r is independent of ϕ, we write $E_x = E_r \cos\phi$. Then, the wave equation reduces to

$$\frac{\partial^2}{\partial r^2}E_r + \frac{1}{r}\frac{\partial E_r}{\partial r} - \frac{E_r}{r^2} + \left(\frac{\omega^2}{c^2}\epsilon_r - k_z^2\right)E_r = 0, \qquad (99)$$

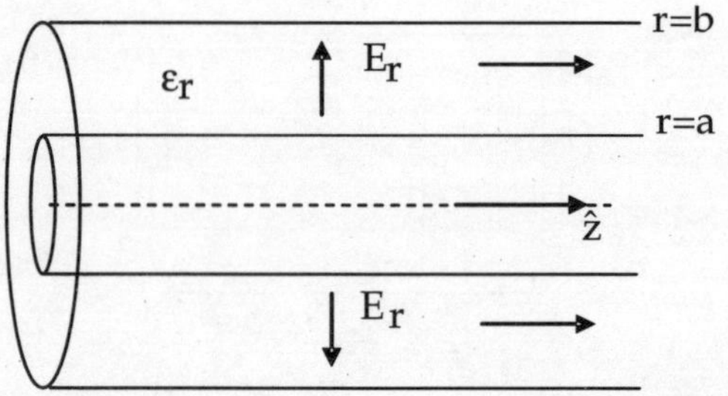

Figure 3.18: Coaxial cable with concentric metallic cylinders of radil a and b. The wave propagates along $\hat{z}$, through the space between the cylinders.

giving a well-behaved solution

$$E_r = \frac{A_0}{r} e^{-i(\omega t - k_z z)}, \tag{100}$$

with

$$k_z = \left(\frac{\omega}{c}\right) \epsilon_r^{\frac{1}{2}} . \tag{101}$$

Eq. (101) is the dispersion relation for TEM mode. It is the same as that for a plane wave in an unbounded dielectric. The magnetic field of the mode is

$$B_\phi = \sqrt{\epsilon_r} \frac{A_0}{cr} e^{-i(\omega t - k_z z)}.$$

Time average Poynting vector is

$$\mathbf{S}_{av} = \hat{z}\sqrt{\epsilon_r} \frac{A_0^2}{2\mu_0 c r^2}$$

and total time average power flow in the cable is

$$P_z = \pi \sqrt{\epsilon_r} \left(\frac{A_0^2}{\mu_0 c}\right) \ln\left(\frac{b}{a}\right).$$

The TEM mode suffers attenuation due to finite conductivity of the inner and outer conductors, as well as losses in the dielectric. The attenuation rate is higher at higher frequencies. Hence, coaxial cables are mostly used below 1 GHz for long distance transmission.

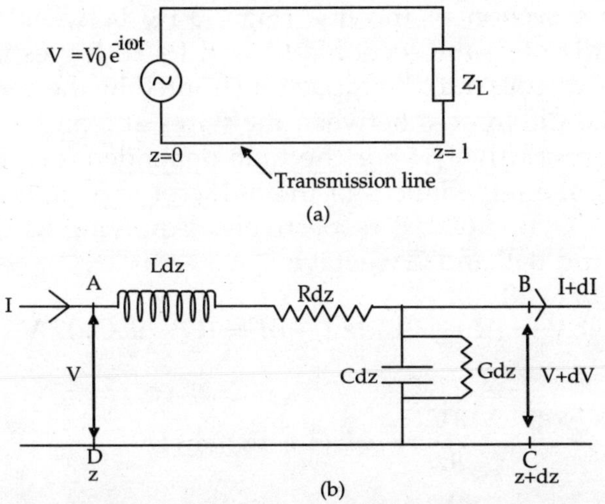

Figure 3.19: (a) Schematic of a transmission line connecting a voltage source at z=0 to a distant load at z=l. (b) Segment of the line between z and z+dz.

3.8 Transmission Line

Point-to-point low frequency power transmission, over long distances, is achieved through transmission lines. A transmission line comprises of two parallel conductors, e.g., a pair of parallel wires, a wire and a parallel conducting plane, a coaxial cable etc. The electromagnetic energy propagates through the medium, in between the conductors (as happens in a coaxial cable) or surrounding them (as in a parallel wire transmission line). The field structure calculation in such lines is usually tedious. However, a distributed circuit analysis called *transmission line theory*, provides much of the relevant information in a simple way. Schematically, one may represent any transmission line by two parallel wires with appropriate inductance per unit length L and capacitance per unit length C (for a coaxial cable, L and C have been calculated in unsolved problems 2 and 3 of unit 1). The line may also have resistance per unit length R (due to finite resistivity of the conducting wires) and conductance per unit length G (due to finite conductivity of the surrounding medium that may leak the current from one wire to the other through the medium).

Consider a section of the line (Fig. 3.19) between z and z + dz. Current I enters the section at A and I + dI leaves it at B. The difference − dI goes into the capacitor shunted by the conductance. The potential differences between the wires at z and z +dz are V and V +dV respectively. For the time dependence of I and V as exp $(-i\,\omega\,t)$, the impedances of the inductor and capacitor are $-i\,\omega\,Ldz$ and $-1/\,(i\,\omega\,(Cdz))$ respectively. Applying Kirchoff's law between A and B, B and C we have

$$-dV = (R - i\omega L)dz\, I, -dI = (G - i\omega C)dz\, V,$$

or

$$\frac{\partial V}{\partial z} = i\omega L\,(1 + iR/\omega L)\, I, \tag{102}$$

$$\frac{\partial I}{\partial z} = i\omega C\,(1 + iG/\omega C)\, V. \tag{103}$$

Eliminating I from these, we get

$$\frac{\partial^2 V}{\partial z^2} + k_z^2 V = 0, \tag{104}$$

giving

$$V = A_1 e^{ik_z Z} + A_2 e^{-ik_z Z}, \tag{105}$$

where $k_z^2 = \omega^2 LC\,(1 + iR/\omega L)\,(1 + iG/\omega C)$ and factor exp $(-i\,\omega t)$ is implied in both the terms. The first term represents the voltage signal propagating along $+\hat{z}$ and the second represents that propagating along $-\hat{z}$. Usually $R << \omega\, L$, $G << \omega C$. Further, if one uses expressions for L and C, one obtains $LC = \epsilon_r /\, c^2$, where c is the velocity of light in free space. Thus

$$k_z = \left(\frac{\omega}{c}\right) \epsilon_r^{\frac{1}{2}} \tag{106}$$

for a loss-free line. This is the same dispersion relation as we get for the TEM waves in an unbounded dielectric or a coaxial cable. Using eq. (106) in eq. (103), we obtain

$$I = \left(\frac{1}{Z_0}\right)\left(A_1 e^{ik_z z} - A_2 e^{-ik_z z}\right),$$

where $Z_0 = \omega L\,(1 + iR/\,\omega L)/\,k_z$. For a low loss line

$$Z_0 = \sqrt{\frac{L}{C}}.$$

Z_0 is known as the **characteristic impedance** of the line. Using the boundary conditions

(i) at z = 0, V = V_0 and

(ii) at z = l, V = Z_L I,

we obtain $A_2 = V_0 - A_1$,

$$A_1 = V_0/\left[1 + \left\{\frac{(Z_L - Z_0)}{(Z_0 + Z_L)}\right\}\exp\,(ik_z l)\right].$$

The input impedance of the line, defined as the ratio of V to I at z = 0, turns out to be

$$Z_{in} = Z_0\,\frac{Z_L - iZ_0\tan\,(k_z l)}{Z_0 - iZ_L\tan\,(k_z l)}.$$

For a lossless line, k_z and Z_0 are real, as given above. The input impedance depends on the length of the line. For instance

$$\begin{aligned} Z_{in} &= Z_L \quad \text{for } k_z l = n\pi \\ &= \frac{Z_0^2}{Z_L} \quad \text{for } k_z l = \left(n + \frac{1}{2}\right)\pi \\ &= Z_0 \quad \text{for } Z_L = Z_0, \end{aligned}$$

where n is an integer. For a line terminated in a load (resistance), equal to the characteristic impedance of the line, the input impedance is always equal to Z_0 irrespective of its length. In this case, $A_2 = 0$ and there is no reflected wave. All the power that goes from the source is transmitted to the load. $Z_L = Z_0$ is called the *matched load*. One may also note that an open ended line ($Z_L \to \infty$) has $Z_i = i\,Z_0 \cot\,(k_z\,l)$ whereas a short-circuit line ($Z_L = 0$) has $Z_{in} = -i\,Z_0 \tan\,(k_z\,l)$. In both cases, input impedance is reactive and length sensitive.

PROBLEM 3.5. Consider a transmission line of length l terminated in a load $Z_L = Z_0$. At a distance $\lambda/4$ from the load, the line is shunted by a resistance Z_0. Obtain the input impedance of the line.

Solution: The input impedance at $z = l - \lambda/4$ due to the right side line and the end load is Z_0. This impedance is in parallel to shunt impedance Z_0. Hence, effective load at $z = l - \lambda/4$ is $Z_L = Z_0/2$, and the input impedance at $z = 0$,

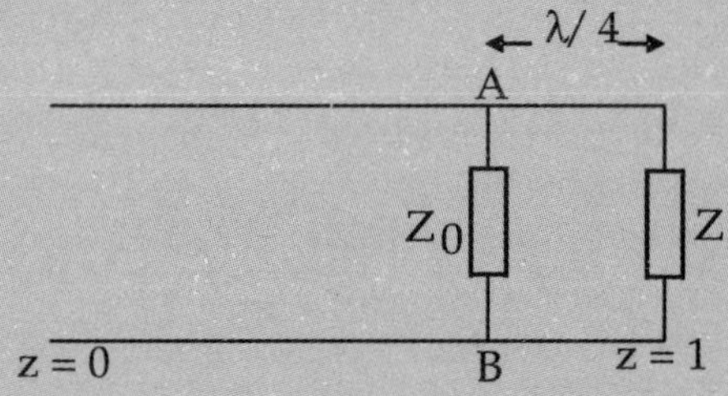

Figure 3.20: Transmission line with load Z_0 at z=l and a shunt load Z_0 at $z = l - \frac{\lambda}{a}$.

$$Z_{in} = Z_0 \frac{Z_0/2 - iZ_0 \tan k_z (l-\lambda/4)}{Z_0 - iZ_0/2 \tan (k_z (l-\lambda/4))}$$

$$= Z_0 \frac{1 + 2i \cot (k_z l)}{2 + i \cot (k_z l)}.$$

3.9 Conclusions

The interface between a conductor and a dielectric supports surface plasma wave, that propagates along the interface and whose amplitude falls off away from the interface in both media. This wave has frequency $\omega < \omega_{sc} = \omega_p / (\in_r + \in_L)^{1/2}$. At optical frequencies, this wave can be excited over metal-free space boundary by a laser, using attenuated total reflection (ATR) geometry, where a thin metal film is deposited over a prism and the laser is launched obliquely onto the prism-metal interface. At a specific angle of incidence $\theta_i = \theta_R$, the laser reflectivity falls down sharply. This property is employed in fabricating high-sensitivity surface plasmon sensors.

Hollow metal waveguides are useful guiding structures at microwave frequencies (1 GHZ < $\omega/ 2\pi$ < 100 GHz). The waves travel in TE or TM modes, with phase velocity v_{ph} greater than c. Since v_{ph} is a function of frequency, a microwave pulse of finite frequency width suffers *dispersion broadening*. Dispersion is an inherent characteristic of waveguide propagation. The phase velocity of waves can be slowed down by using sheath helix or periodic structures. Such structures are used in travelling wave tube and linear electron accelerator. In these devices, an electron beam is launched through the waveguide with velocity slightly higher or lower than the phase velocity of the wave. In the former, the beam resonantly gives energy to the wave, while in the latter it takes energy from the wave and gets accelerated. At millimetre and sub-millimetre wavelengths waveguide modes become too crowded and have large absorption, hence strip lines are more preferred guiding structures. At optical frequencies, dielectric waveguides or optical fibres are used that offer low-loss transmission upto thousands of kilometers.

Transmission lines provide low-loss transmission at lower frequencies. Coaxial cable is the most versatile transmission line used in communication and signal processing.

UNSOLVED PROBLEMS

1. A medium wave of frequency 1 MHz propagates on earth's surface as a surface wave. The earth is characterised by $\sigma = 1$ MKs and $\epsilon_r = 4$. Estimate the wavelength and attenuation length of the surface wave. If the field amplitude at the earth surface is 1 V/m, estimate its value at a height of 200 m.

   [**ANS.** $\lambda \approx 300$ m, attenuation length $\approx 1.7 \times 10^3$ km
   $|E| \approx 0.98$ V/m]

2. In a rectangular waveguide of x dimension a = 4 cm, a TM mode has

$$E_z = A\sin(2\pi x/a)\cos(\omega t - k_z z).$$

Obtain the phase velocity at $\omega/2\pi = 30\,\text{GHz}$.

[ANS. $4c/\sqrt{15}$.]

3. An open resonator, comprising of two plane metallic mirrors of large but finite conductivity σ at $z = 0, z = L$, has electromagnetic field inside it. To the zeroth order, $\mathbf{E} = \hat{x}A\sin(4\pi z/L)\exp(-i\omega t)$. What is the value of ω?. Estimate the amplitude of the electric field at a depth of $d = L/100$ inside the mirror, when $\sigma/(\epsilon_0\,\pi c/L) = 10^4$.

[HINT: *Obtain* H_y *at* $z = 0$. *Inside the mirror* $H_y(z) \approx H_y(0)$ $\exp(-i\,kz), k \approx (\omega/c)\,\epsilon_{eff}^{1/2}$. *Hence deduce* E_x.]

[ANS. $\omega = 4\pi c/L, |E_x| \approx 2 \times 10^{-4}A$.]

4. In a rectangular waveguide, filled with a non-dispersive dielectric of relative permittivity ϵ_r, the dispersion relation of a TM_{mn} mode is

$$\omega^2 = \left(k_z^2 + m^2\pi^2/a^2 + n^2\pi^2/b^2\right)c^2/\,\epsilon_r,$$

where a and b are the x and y dimensions of the waveguide. Show that

$$v_{ph}v_g = c^2/\,\epsilon_r,$$

where $v_{ph} = \omega/\,k_z$ and $v_g = \partial\omega/\partial\,k_z$ are the phase and group velocities respectively.

5. A cylindrical waveguide of radius a is filled with a plasma of effective permittivity $\epsilon_{eff} = 1 - \omega_p^2/\omega^2$. Deduce the dispersion relation for TM_{lm} mode. Hence show that the cutoff frequency is $\omega_{lm} = \left(\omega_p^2 + \chi_{lm}^2 c^2/a^2\right)^{1/2}$ and $v_{ph}v_g = c^2$, where v_{ph} and v_g are the phase and group velocities respectively.

6. An amplitude-modulated microwave signal propagates through a cylindrical waveguide of radius a in TM_{01} mode. At $z = 0$, $E_z = A\, J_0\,(\chi_{01}\, r/a)\cos(\Omega t)\cos(\omega t)$, where $\Omega << \omega$ and ω is twice the cutoff frequency of the mode. Show that

$$E_z\,(z,t) = AJ_0\,(\chi_{01} r/a)\cos\left[\Omega\left(t - 2Z/\sqrt{3}c\right)\right]\cos\left[\omega\left(t - \sqrt{3}Z/2c\right)\right].$$

7. A 50 Ω transmission line, connected to an ac voltage source at one end, is terminated into a load $Z_L = 50 + 100i$ ohm at the other. Estimate

(i) the voltage standing wave ratio
(ii) the distance (in terms of λ), from the load, where the voltage minimum occurs.

[ANS. (i) 5.6 (ii) 3 λ/ 16]

8. A surface plasma wave is propagating along $\hat{z}$ direction, on a vacuum-plasma interface ($x = 0$). On the vacuum side ($x > 0$), $E_z = A\exp(-\alpha x)\exp[-i(\omega t - \sqrt{3}\omega z/c]$. Estimate (i) α, (ii) ω_P/ω in the plasma, (iii) **E** in the plasma. Neglect collisions and take $\varepsilon_{eff} = 1 - \omega_p^2/\omega^2$ for the plasma.

[ANS. (i) $\alpha = \sqrt{2}\omega/c$, (ii) $\omega_P/\omega = \sqrt{5/2}$,
(iii) $\mathbf{E} = \mathbf{A}(\hat{z} - \hat{x}i\sqrt{3}\omega/c\alpha')e^{\alpha' x}e^{-i(\omega t - \sqrt{3}\omega z/c)}$, $\alpha' = 3\,\omega/\sqrt{2}c$.]

9. An electromagnetic wave propagating through a cylindrical waveguide in TM_{01} mode at twice the cutoff frequency suffers 2 dB power loss over a length of 20 cm. If the frequency is doubled, what would be the power loss over the same length?

[ANS. 2.53 dB]

10. The amplitude of axial component of electric field $|E_z|$ of the TM_{01} mode, at $r = 0$ in a cylindrical waveguide at twice the cutoff frequency, is 1V/m. If the same power is carried at twice this frequency, what would be $|E_z|_{r=0}$?

[ANS. 0.46 V/m.]

11. A microwave pulse of carrier frequency $\omega = 1.5\,\omega_{01}$ takes 1 ns to propagate through a cylindrical waveguide in TM_{01} mode, where ω_{01} is the mode cutoff frequency. If $\omega = 2\,\omega_{01}$, what would be the time of travel for the pulse?

[ANS. 0.86 ns.]

12. A 50Ω coaxial cable has dielectric with $\varepsilon_r = 2$. If ε_r is 18, what would be the characteristic impedance?

[ANS. 16.6Ω]

13. A transmission line of length $\lambda/2$ and characteristic impedance Z_0 is terminated in a resistive load of value Z_0 at the end. In the middle, it is shunted by another resister Z_0. Obtain time average power dissipation, when the voltage at the input end is $V_0 \cos \omega t$.

[ANS. $V_0^2/4Z_0$.]

14. Two parallel conducting wires, each of radius a and axis-to-axis separation b constitute a transmission line. Obtain the capacitance per unit length of the line. Take $\varepsilon = \varepsilon_0$ for the medium surrounding the wires and $b >> a$.

[ANS. $C = \dfrac{\pi \varepsilon_0}{\ln b/a}$.]

References and Suggested Reading

[1] Jordan E. C. and Balmain, K. G., *Electromagnetic waves and radiating systems,* N. J., USA, Prentice Hall, 1970.
[2] Chatterjee, R. *Elements of microwave engineering,* New Delhi, India, Affiliated East-West Press Pvt. Ltd., 1994.

Unit 4

Radiation

Introduction

A current-carrying wire produces a magnetic field **B** in the region surrounding it (Fig. 4.1). If the current is periodic in time, it will produce a magnetic field–also periodic in time. According to Faraday's law of electromagnetic induction ($\nabla \times \mathbf{E} = -\partial \mathbf{B}/\partial t$), this magnetic field is accompanied by a time-varying electric field **E**. The Poynting vector due to these **E** *and* **B** fields turns out to be finite, pointing away from the wire. **Thus a wire carrying periodic current is a source of electromagnetic radiation**. If the wire were surrounded by a conducting cylinder carrying an equal and opposite current, eg., a coaxial cable, there would be no magnetic field outside the cylinder. Hence, only the exposed portion of the wire (not

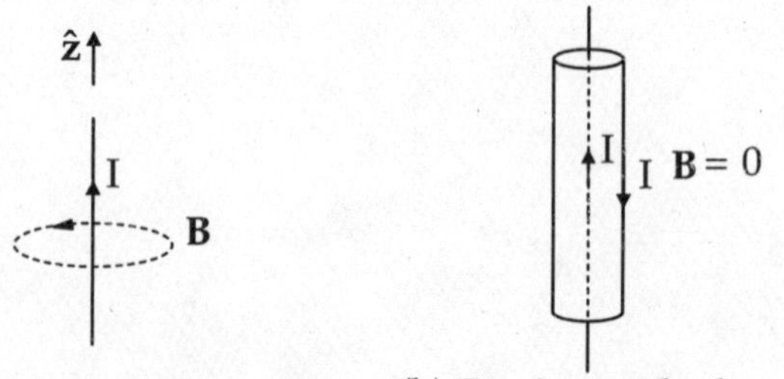

(a) **B** = 0 outside the wire (b) **B** = 0 outside the cable

Figure 4.1: Magnetic field due to (a) a current carrying wire and (b) a current carrying coaxial cable.

surrounded by a conducting cylinder) is the source of electromagnetic waves. This exposed portion of the wire is called an **antenna**.

The simplest antenna is a finite length straight wire with a cut in the middle. In fig. 4.2, wire AB is an antenna. It is cut in the middle and points C and D are connected to inner and outer conductors of a coaxial cable respectively, to feed current from a current source. Such a wire is known as *centre-fed dipole antenna*. The antenna acts as a resistive load for the transmission line (coaxial cable). When load impedance between C and D matches with the characteristic impedance of the line, there is no power reflection from the antenna into the line.

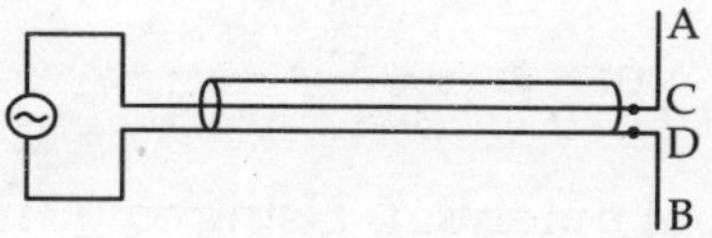

Figure 4.2: A centre-fed dipole antenna connected to a power source through a coaxial cable.

An important aspect of the fields produced by an antenna may be visualised intuitively. The fields produced at a distance d away from an antenna element at time t, would depend on the current in the element at time t - d/c i.e. time preceded by that required for propagation of the signal (fields), where c is the velocity of light in vacuum.

The starting point of an antenna problem is to obtain the vector potential, due to the current distribution in the antenna. From the vector potential, electric and magnetic fields, Poynting vector and total power radiated by the antenna are deduced. We begin with a general discussion of vector and scalar potentials for time dependent fields.

4.1 Vector and Scalar Potentials

From the second Maxwell's equation, $\nabla . \mathbf{B} = 0$. Thus, we may write **B** as curl of a vector (as divergence of curl of a vector always vanishes)

$$\mathbf{B} = \nabla \times \mathbf{A}. \qquad (1)$$

B is a physical observable – an actual quantity, while **A (r, t)** is a suitable mathematical function, to be chosen such that its curl produces the actual **B** . **A** is called the vector potential.

Using eq. (1) in the third Maxwell's equation, $\nabla\times\mathbf{E} = -\partial\mathbf{B}/\partial t$, we get

$$\nabla\times\left(\mathbf{E}+\frac{\partial\mathbf{A}}{\partial t}\right) = 0. \tag{2}$$

The quantity within the parenthesis can be written as the gradient of a scalar,

$$\mathbf{E}+\frac{\partial\mathbf{A}}{\partial t} = -\nabla\phi$$

$$\text{or}\quad \mathbf{E} = -\nabla\phi - \frac{\partial\mathbf{A}}{\partial t}. \tag{3}$$

ϕ (**r**,t) is called the **scalar potential**. Using eq. (3) in the first Maxwell's equation (*Poisson's equation*), $\nabla.\mathbf{D} = \rho$, we get

$$\nabla^2\phi + \frac{\partial}{\partial t}\nabla.\mathbf{A} = -\rho/\varepsilon_0, \tag{4}$$

where we have taken $\mathbf{D} = \epsilon_0\ \mathbf{E}$. If we use eqs. (1) and (3) in the fourth Maxwell's equation, $\nabla\times\mathbf{H} = \mathbf{J} + \partial\mathbf{D}/\partial t$ and use the vector identity $\nabla\times(\nabla\times\mathbf{A}) = -\nabla^2\mathbf{A} + \nabla(\nabla.\mathbf{A})$, we obtain

$$\nabla^2\mathbf{A} - \frac{1}{c^2}\frac{\partial^2\mathbf{A}}{\partial t^2} - \frac{1}{c^2}\nabla\frac{\partial\phi}{\partial t} - \nabla(\nabla.\mathbf{A}) = -\mu_0\mathbf{J}. \tag{5}$$

Thus, we have reduced the set of four Maxwell's equations into two equations (4 and 5). However, they are still coupled. We can decouple them by exploiting the liberty in the definitions of the potentials. **B** is defined as $\nabla\times\mathbf{A}$. One may add gradient of any scalar function to **A. B** will still remain the same (as curl of a gradient is always zero). Thus, **B** remains unchanged by the transformation

$$\mathbf{A}\rightarrow\mathbf{A}' = \mathbf{A} + \nabla\psi. \tag{6}$$

In order that the electric field also remains unchanged, ϕ must be transformed to

$$\phi\rightarrow\phi' = \phi - \frac{\partial\psi}{\partial t}. \tag{7}$$

The freedom implied in the definitions of **A** and ϕ can thus be used to choose **A** and ϕ, such that

$$\nabla.\mathbf{A} + \frac{1}{c^2}\frac{\partial\phi}{\partial t} = 0. \tag{8}$$

Then eqs. (4) and (5) decouple to

$$\nabla^2\phi - \frac{1}{c^2}\frac{\partial^2\phi}{\partial t^2} = -\rho/\epsilon_0, \tag{9}$$

$$\nabla^2\mathrm{A} - \frac{1}{c^2}\frac{\partial^2\mathbf{A}}{\partial t^2} = -\mu_0\mathbf{J}. \tag{10}$$

Condition (8) is called the *Lorentz gauge*. The transformations of potentials, as given by eqs. (6) and (7), are called **gauge transformations**. The invariance of fields under gauge transformation is called **gauge invariance**.

PROBLEM 4.1. $\mathbf{A}(\mathbf{r}, t)$ and $\phi(\mathbf{r}, t)$ are the scalar and vector potentials for which $\nabla.\mathbf{A} + \frac{1}{c^2}\frac{\partial\phi}{\partial t} \neq 0$. Show that, by a gauge transformation, it is always possible to obtain $\mathbf{A}'(\mathbf{r}, t)$ and $\phi'(\mathbf{r}, t)$ (representing the same electric and magnetic fields) which satisfy the Lorentz gauge.

Solution: Let $\nabla.\mathbf{A} + \frac{1}{c^2}\frac{\partial\phi}{\partial t} = F(\mathbf{r}, t)$.
We transform **A** and ϕ to $\mathbf{A}' = \mathbf{A} + \nabla\psi, \phi' = \phi - \partial\psi/\partial t$. Demanding

$$\nabla.\mathbf{A}' + \frac{1}{c^2}\frac{\partial\phi'}{\partial t} = 0,$$

we get

$$\nabla^2\psi - \frac{1}{c^2}\frac{\partial^2\psi}{\partial t^2} = -F.$$

Since ψ satisfying this equation can always be found, a set of vector and scalar potentials satisfying the Lorentz gauge always exists. In fact, the potentials satisfying the Lorentz gauge too, have an arbitrariness as the **restricted gauge** transformation

$$\mathbf{A} \rightarrow \mathbf{A} + \nabla\psi, \phi \rightarrow \phi - \frac{\partial \psi}{\partial t},$$

where $\nabla^2\psi - \frac{1}{c^2}\frac{\partial^2\psi}{\partial t^2} = 0$, preserves the Lorentz condition (provided $\mathbf{A}$ and ϕ satisfy it initially).

4.2 Vector Potential due to a Current Distribution

The vector potential $\mathbf{A}(\mathbf{r}, t)$ due to a current distribution $\mathbf{J}(\mathbf{r}', t)$ can be obtained by solving eq. (10), using Green's function technique. Alternatively we write down a vector potential intuitively, as a generalisation of the static case, to account for time delay between the cause and effect at a distance

$$\mathbf{A}(\mathbf{r}, t) = \frac{\mu_0}{4\pi}\int \frac{\mathbf{J}(\mathbf{r}', t - R/c)}{R} dV' \tag{11}$$

where $\mathbf{R} = \mathbf{r} - \mathbf{r}'$ and the integral is to be carried out over the entire volume of the source. One may notice that the vector potential at $\mathbf{r}, t$ due to a current element at $\mathbf{r}'$ depends on the value of current density at time t - R/ c, as R/ c is the time of flight of the effect **(A)** from the cause (current element).

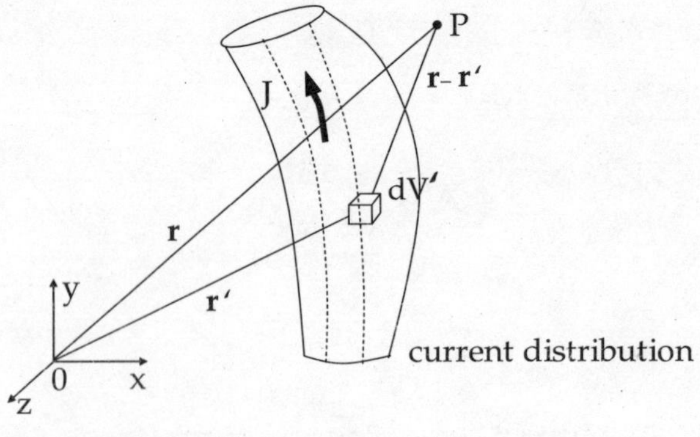

Figure 4.3: Locations of point of observation P(**r**) and current element of a large current distribution.

PROBLEM 4.2. Show that the vector potential given by eq. (11) satisfies eq. (10).

Solution: From eq. (11),

$$\frac{\partial^2 \mathbf{A}}{\partial t^2} = \frac{\mu_0}{4\pi}\int \frac{1}{R}\frac{\partial^2}{\partial t^2}\mathbf{J}\left(\mathbf{r}',t-R/c\right)dV',$$

$$\nabla^2 \mathbf{A} = \frac{\mu_0}{4\pi}\int \nabla_R^2\left[\frac{\mathbf{J}\left(\mathbf{r}',t-R/c\right)}{R}\right]dV',$$

where $\mathbf{R} = \mathbf{r} - \mathbf{r}'$ and $\mathbf{r}'$ is a constant as far as ∇ operation is concerned. Hence $\nabla = \nabla_R = \hat{x}\partial/\partial R_x + \hat{y}\partial/\partial R_y + \hat{z}\partial/\partial R_z$,

$$\nabla_R^2\left[\frac{\mathbf{J}\left(\mathbf{r}',t-R/c\right)}{R}\right] = \frac{1}{R^2}\frac{\partial}{\partial R}\left\{R^2\frac{\partial}{\partial R}\left[\frac{\mathbf{J}\left(\mathbf{r}',t-R/c\right)}{R}\right]\right\}$$

$$= -\mathbf{J}(\mathbf{r}',t-R/c)4\pi\delta^3(\mathbf{R}) + \frac{1}{R}\frac{\partial^2}{\partial R^2}\left[\mathbf{J}\left(\mathbf{r}',t-R/c\right)\right].$$

Here we have employed $\nabla^2\left(\frac{1}{R}\right) = -4\pi\delta^3(\mathbf{R})$. Using the above expressions, we obtain

$$\nabla^2 \mathbf{A} - \frac{1}{c^2}\frac{\partial^2 \mathbf{A}}{\partial t^2} = -\mu_0 \mathbf{J}(\mathbf{r},t) + \frac{\mu_0}{4\pi}\int\frac{1}{R}\left(\frac{\partial^2}{\partial R^2} - \frac{1}{c^2}\frac{\partial^2}{\partial t^2}\right)\mathbf{J}\left(\mathbf{r}',t-R/c\right)dV'$$

$$= -\mu_0 \mathbf{J}(\mathbf{r},t)$$

as $\frac{\partial^2 \mathbf{J}}{\partial t^2} = \frac{\partial^2 \mathbf{J}}{\partial \xi^2}, \frac{\partial^2 \mathbf{J}}{\partial R^2} = \frac{1}{c^2}\frac{\partial^2 \mathbf{J}}{\partial \xi^2}$, with $\xi = t - R/c$.

4.3 Short Antenna

Consider a small antenna of length $\hat{\mathbf{z}}\,dl$ carrying a current $I = I_0 e^{-i\omega t}$ and placed at $\mathbf{r}' = 0$.

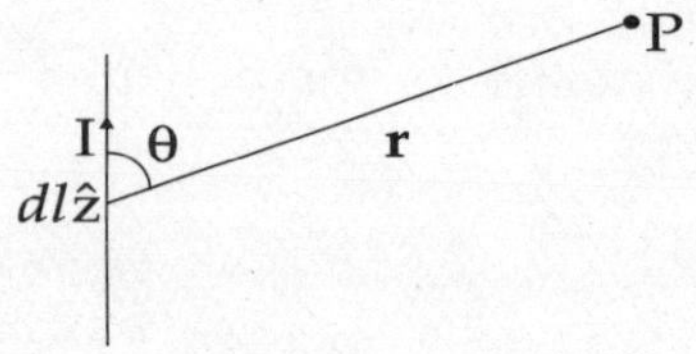

Figure 4.4: Schematic of a small antenna of length dl $\hat{z}$.

The vector potential at $(\mathbf{r}, t)$ due to the antenna is

$$\mathbf{A}(\mathbf{r}, t) = \hat{\mathbf{z}} \frac{\mu_0 I_0 dl}{4\pi r} e^{-i(\omega t - \frac{\omega}{c} r)}. \tag{12}$$

The magnetic field, $\mathbf{B} = \nabla \times \mathbf{A}$, is

$$\mathbf{B}(\mathbf{r}, t) = \frac{\mu_0 I_0 dl}{4\pi r} \left(\frac{i\omega}{c} - \frac{1}{r} \right) \hat{r} \times \hat{z} e^{-i(\omega t - \frac{\omega}{c} r)}, \tag{13}$$

where r refers to spherical polar coordinates. **B** has two terms

(i) one whose amplitude goes as $1/r$ is called the **radiation field**

(ii) the other with $1/r^2$ dependence is called the **induction field**.

At $r = c/\omega \approx \lambda/6$ (λ being the wavelength of the wave), the two terms are of equal magnitude. For $r >> c/\omega$, the induction field can be neglected and one may write

$$\mathbf{B} = \frac{\mu_0 I_0 dl}{4\pi r} \frac{i\omega}{c} (\hat{\mathbf{r}} \times \hat{\mathbf{z}}) e^{-i(\omega t - \frac{\omega}{c} r)}. \tag{14}$$

Using the fourth Maxwell's equation, the electric field can be written as

$$\mathbf{E} = \frac{i}{\omega \epsilon_0} \nabla \times \mathbf{H}. \tag{15}$$

At $r >> \lambda$, one may replace ∇ by $(i\omega/c)\,\hat{r}$. Assuming the r-variation of phase much stronger than that of the amplitude, we get

$$\mathbf{E} = -\frac{I_0 dl}{4\pi r \epsilon_0} \frac{i\omega}{c^2} (\hat{\mathbf{r}} \times (\hat{\mathbf{r}} \times \hat{\mathbf{z}})) e^{-i(\omega t - \frac{\omega}{c} r)}. \tag{16}$$

If need arises, we may write $\hat{\mathrm{r}}$ as $\hat{\mathrm{r}} = \sin\theta\cos\phi\hat{\mathrm{x}} + \sin\theta\sin\phi\hat{\mathrm{y}} + \cos\theta\hat{\mathrm{z}}$. The time average Poynting vector (intensity) is

$$\mathbf{S}_{\mathrm{av}} = \frac{1}{2}\mathrm{Re}\left(\mathbf{E}\times\mathbf{H}^*\right) = \hat{\mathrm{r}}\frac{|\mathrm{H}|^2}{2\,\epsilon_0\,c} = \left(\frac{\mathrm{I}_0 dl}{4\pi \mathrm{r}}\right)^2\frac{\omega^2}{2c^3\,\epsilon_0}\sin^2\theta\hat{\mathbf{r}}. \qquad (17)$$

The intensity scales as the square of wave frequency, inversely as the square of distance and varies as $\sin^2\theta$, where θ is the angle between the direction of observation and the direction of current flow in the antenna. It is maximum in the direction transverse to the direction of current flow ($\theta = \pi/2$) and vanishes in the end-on side ($\theta = 0$) of the antenna. Had we retained the induction field, its contribution to time average Poynting vector would vanish.

4.3.1 Radiation Pattern

It is useful to plot polar lines from the centre of an antenna in different directions, such that, the length of a line is proportional to intensity S_{av} of radiation in that direction. The locus of tips of these lines gives the radiation pattern of the antenna. In the case of a dipole antenna, $\mathrm{S}_{\mathrm{av}}\alpha\sin^2\theta$. The radiation pattern in the x-z (or y-z) plane is shown in fig. 4.5. It has maxima at $\theta = \pi/2$. In the x-y plane, the radiation pattern (showing variation with azimuthal polar angle ϕ) is a circle as there is no ϕ dependence.

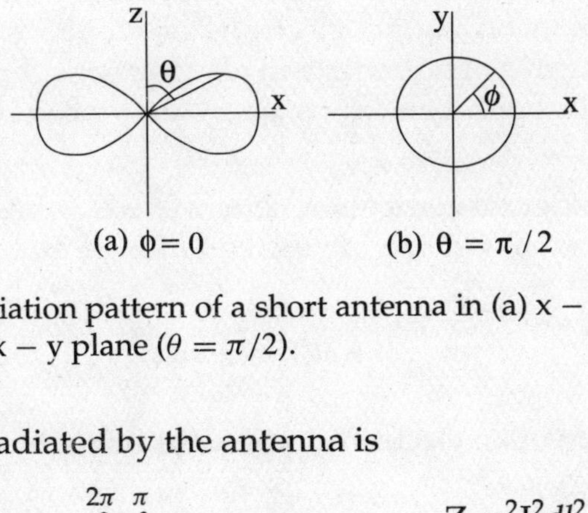

Figure 4.5: Radiation pattern of a short antenna in (a) x − z plane ($\phi = 0$), (b) x − y plane ($\theta = \pi/2$).

Total power radiated by the antenna is

$$\mathrm{W} = \int_0^{2\pi}\int_0^{\pi} \mathrm{S}_{\mathrm{av}}\mathrm{r}^2\sin\theta d\theta d\phi = \frac{\mathrm{Z}_0\omega^2\mathrm{I}_0^2 dl^2}{12\pi c^2}, \qquad (18)$$

where $Z_0 = (\mu_0/\epsilon_0)^{1/2} \approx 377$ ohm. Expressing W as $\mathbf{I}_0^2\mathbf{R}_{rad}/2$, one obtains the radiation resistance of the dipole antenna,

$$\mathbf{R}_{rad} = 80\pi^2 \left(\frac{dl}{\lambda}\right)^2. \tag{19}$$

In the above treatment, we have assumed the current distribution along the length of the antenna to be uniform. However, in actual centre-fed dipole antennae, the current is maximum at the centre and vanishes at the ends. With this, the field strength at every point is reduced to one half and the total power radiated by the antenna becomes one fourth of that given in eq. (18). The radiation resistance also becomes one fourth.

A useful quantity in antenna theory is **antenna gain**. It is defined as the ratio of maximum intensity S_{avm} obtained in certain direction by the antenna, to the intensity S_{av0} one would obtain from an isotropic radiator of the same power. For an antenna of power W, S_{av0} at a distance r is $S_{av0} = W/4\pi r^2$. Hence the gain is

$$G = \frac{S_{avm}}{W/4\pi r^2}.$$

For a dipole antenna, G = 3/2.

PROBLEM 4.3. Rayleigh scattering is a process in which an electromagnetic wave of wavelength λ, incident on a gas, induces oscillatory motion on bound electrons of atoms that behave as small dipole antennae and radiate in all directions. Employing the spring model for bound electrons, show that the scattered power scales as $1/\lambda^4$.

Solution: Consider an electromagnetic wave $\mathbf{E} = \hat{x}A\cos(\omega t - kz)$ incident on an atom. On solving the equation of motion,

$$m\frac{\partial^2 x}{\partial t^2} + \omega_0^2 x = -eE_x$$

the displacement of an electron cloud turns out to be

$$x = \left(\frac{eA}{m\left(\omega^2 - \omega_0^2\right)}\right)\cos(\omega t - kz),$$

where $-e$ and m are the electron charge and mass respectively and ω_0 is the natural frequency of oscillation of the mass attached to the spring .The oscillatory electron velocity can be written as

$$v_x = \left[\frac{eA\omega}{m\left(\omega_0^2 - \omega^2\right)} \right] \sin(\omega t - kz) .$$

For $\omega << \omega_0$, v_x scales as ω. Hence, the oscillatory electron current I also scales as ω. Since the radiated power scales as $I^2\omega^2$, it goes as ω^4 or λ^{-4}.

PROBLEM 4.4. Two identical dipoles, each of length $dl\hat{z}$, and carrying a current $I_0 \exp(-i\omega t)$ are placed at (0,0,0) and (d,0,0). Deduce the radiation pattern in

(i) $\theta = \pi/2$ plane and

(ii) $\phi = 0$ plane for $\omega d / c = \pi$.

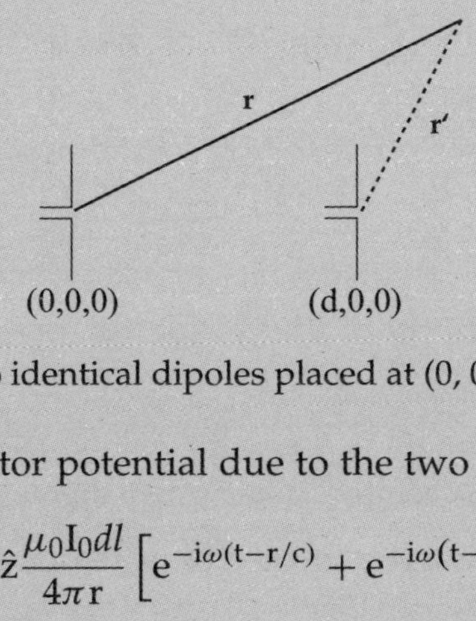

Figure 4.6: Two identical dipoles placed at (0, 0, 0) and (d, 0, 0).

Solution: The vector potential due to the two antennae at $r >> d$ is

$$A = \hat{z}\frac{\mu_0 I_0 dl}{4\pi r}\left[e^{-i\omega(t-r/c)} + e^{-i\omega(t-r'/c)}\right],$$

where $r' = \left|\hat{r} - d\hat{x}\right| = r - dx/r = r - d\sin\theta\cos\phi$.
It simplifies to give

$$A = \hat{z}\frac{\mu_0 I_0 dl}{4\pi r} e^{-i\left(\omega t - \omega r/c\right)}\left(1 + e^{-i\psi}\right),$$

where $\psi = (\omega d/c)\sin\theta\cos\phi$. Using $\mathbf{B} = \nabla \times \mathbf{A}$, $\mathbf{E} = -\nabla \times \mathbf{H}/i\omega\varepsilon_0$ and replacing ∇ by $i(\omega/c)\hat{\mathbf{r}}$, we obtain

$$\mathbf{B} = i(\omega A_z/c)\hat{\mathbf{r}} \times \hat{\mathbf{z}},$$

$$\mathbf{S}_{av} = \frac{1}{2}\mathrm{Re}(\mathbf{E}^* \times \mathbf{H}) = \frac{\hat{\mathbf{r}}|H|^2}{2\,\epsilon_0\,c}$$

$$= \hat{\mathbf{r}}\,S_0 \sin^2\theta \cos^2(\psi/2); \quad S_0 = \frac{2\omega^2\mu_0}{c}\left(\frac{I_0 dl}{4\pi r}\right)^2.$$

For $\theta = \pi/2$, $S_{av} = S_0 \cos^2((\pi/2)\cos\phi)$.
For $\phi = 0$, $S_{av} = S_0 \sin^2\theta \cos^2((\pi/2)\sin\theta)$.

4.4 Half-Wave Dipole

Consider a centre-fed dipole antenna of length $2l = \lambda / 2$ (Fig. 4.7). The current distribution in the antenna may be taken as

$$I = I_0 \sin\frac{\omega}{c}(l - z)e^{-i\omega t} \qquad z > 0$$
$$= I_0 \sin\frac{\omega}{c}(l + z)e^{-i\omega t} \qquad z < 0,$$

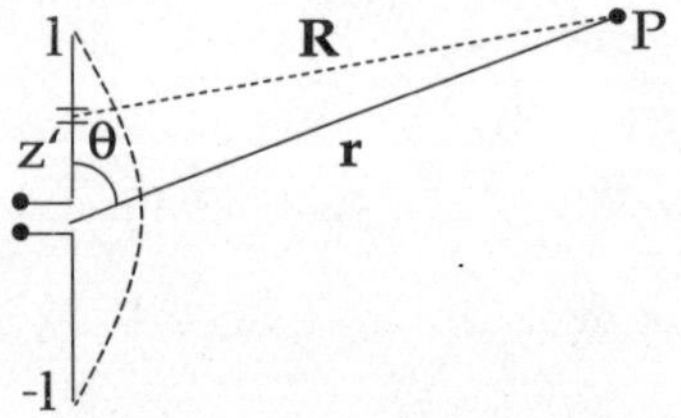

Figure 4.7: A centre-fed half-wave dipole antenna.

From eq. (11) we may write the vector potential as

$$\mathbf{A}(\mathbf{r},t) = \hat{\mathbf{z}}\frac{\mu_0 I_0}{4\pi}\left[\int_0^{\lambda/4} \frac{\sin\left(\frac{\pi}{2} - \frac{\omega}{c}z'\right)}{R} e^{-i\left(\omega t - \frac{\omega}{c}R\right)} dz' + \int_{-\lambda/4}^{0} \frac{\sin\left(\frac{\pi}{2} + \frac{\omega}{c}z'\right)}{R} e^{-i\left(\omega t - \frac{\omega}{c}R\right)} dz'\right]. \qquad (20)$$

At large distances, $R \approx r - \mathbf{z}' \cos\theta$ and eq. (20) simplifies to

$$\mathbf{A}(\mathbf{r}, t) = \hat{\mathbf{z}} \frac{\mu_0 I_0 e^{-i\left(\omega t - \frac{\omega}{c} r\right)}}{4\pi r} \int_0^{\lambda/4} \left[\cos\left(\tfrac{\omega}{c} z'(1 + \cos\theta)\right) + \cos\left(\tfrac{\omega}{c} z'(1 - \cos\theta)\right)\right] dz'$$

$$= \hat{\mathbf{z}} F e^{-i\left(\omega t - \frac{\omega}{c} r\right)}, \tag{21}$$

$$\text{where} \quad F = \frac{c\mu_0 I_0}{4\pi\omega r} \frac{\cos\left[(\pi\cos\theta)/2\right]}{\sin^2\theta}. \tag{22}$$

The magnetic and electric fields and the average Poynting vector are

$$\mathbf{B} = \nabla \times \mathbf{A} = i\frac{\omega}{c}(\hat{\mathbf{r}} \times \hat{\mathbf{z}}) F e^{-i\left(\omega t - \frac{\omega}{c} r\right)},$$

$$\mathbf{E} = i\nabla \times \mathbf{H}/\omega \in_0 = -i\frac{\omega}{c^2 \in_0} \hat{\mathbf{r}} \times (\hat{\mathbf{r}} \times \hat{\mathbf{z}}) F e^{-i\left(\omega t - \frac{\omega}{c} r\right)},$$

$$\mathbf{S_{av}} = \hat{\mathbf{r}} \frac{\eta I_0^2}{8\pi^2 r^2} \frac{\cos^2\left(\frac{\pi}{2}\cos\theta\right)}{\sin^2\theta}; \quad \eta = \frac{1}{\in_0 c}. \tag{23}$$

The total power radiated by the antenna, through the hemispherical surface of radius r, is

$$\frac{W}{2} = \frac{\eta I_0^2}{4\pi} \int_0^{\Pi/2} \frac{\cos^2\left(\frac{\pi}{2}\cos\theta\right)}{\sin^2\theta} d\theta$$

$$\approx 0.609 \eta I_0^2 / 4\pi. \tag{24}$$

The radiation pattern of a half-wave dipole is a circle in the x-y plane ($\theta = \pi / 2$ plane). In the x-z plane, it is similar to a short dipole but more complicated.

4.5 Conclusions

Current-fed antennae are the main source of electromagnetic radiation for frequencies $\omega/2\pi < 1\,GHz$. The radiated power of an antenna scales as the square of current and square of radiation frequency. The low frequency antennae, ($\omega/2\pi \leq 100\,Hz$) thus radiate very weakly. To achieve substantial power at these frequencies, the length of the antenna needs to be increased.

Radiation from an antenna is not the same in all directions. For a dipole antenna, there is no radiation in the end-on direction. The maximum radiation is in the equatorial plane. The antenna gain for a short wave dipole is, however, very modest. It can be increased by orders of magnitude (i.e. the radiation can be sharply directed in a given direction) by using antenna array, as discussed in unit 5.

It should be mentioned that a transmitting antenna can also be employed as a receiving antenna. The electromagnetic waves, incident on it, induce an electrical current that can be amplified and processed to retrieve relevant information from the wave. It can be shown that the directional (radiation) pattern of a receiving antenna is identical with that for a transmitting antenna.

UNSOLVED PROBLEMS

1. A dipole antenna of length dl =1m carries a current of $I = I_0 \exp(-i\omega t)$ with $I_0 = 10$ A and $\omega/2\pi = 1$ MHz. Estimate the radiation intensity at a distance of 20 km in the equatorial plane, i.e. at 90° to the direction of antenna length.

[ANS. $(\pi/24)\,10^{-13}$ W/cm^2.]

2. Estimate the amplitude of electric field, at a distance of 40,000 km from a 1kW transmitter, in the direction where antenna gain is 30 dB.

[ANS. 1.9×10^{-4}V/m.]

3. Two dipoles with current length products $\hat{x}I_0\, dl\, \exp(-i\omega t)$ and $\hat{y}iI_0\, dl\, \exp(-i\omega t)$ are placed at (0,0,0) and (d,0,0). Obtain S_{av} at large distances from the antenna. Plot the antenna pattern in $\phi = 0$ plane.

$$\left[\textbf{Ans. } \hat{r}\left(\frac{I_0 dl}{4\pi r}\right)\frac{w^2}{2\epsilon_0 c^3}\left[1+\cos^2\theta-2\sin^2\theta\cos\phi\sin\phi\sin\Psi\right]\right]$$

4. Two dipole antennae of length $dl\hat{z}$ and currents $I_0 \exp(-i\omega t)$ and $I_0 \exp[-i(\omega t+\pi/2)]$ are mounted at (0,0,h) and (0,0,–h). Obtain the vector potential at a far point.

$$\left[\textbf{Ans. } \mathbf{A}(\mathbf{r},t)=\hat{z}\frac{2\mu_0 I_0 dl}{4\pi r}\sin\left(\frac{\omega}{c}h\cos\theta\right)e^{-t(\omega t-\frac{\omega}{c}r+\frac{\pi}{2})}.\right]$$

5. A ground-based radiowave transmitter of frequency $\omega/2\pi = 5\,\text{MHz}$, power $W = 100$ kW and antenna gain 20 dB is employed to heat the electrons, at a height $h = 120$ Km in the ionosphere, where electron density is $10^{11}\,\text{m}^{-3}$, collision frequency $\nu = 10^5\,\text{s}^{-1}$ and plasma temperature $T_0 = 0.2\,\text{eV}$. Neglecting the attenuation and refraction of the wave upto that height, estimate the time average heating rate R, viz., the energy absorbed per second by an electron. If an electron losses energy via collisions at a rate $R' = (3/2)\delta\nu(T_e - T_0)$, where $\delta = 3\times 10^{-4}$ is the mean fraction of the excess energy lost in a collision, estimate the rise in the electron temperature $T_e - T_0$ in the steady state.

$$\left[\textbf{Ans. } R=\mu_0 cWG\nu e^2/(4\pi h^2 m\omega^2)\approx 0.4\,\text{eV/s},\right.$$
$$\left.T_e - T_0 = 9\times 10^{-3}\,\text{eV}.\right]$$

6. Obtain the vector potential due to a travelling wave current distribution $I = I_0 \exp[-i(\omega t - \omega z/c)]$ of length $L\hat{Z}$ and show that at a far point

$$|E_\theta| = \frac{30 I_0 \sin\theta}{r(1-\cos\theta)}\sqrt{2-2\cos\left[\frac{\omega}{c}(1-\cos\theta)L\right]}.$$

7. Given $\mathbf{A} = (\hat{x}+\hat{z})\frac{A_0}{r}e^{-i(\omega t-kr)}$, $k=\omega/c$, obtain $\mathbf{S_{av}}$ at $r >> c/\omega$.

$$\left[\textbf{Ans. } \mathbf{S_{av}}=\hat{r}\left(\frac{\omega^2 A_0^2}{2\mu_0 c r^2}\right)\left\{2\sin^2\theta\sin^2\phi+(\sin\theta\cos\phi-\cos\theta)^2\right\}\right]$$

8. An electromagnetic wave has $\mathbf{B} = \hat{\mathbf{y}}\mathbf{b_0} \exp[-i\{\omega t - \omega(2x + z)/c\}]$. Construct a suitable vector potential $\mathbf{A}$ (that has $\nabla . \mathbf{A} = \mathbf{0}$) for it.

[HINT: *Presume a vector potential with the same phase variation as* $\mathbf{B}$, $\mathbf{A} = \mathbf{a} \exp[-i(\omega t - \mathbf{k} \cdot \mathbf{r})]$, *where* $\mathbf{k} = (2\omega/c)\hat{x} + (\omega/c)\hat{z}$, *and then employ* $\mathbf{k} \cdot \mathbf{a} = 0$ and $\mathbf{B} = \nabla \times \mathbf{A} = i\mathbf{k} \times \mathbf{A}$ *to obtain* $\mathbf{a}$.]

[ANS. $\mathbf{A} = -i(cb_0/5\omega)(\hat{x} - 2\hat{z}) \exp[-i\{\omega t - \omega(2x + z)/c\}]$.]

9. Construct suitable vector potentials (that have $\nabla \cdot \mathbf{A} = \mathbf{0}$) for the following electromagnetic waves

(i) $\mathbf{E} = (\hat{x} + i\hat{y})E_0 \exp[-i(\omega t - \omega z/c)]$,

(ii) $\mathbf{B} = (\hat{x} - \hat{z})b_0 \exp[-i(\omega t - 2\omega y/c)]$.

$$\left[\textbf{ANS.}\quad \begin{aligned} &\text{(i) } \mathbf{A} = \mathbf{E}/i\omega, \\ &\text{(ii) } \mathbf{A} = (\hat{x} + \hat{z})(b_0 c/2i\omega) \exp[-i(\omega t - 2\omega y/c)] \end{aligned} .\right]$$

10. A short antenna of length $dl\hat{z}$ carries a current $I_0[\exp(-i\omega t) + 3\exp(-2i\omega t)]$. Estimate the ratio of intensities at large distances, for the frequencies ω and 2ω.

[ANS. 1:36]

REFERENCES AND SUGGESTED READING

[1] Jordan E. C. and Balmain K. G., *Electromagnetic waves and Radiating systems*, second edition, New Delhi, Prentice Hall of India Pvt. Ltd. (1991).

[2] Ramo S., Whinnery J. R., and Duzer T. V., *Fields and waves in communication electronics*, second edition, New York, John Wiley & Sons, (1984).

Unit 5

Radio Communication and Radar

Introduction

One of the major applications of electromagnetic waves is in the area of transmission of information and control signals. The frequency or amplitude of a carrier electromagnetic wave is modulated by the information we wish to transmit. The carrier wave travels through free space or a guiding system to the receiver, where information is extracted from the carrier wave. Information can also be sent in bits and pieces by a succession of short pulses and the information can be reconstructed at the receiver end.

Electromagnetic waves of different frequencies are used for different kinds of communication. The ones below the frequency of 3 GHz are commonly called radio waves and are divided into the following frequency bands:

0-30 Hz	ULF	(ultra low frequency waves)
30-300 Hz	ELF	(extremely low frequency waves)
300 Hz-30 KHz	VLF	(very low frequency waves)
30 KHz-300 KHz	LF	(low frequency waves)
300 KHz-3 MHz	MF	(medium frequency waves)
3 MHz-30 MHz	HF	(high frequency waves or short waves)
30 MHz-300 MHz	VHF	(very high frequency waves)
300 MHz-3GHz	UHF	(ultra high frequency waves)

Each band has its specific utility. ULF waves, for instance, are used for underground explorations and geoseismic survey. ELF waves are suited for submarine communications. VLF waves are useful for ionospheric and magnetospheric explorations. Medium frequency waves and short waves are employed for radio transmission. VHF waves are used for television transmission and telephone links. UHF band finds use in telephone and radar. In this chapter, we discuss four specific applications:

a. medium wave communication
b. short wave radio communication
c. cellular communication.
d. radar.

The Earth and the ionosphere play important roles in medium wave and short wave communication. Hence, we begin with an analysis of the electrical properties of these media.

5.1 Effective Relative Permittivity of the Earth

At medium and short wave frequencies, the effective relative permittivity of the Earth is

$$\epsilon_{eff} = \epsilon_r + i\frac{\sigma}{\omega\,\epsilon_0} \approx i\frac{\sigma}{\omega\,\epsilon_0}, \tag{1}$$

where ω is the frequency of the radio wave, ϵ_r is the relative permittivity of the earth, ϵ_0 is the free space permittivity and σ is the electrical conductivity of the Earth. σ varies form 5×10^{-3} mho/m to 1 mho/m and ϵ_r from 2 to 80, as one goes from dry soil to wet soil and sea water. The Earth acts as a good reflector for medium and short waves. It also allows their propagation as surface waves, over the Earth-atmosphere boundary. The propagation constant of a surface wave of frequency ω (as derived in unit 3) is

$$k = \frac{\omega}{c}\left(\frac{\epsilon_{eff}}{1+\epsilon_{eff}}\right)^{1/2} \approx \frac{\omega}{c}\left(1 + i\frac{\omega\,\epsilon_0}{2\sigma}\right). \tag{2}$$

Its imaginary part, k_i, goes as square of frequency and attenuation becomes severe at short waves. Thus, only medium waves can propagate as surface waves. Typically at the frequency of 1 MHz, if one takes $\sigma \approx 10^{-2}$ mho/m, attenuation constant $k_i \approx 5 \times 10^{-5}\,m^{-1}$. On wet soil and sea water, surface waves can propagate with much less attenuation.

5.2 Characteristics of the Ionosphere

The ionosphere is the ionised region of upper atmosphere, ranging from a height of 65 km above the ground to over a thousand kilometers (Fig. 5.1). The electron density peaks at a height of about 300 km. The medium and short waves, launched from a ground-based antenna, penetrate into the ionosphere and suffer reflection somewhere in the interior.

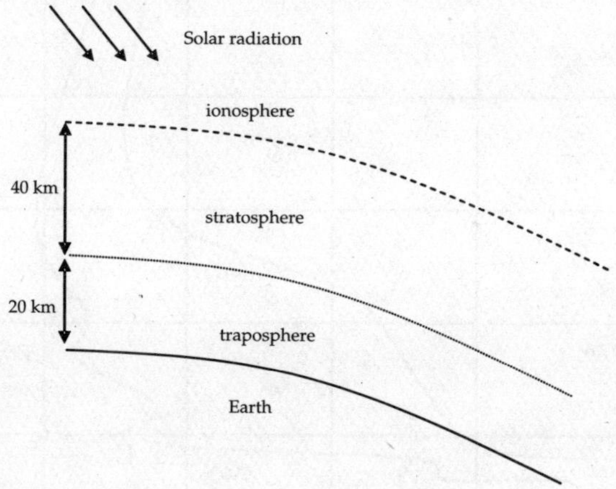

Figure 5.1: Different regions of Earth's atmosphere.

Ionosphere is the result of ionisation of air, primarily nitrogen and oxygen of the Earth's atmosphere, due to the ultraviolet part of the solar spectrum. The rate of ionisation R_i (the number of free electrons produced per unit volume per second) is proportional to intensity of ultraviolet radiation I_u and air pressure p. If one denotes height above the ground as h, then p decreases almost exponentially with h while I_u increases with h (as the source of ultraviolet radiation is the Sun and its intensity diminishes as it approaches the Earth's surface). Thus, R_i maximises at a certain height. The process of ionisation is countered by recombination and attachment of free electrons with ions and neutral atoms. Let the electron loss rate per unit volume per second be R_c. A steady state is realised when

$$R_i = R_c.$$

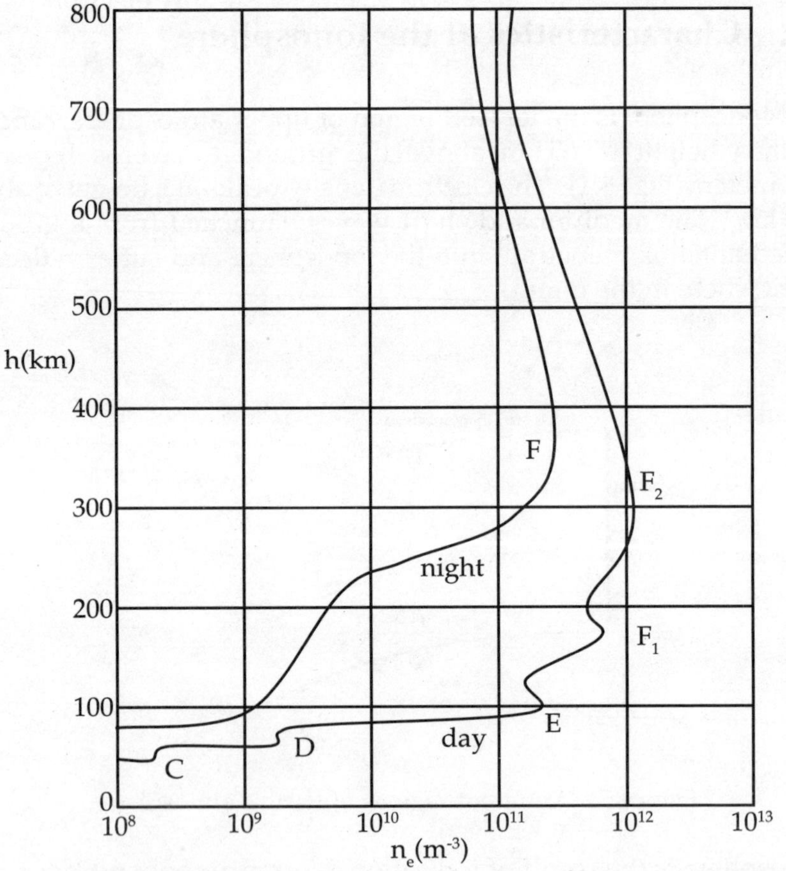

Figure 5.2: Typical variation of free electron density, with height above the ground.

This equation gives the free electron density profile, n_e (h), of the ionosphere, as shown in fig. 5.2. We have also shown the variation of electron collision frequency with height in fig. 5.3. The electron density profile changes with the time of the day and the season, as it depends on the orientation of solar rays and the distance of the Earth from the Sun. Sunspots (events in which the bulk of solar plasma is ejected from the Sun) also cause very significant variations in the ionospheric electron density profile. The lower ionosphere is divided into D, E and F layers. D layer exists primarily during the daytime, from h = 65 km to 90 km. It is a low electron density ($n_e \leq 2 \times 10^3\,cm^{-3}$), collisional plasma with high electron-neutral collision frequency ν ($5 \times 10^5\,s^{-1} < \nu < 10^7\,s^{-1}$). During the night, width of the D region shrinks and electron density diminishes by two orders of magnitude. E layer ranges from h = 90 km to 150 km, has higher electron density ($2 \times 10^3\,cm^{-3} < n_e < 2 \times 10^5\,cm^{-3}$) and lower collision frequency ($5 \times 10^3\,s^{-1} < \nu < 5 \times 10^5\,s^{-1}$); collisions are still dominated by electron- neutral collisions. F region has large electron density ($n_e \leq 10^6\,cm^{-3}$) and weak collisionality, dominated by electron-ion collisions ($\nu \sim 10^3\,s^{-1}$). The electron density profile has four prominent maxima, called D, E, F_1 and F_2 peaks. The values of plasma frequency $\omega_p = (n_e e^2/m\,\epsilon_0)^{1/2}$ corresponding to these peaks (expressed as $f_p = \omega_p/2\pi$), during daytime, are:

$$f_{pD} \cong 0.34\,\text{MHz},\ f_{pE} \cong 3.4\,\text{MHz},\ f_{pF_1} \cong 4.3\,\text{MHz},\ f_{pF_2} \cong 8\,\text{MHz}.$$

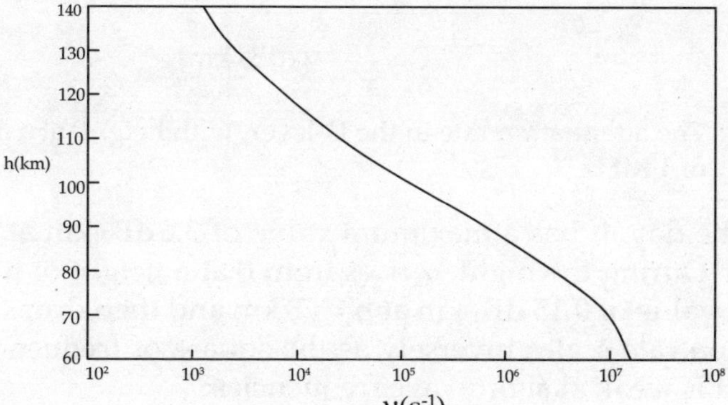

Figure 5.3: The variation of electron collision frequency with height above the ground.

The refractive index of the ionosphere, ignoring the role of Earth's magnetic field, can be written as

$$\eta = \eta_r + i\eta_i = \sqrt{\left(1 - \frac{\omega_p^2}{\omega\,(\omega + i\nu)}\right)}, \tag{3}$$

where ω is the frequency of the radio wave. At medium frequencies, the attenuation constant $k_i = (\omega\eta_i/c)$, in D layer is high during the daytime and weak at night. Fig.5.4 shows the variation of attenuation rate $\alpha = 2 \times 10^4\, k_i \log_{10} e$ in dB/km for a radio wave of frequency 1 MHz in the D region.

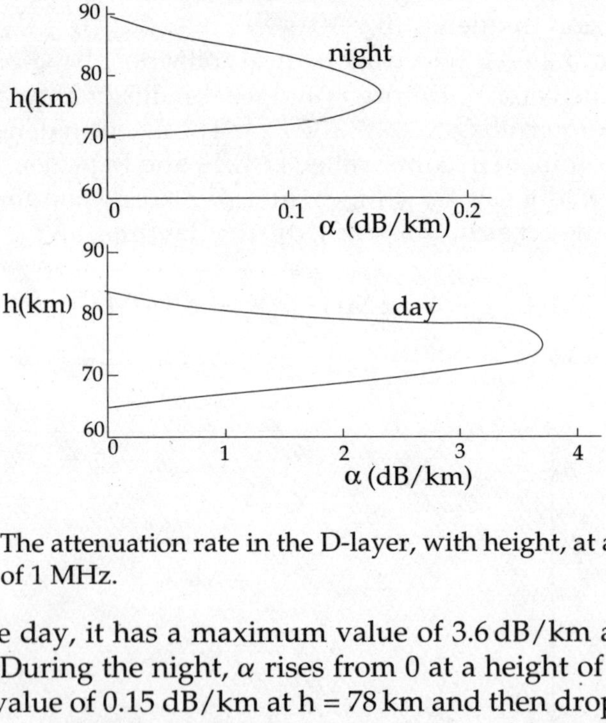

Figure 5.4: The attenuation rate in the D-layer, with height, at a frequency of 1 MHz.

During the day, it has a maximum value of 3.6 dB/km at a height of 74 km. During the night, α rises from 0 at a height of h = 70 km to a peak value of 0.15 dB/km at h = 78 km and then drops off. The attenuation rate scales inversely as the square of frequency of the wave and is weak at short wave frequencies.

Besides absorption, electromagnetic waves undergo refraction in the ionosphere. Since the refractive index decreases with height (below the F peak), the rays move away from the vertical and bend, suffering total reflection from a layer where $\omega_p = \omega \cos i_0$ (i_0 is the angle of incidence at the entry point of the ionosphere).

The earth's magnetic field $\mathbf{B_s}$ plays a significant role in the propagation of radio waves through the ionosphere. The magnetic field influences electrons' response to electric field of the wave, making electrical conductivity a tensor and plasma an anisotropic medium. There are two modes of electromagnetic wave propagation, known as *ordinary* (O) and *extraordinary* (X) modes. These modes, in general, are elliptically polarised and their phase velocities depend on the angle wave vector $\mathbf{k}$ makes with $\mathbf{B_s}$. For $\mathbf{k} \parallel \mathbf{B_s}$, the O and X modes are, respectively, left and right circularly polarised with propagation constants

$$k_{L,R} = \frac{\omega}{c}\sqrt{\left(1 - \frac{\omega_p^2}{\omega\,(\omega \pm \omega_c + i\nu)}\right)}, \tag{4}$$

where $\omega_c = eB_s/m$ is the electron cyclotron frequency ($\approx 5 \times 10^6$ rad / s) and +, − signs refer to O and X modes respectively. The X-mode is sensitive to cyclotron resonance. In this mode, waves with $\omega < \omega_c$ can propagate even in dense plasma ($\omega_p > \omega$). In fact, VLF waves propagate in this mode to thousands of kilometres through the ionosphere and magnetosphere with low attenuation rate.

5.3 Medium Wave Communication

Medium frequency radio communication is a fairly noise-free mode of audio information transmission. The audio signal is converted into an electrical signal, using a microphone or a transducer. The electrical signal modulates the medium frequency (rf) current, that feeds into a transmitting antenna. Two modes of modulation are employed:

a. Amplitude Modulation (AM)
b. Frequency Modulation (FM).

In AM, the amplitude of the rf current is modulated by the electrical signal f(t)

$$I = I_0 \left(1 + \mu f(t)\right) e^{-i\omega_0 t} \tag{5}$$

where ω_0 is the carrier frequency and μ characterises the level of modulation (Fig. 5.5). If one expresses f(t) as a Fourier integral (Appendix A1),

$$f(t) = \frac{1}{\sqrt{2\pi}} \int_{-\infty}^{\infty} g(\Omega)\, e^{-i\Omega t} d\Omega,$$

then the current can be written as

$$I = \frac{1}{\sqrt{2\pi}} \int_{-\infty}^{+\infty} I_\omega e^{-i\omega t} d\omega, \tag{6}$$

where $I_\omega = \left[\sqrt{2\pi}\,\delta(\omega - \omega_0) + g(\omega - \omega_0)\right] I_0$. The second term in I_ω is the same as the Fourier transform of the audio signal, shifted by the rf carrier frequency.

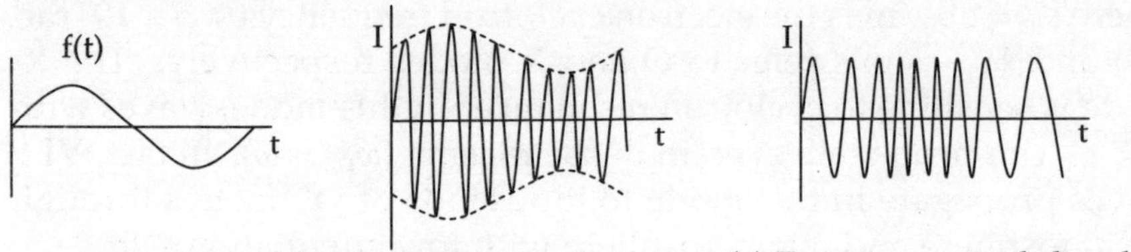

(a) Signal (b) Amplitude-modulated current (c) Frequency-modulated current

Figure 5.5: Amplitude and frequency modulation.

In frequency modulation, the rf frequency is shifted in proportion to the instantaneous magnitude of the modulating signal, while the amplitude of the rf remains constant.

$$I = I_0 e^{-i\phi(t)},$$

$$\omega = \frac{\partial \phi}{\partial \mathbf{t}} = \omega_0 \left(1 + \mu' f(t)\right). \tag{7}$$

In a specific case of f (t) = A cos Ω t,

$$\omega = \omega_0 \left(1 + \mu' A \cos \Omega t\right),$$

$$\phi = \omega_0 \mathbf{t} + \frac{\omega_0\, \mu' A}{\Omega} \sin \Omega t. \tag{8}$$

$\omega_0 \mu' A/\Omega$ is known as the *index of frequency modulation*. One may rewrite I, employing the Bessel function identity

$$e^{i\alpha \sin\theta} = \sum_n J_n(\alpha) e^{in\theta},$$

as

$$I = \sum_n I_0 J_n \left(\frac{\omega_0 \mu' A}{\Omega} \right) e^{-i(\omega_0 + n\Omega)t}. \quad (9)$$

It has frequency components shifted from the carrier frequency, by the harmonics of signal frequency. One may define the *bandwidth* of an FM wave as the width of the frequency spectrum, that contains all components having an amplitude equal to or greater than 1% of the amplitude of the unmodulated wave. Thus, $J_n\left(\frac{\omega_0 \mu' A}{\Omega}\right) \le 0.01$ gives the maximum value of n and the bandwidth is BW = 2n Ω. FM systems, designed for speech communication, usually keep the modulation index low. When $\omega_0 \mu' A/\Omega < 1$, the frequency components beyond the first order sidebands are relatively small. Hence, a bandwidth equal to twice the signal frequency is sufficient to transmit the intelligence. This is the same as the bandwidth required for the transmission of amplitude-modulated wave. At higher *frequency modulation index*, higher order sidebands are important and larger bandwidth is required. This, of course, improves the quality of reception.

The transmitting antenna for medium frequency waves is usually a vertical monopole or a dipole, mounted near the surface of the Earth (Fig. 5.6). The antenna excites a surface plasma wave on the Earth's surface, with amplitude decaying exponentially as one moves away from the ground either into the Earth or into the atmosphere. The antenna also emits waves, that may reach the receiving station via reflection from the ionosphere or directly along the line of sight or after reflection from the Earth. The ionospheric wave is called the *sky wave* whereas the latter two are called *space waves*. During the day, the sky wave is mostly absorbed in the D-layer, hence the signals received at the receiver are the sum of signals from the surface plasma wave and the space waves. If the transmitting and receiving antennae are placed close to the ground, the space wave reflected from the ground is nearly π out of phase with the direct space wave and the sum of the two fields is small. Thus, the surface wave is the main mode of transmission.

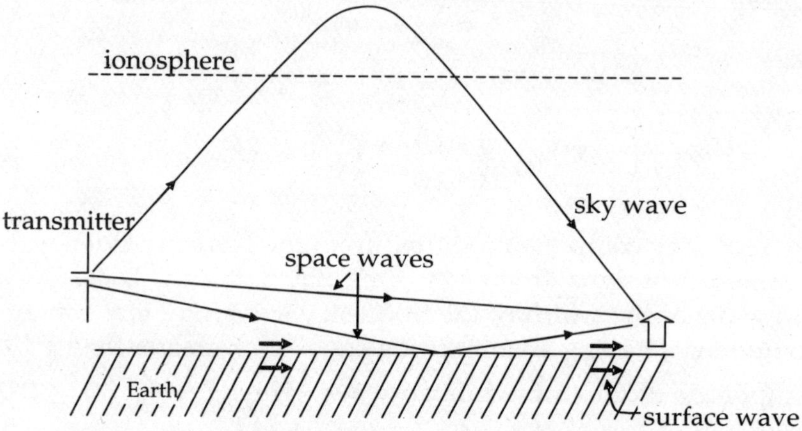

Figure 5.6: Different modes of medium wave propagation.

The excitation of waves by a small vertical dipole, mounted above a planar earth, has been analysed by several researchers, starting with *Sommerfeld* in 1906. *Norton* simplified the field expressions and put them in usable forms. Denoting $\hat{x}$ as the vertical direction, the vertical component of field on the ground, at a distance R from the vertical dipole of length dl carrying current $I = I_\omega e^{-i\omega t}$, can be written as

$$E_x \approx i60\frac{\omega}{c}\frac{I_\omega dl}{R}Fe^{-i(\omega t-\frac{\omega}{c}R)}, \tag{10}$$

where F is the attenuation factor. $|F|$ depends on two quantities, $p = \omega R/(2\chi)$ (called numerical distance) and $b = \tan^{-1}[(\epsilon_r + 1)/\chi]$, where $\chi = \sigma/\omega\epsilon_0$. For $b < 0.1$ radian,

$$\begin{aligned} |F| &\approx e^{-0.43p+0.01p^2} && \text{for } p < 4.5 \\ &\approx \frac{1}{2p-3.7} && \text{for } p > 4.5. \end{aligned} \tag{11}$$

The field decays more rapidly than R^{-1}, with the distance of propagation. At long distances, it goes as R^{-2}. During night, the sky wave reaching the receiving antenna has finite amplitude and interferes with the surface wave signal.

The typical antenna powers, employed for MW broadcast, are tens of kilowatt to ensure high signal to noise ratio at the receiver sites hundreds of kilometres away.

PROBLEM 5.1. A 1 MHz radio wave propagates along $\hat{z}$ as a surface wave, over the surface ($x = 0$) of the Earth of $\sigma = 0.1$mho/m, $\varepsilon_r = 4$. If the amplitude of the horizontal component of the electric field on the surface of the Earth is 10^{-2} V/m, estimate its value at the height of 1 km.

Solution: The horizontal component of the electric field of a (non-diverging) surface wave, in free space above the Earth, goes as

$$E_z = A_0 e^{-\alpha x} e^{-i(\omega t - k_z z)},$$

$$\alpha = (\omega/c)\left[-(1+\varepsilon_{eff})\right]^{-\frac{1}{2}} \approx \sqrt{\frac{\omega\varepsilon_0}{2\sigma}}\left(\frac{\omega}{c}\right)(1+i).$$

Hence the amplitude of E_z at the height h above the Earth is

$$A = A_0 \exp[-(\omega h/c)(\omega\varepsilon_0/2\sigma)^{1/2}] \cong 7\text{mV/m}.$$

5.4 Short Wave Communication

Short waves have wavelength in the range 10-100 m. These are transmitted by antennae, mounted on the ground or at a certain height above the ground. They propagate to long distances as sky waves, suffering reflection in the ionosphere.

5.4.1 Ray Path

In calculating the ray path of the short wave inside the ionosphere, one must take into account the curvature of the Earth-ionosphere system. Let the radius of the Earth be R_E. The ionosphere begins at a height h_0 above the ground. As height increases, refractive index $\eta(h)$ decreases. We may model the ionosphere as a succession of spherically stratified layers of widths h_1-h_0, h_2-h_1, h_3-h_2, ...having refractive indices η_1, η_2, η_3, ...respectively (Fig. 5.7).

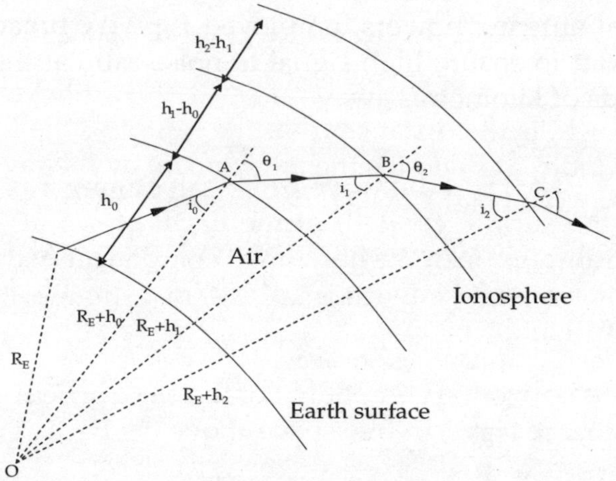

Figure 5.7: Schematic of a ray path in a radially stratified ionosphere.

A ray, launched from the ground transmitter, travels in straight line till it hits the ionosphere. Let the angle of incidence (the angle ray makes with radial direction OA at point A) be i_0. As the ray penetrates into the layer, it bends away from the normal (as $\eta_1 < 1$). However, in each layer, ray travels in straight line. For instance, AB is a straight path in layer 1. At A, the ray makes angle θ_1 with layer normal OA whereas at B it makes angle i_1 with layer normal OB. If one extrapolates line AB and draws a perpendicular on it from O, one gets

$$(R_E + h_0)\sin\theta_1 = (R_E + h_1)\sin i_1.$$

Similarly, for the line segment BC

$$(R_E + h_1)\sin\theta_2 = (R_E + h_2)\sin i_2.$$

Applying the Snell's law at A and B, we get

$$\eta_1 \sin\theta_1 = \sin i_0,$$
$$\eta_2 \sin\theta_2 = \eta_1 \sin i_1.$$

These equations give

$$(R_E + h_0)\sin i_0 = \eta_1(R_E + h_1)\sin i_1 = \eta_2(R_E + h_2)\sin i_2.$$

In general

$$\eta r \sin i = \text{constant}, \tag{12}$$

where r = R_E + h, is the distance from the centre of the Earth, η is the refractive index at r and i is the angle made by the ray with the radius vector at that height.

If one considers a line segment PQ of a ray, characterising points P and Q by polar coordinates (r, θ) and (r + Δ r, θ + Δ θ), (Fig. 5.8), one may deduce

$$\sin i = \frac{1}{\left[1 + \frac{1}{r^2}\left(\frac{dr}{d\theta}\right)^2\right]^{1/2}} \tag{13}$$

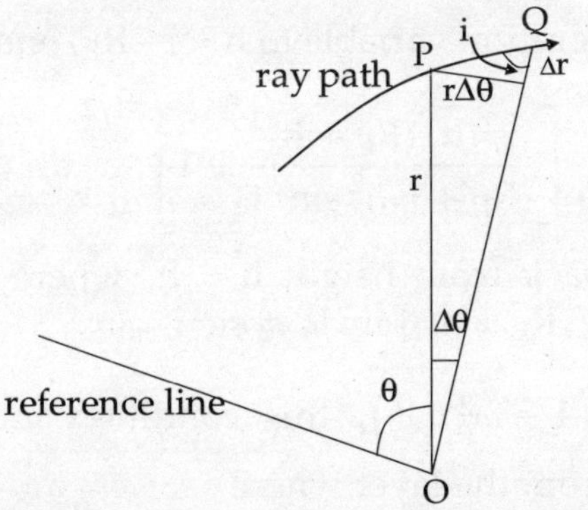

Figure 5.8: Segment of a ray path PQ. OP makes an angle θ with a reference line. tan i = $r\Delta\theta/\Delta r$.

Using eq.(13) in eq.(12), one obtains the equation for the ray trajectory,

$$\frac{\eta^2 r^2}{1 + \frac{1}{r^2}\left(\frac{dr}{d\theta}\right)^2} = (R_E + h_0)^2 \sin^2 i_0$$

or

$$\int_{R_E+h_0}^{R_E+h} \frac{1}{r}\left[\frac{\eta^2 r^2}{(R_E + h_0)^2 \sin^2 i_0} - 1\right]^{-\frac{1}{2}} dr = \theta - \theta_0, \tag{14}$$

where θ_0 is the polar angle of point A.

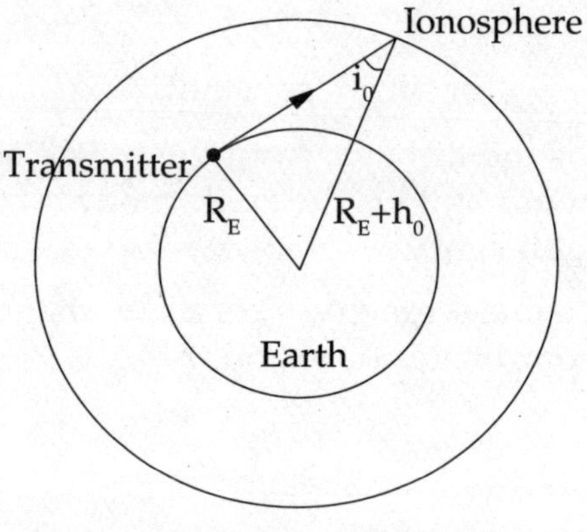

Figure 5.9: Estimation of the maximum angle of incidence of a ray, launched from a ground-based transmitter into the ionosphere.

Changing the integration variable to $h = r - R_E$, one may write

$$\int_{h0}^{h} \frac{1}{R_E + h}\left[\frac{\eta^2(h)\,(R_E + h)^2}{(R_E + h_0)^2 \sin^2 i_0} - 1\right]^{-1/2} dh = \theta - \theta_0.$$

The ray reflects back from height $h = h_r$ where $dr/d\theta = 0$, i.e., $\eta\,(h_r) = [(R_E + h_0)/(R_E + h_r)] \sin i_0 \approx \sin i_{rm0}$.

Since $\eta \cong \sqrt{\left(1 - \omega_p^2/\omega^2\right)}$, this condition indicates that the reflection occurs from the layer where

$$\omega_p(h_r) = \omega \cos i_0. \tag{15}$$

For a radially stratified ionosphere, the largest value of i_0 corresponds to a ray grazing the Earth. From fig. 5.9, we have

$$\sin i_0 = \frac{R_E}{R_E + h_0}.$$

Hence, reflection occurs only when

$$\omega_p > \omega\sqrt{\frac{2h_0}{R_E}} \sim \frac{\omega}{4}.$$

Since the maximum value of plasma frequency in the ionosphere is $\sim$ 8 MHz, the maximum usable frequency (MUF) via ionospheric reflection is ~32 MHz. Frequencies, higher than MUF, escape through the ionosphere.

5.4.2 Attenuation in the ionosphere

The refractive index of the ionosphere is a complex quantity. From eq. (3), the attenuation constant k_i is

$$k_i = \frac{\omega}{2\eta_r}\frac{\omega_p^2}{\omega^2}\frac{\nu}{\omega}, \tag{16}$$

where $\eta_r = \sqrt{\left(1 - \omega_p^2/\omega^2\right)}$. As the short wave travels through the ionosphere, its amplitude is diminished by a factor $A = \exp\left(-\int k_i ds\right)$, where integration is over the entire path from the entry into the ionosphere to the exit. Since

$$ds = dr\sqrt{\left[1 + r^2\left(\frac{d\theta}{dr}\right)^2\right]} = \frac{dr}{\left[1 - \frac{(R_E+h_0)^2\left(\sin^2 i_0\right)}{\eta_r^2 r^2}\right]^{\frac{1}{2}}}$$

and ray path is symmetrical about the turning point $h = h_r$, one may write

$$A = \exp\left[-\int_{h_0}^{h_r}\frac{\nu}{c}\frac{\omega_p^2}{\omega^2}\frac{dh}{\left[\eta^2 - \frac{(R_E+h_0)^2\sin^2 i_0}{(R_E+h)^2}\right]^{\frac{1}{2}}}\right]$$

$$= \exp\left(-\int_{h_0}^{h_r}\frac{dh\,(\nu/c)\,\omega_p^2(h)/\omega^2}{\left(\cos^2 i_0 - \omega_p^2(h)/\omega^2\right)^{\frac{1}{2}}}\right) \tag{17}$$

This integral can be evaluated numerically, for the actual density and collision frequency profiles in the ionosphere.

PROBLEM 5.2. Assuming

$$\omega_p^2 = \begin{cases} \omega_{p0}^2 \frac{(h-h_0)}{L_n}; & h > h_0 \\ 0 & h < h_0 \end{cases}$$

and ν to be constant, evaluate the attenuation factor A given in eq.(17).

Solution: The turning point, for the wave of frequency ω occurs at

$$h_r \cong h_0 + \left(\omega^2 \cos^2 i_0 / \omega_{p0}^2\right) L_n$$

and

$$A \cong \exp\left[-\frac{\nu}{\omega}\frac{\omega_{p0}^2}{\omega^2 L_n}\int_{h_0}^{h_r} \frac{(h-h_0)\,dh}{\left(\cos^2 i_0 - \frac{\omega_{p0}^2}{\omega^2}\frac{h-h_0}{L_n}\right)^{\frac{1}{2}}}\right]$$

$$= \exp\left[-\frac{4}{3}\frac{\nu}{c}\frac{L_n\omega^2}{\omega_{p0}^2}\cos^3 i_0\right].$$

5.5 Cellular Communication

Wireless telephone communication is emerging as a dominant mode of personal communication. In North America, cellular communication is assigned a 25 MHz band between 824 and 849 MHz for mobile unit to base transmission and a similar band between 869 and 894 MHz has been allotted for transmission from the base to mobile.

A cellular system comprises a mobile telephone switching office (MTSO), the cell sites and mobile units (Fig.5.10). The MTSO is connected by a trunk group to a nearby telephone exchange, providing an interface to and connectivity with, the public switching telephone network (PSTN).

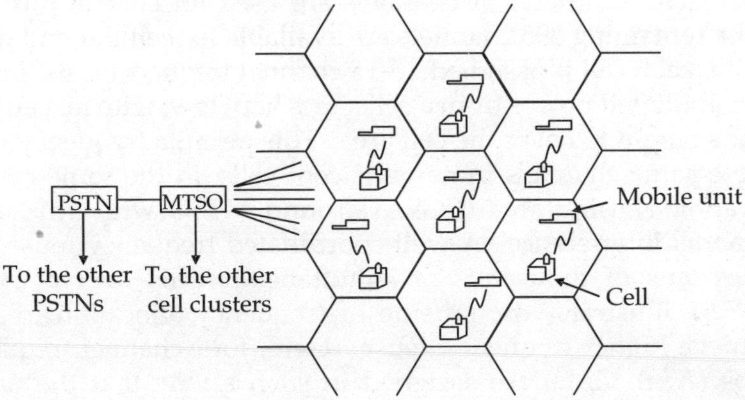

Figure 5.10: Layout of a cellular system. Mobile telephone switching office (MTSO), connecting to various cells in its area, is connected to public switching telephone network (PSTN).

The area around a MTSO is divided into small geographic cells, which are usually hexagonal and are situated 6-12 km apart. Each cell has a radio facility, housed in a building. The radio facility can connect and control any mobile unit within its zone or responsible geographical area. The mobile units may be hand-held or vehicle-mounted terminals. The radio transmitters, located at the cell site, have a maximum effective radiated power (ERP) of 100 W. The MTSO provides switching and control function for the cell sites. It also controls cell site activities and mobile action through command and control data channels.

A mobile unit comprises a control unit, a radio transceiver and an antenna. The control unit has a telephone handset, a push-button keypad to enter commands into the cellular/telephone network and audio and visual indications for customer alerting and cell progress. The transceiver permits full duplex communication. i.e. simultaneous bidirectional transmission between mobile and cell sites. Its ERP is around 6 W. Hand-held terminals combine all functions into one small package, that can be easily held in one hand. The ERP of a hand-held terminal is 0.6 W.

Each cellular band is split into two, to permit competition. Thus, each cellular operator has a bandwidth of 12.5 MHz for each direction of transmission. This is divided into 416 channels of width

30kHz each. Of these, 21 channels are used for control purposes and the remaining 395 channels are available for cellular end users. Usually, each cell is assigned 10-15 channel frequencies for mobile traffic. Radiated power from a cell site is kept low, with just enough antenna height to cover the cell area. This permits frequency reuse of these same channels in non-adjacent cells, in the same cellular geographic sensing area (CGSA) around MTSO with little or no co-channel interference. A well-coordinated frequency reuse plan enables tens of thousands of simultaneous cells over a CGSA. Fig. 5.11 illustrates the scheme of frequency allocation to cells, to achieve higher frequency reuse. Here, four channel frequency groups (A, B, C, D) are assigned in such a way that the use of same frequency in adjacent cells is avoided. Connected cells may further be divided into smaller cells, each with its own base station. The smaller cells would use lower transmitter power and smaller antenna length, thus permitting greater frequency reuse. However, there is a practical limit to cell splitting $\geq$ 1.6 km.

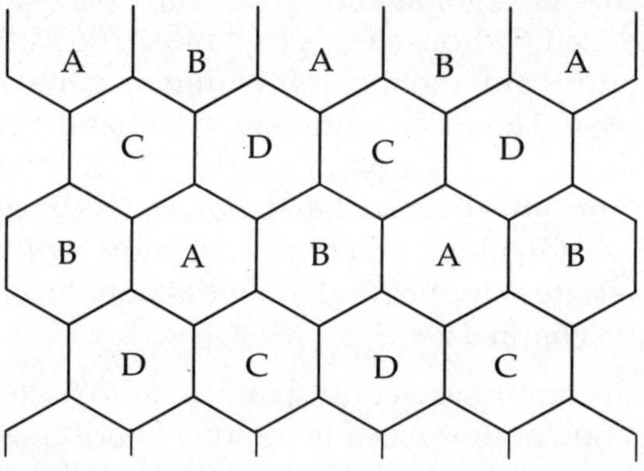

Figure 5.11: Cell separation in a cellular geographic service area with four channel frequency groups A, B, C, D.

5.5.1 Path Loss

Wireless signals suffer heavy losses due to the presence of the Earth, trees, buildings and hills in the transmission path or its neighbouring areas. Theoretical predictions of these losses are complicated.

However, empirical models have been developed which can account for the observed losses with reasonable accuracy. We mention one such model.

5.5.1.1 *The Okumura Model*

Based on the data from Tokyo and neighborhood for mobile terminals, the transmission loss in dB is

$$L_{dB} = 69.55 + 26.16 \log f - 13.82 \log h_t - A(h_r) + (44.9 - 6.55 \log h_t) \log d, \tag{18}$$

where f is the frequency in MHz (in the range 150- 1500 MHz), h_t is the transmitting antenna height above the Earth in metres (between 30 and 300 m), d is the transmitter - receiver distance in km (between 1 and 20 km), h_r is mobile antenna height in metres (between 1 and 10 m) and A (h_r) is the correction factor.

$A(h_r) = 3.2[\log(11.75h_r)]^2 - 4.97$, for a large city when $f > 400$ MHz.

In a typical case of f=900 MHz, h_t=40 m, h_r=5 m, d=10 km, one finds A $(h_r) \approx 146$ dB. Inside the multistoreyed buildings, there exist additional losses due to the floors and walls that come in the way of transmission path.

5.6 Radar

Radar is an abbreviation for **R**adio **D**etection and **R**anging. It is a device that sends an electromagnetic signal towards an object, receives it back after reflection and infers distance and velocity of the object (Fig. 5.12). The time delay τ, after which the reflected

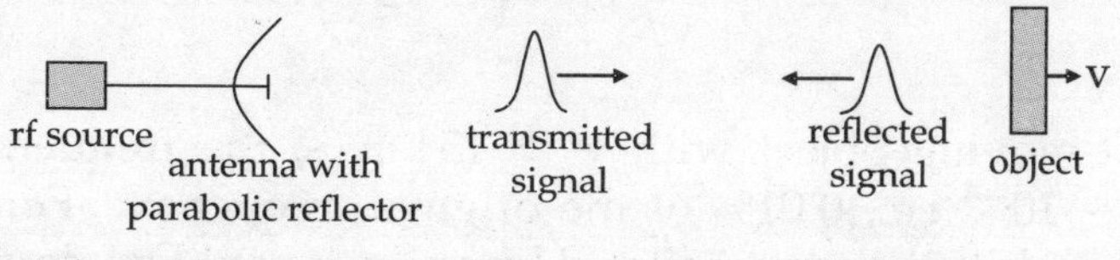

Figure 5.12: Schematic of a radar reflection.

signal is received back by the radar, gives the distance $d = c\tau/2$ at which the object is located at the time of reflection. The Doppler shift in frequency Δf provides the velocity v of the object.

5.6.1 The Doppler Shift

Consider a radar located at $z = 0$ and an object moving along z axis with velocity $v\hat{z}$ (Fig. 5.13). At $t = 0$, the object is at $z = d$. The radar sends a rf signal to detect the object. Let it release two successive crests at $t=0$ and $t = T = 2\pi/\omega$, that hit the object at times t_1 and t_2 and return back to the radar at t_{1r} and t_{2r} respectively. If the velocity of the signal is c, then

$$ct_1 = vt_1 + d, \quad c(t_2 - T) = vt_2 + d,$$
$$t_{1r} = 2t_1, \quad t_{2r} = T + 2(t_2 - T).$$

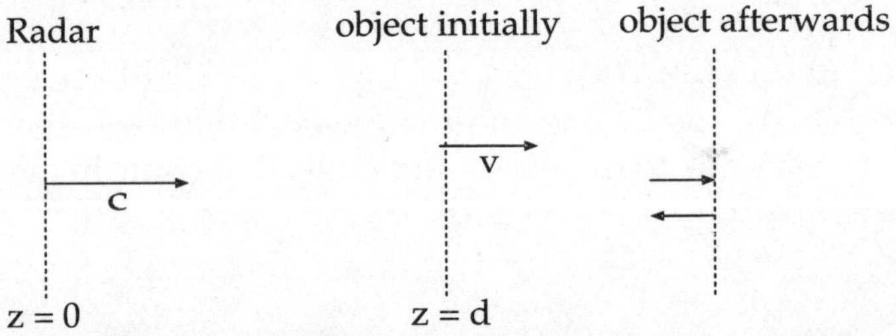

Figure 5.13: Reflection of radar signal from a moving object.

These equations give the time period of the reflected signal $T' = t_{2r} - t_{1r} = (c+v)T/(c-v)$ and its Doppler shifted frequency as

$$\omega' = \frac{1 - v/c}{1 + v/c}\omega \approx \left(1 - 2\frac{v}{c}\right)\omega. \tag{19}$$

For a fast-moving object with $v \approx 15\,\text{km/s}$, the frequency shift $\Delta\omega/\omega \sim -10^{-4}$ i.e., 0.01% of the original frequency. For slower objects, this shift is even smaller. However, it could be detected by mixing the reflected signal with the launched one in a mixer.

5.6.2 Received Power to Transmitted Power Ratio

An estimate of the ratio of the received power to the transmitted power of a radar can be obtained as follows. Let P_t be the transmitted power, A_t be the aperture of the transmitting antenna (or parabolic reflector), r be the distance of the object from the radar and σ be the scattering cross-section of the object (assumed to be an isotropic reflector). The Poynting flux at the object is $P_t\, G/4\pi r^2$ (where G is the antenna gain related to A_t and radiation wavelength λ, as $G = A_t\, 4\,\pi\,/\lambda^2$) and the power intercepted by the object is

$$P_i = \frac{P_t A_t \sigma}{r^2 \lambda^2}.$$

The power received by the radar is

$$P_r = \frac{P_i A_t}{4\pi r^2} = \frac{P_t A_t^2 \sigma}{4\pi r^4 \lambda^2}. \tag{20}$$

The power ratio P_r/P_t scales inversely as the fourth power of distance of the object and square of the wavelength. Hence, shorter wavelengths are preferred for radar communication.

5.6.3 Beyond-the-Line-of-Sight Detection

Early radars employed UHF waves and detected only line-of-sight objects. This is a severe limitation, as the Earth has a curvature and low-lying objects beyond a few tens of kilometers cannot be detected by this technique. Consider a radar mounted at a height h above the ground (Fig. 5.14). The direct line of sight distance d can be calculated as follows:

$$(R + h)^2 = d^2 + r^2$$

$$d \approx \sqrt{2Rh}.$$

Conversely, if a plane is flying at a height h above the ground, then it can be detected by a ground-based radar if the latter is placed within a distance $(2\,Rh)^{1/2}$. This may not leave enough time for responding to the threat of the bomber spacecraft.

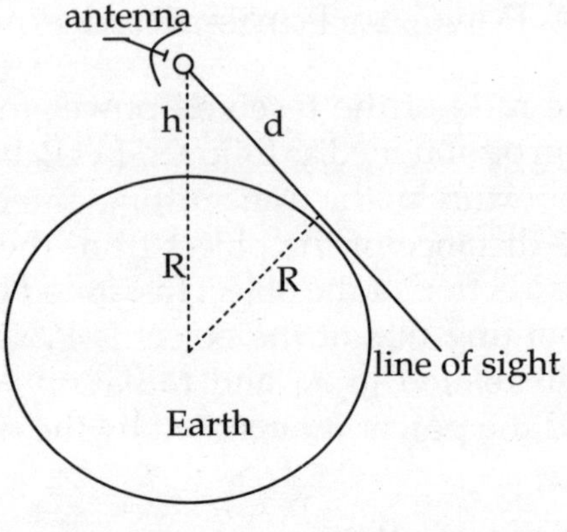

Figure 5.14: Line-of-sight distance for an antenna, located at a height h above the Earth's surface, of radius R.

In this scenario, radar detection via satellite emerged as a tremendous breakthrough. A satellite-borne antenna, with a parabolic reflector, may scan the objects on the ground over a very wide range and the information may be transmitted to the ground-based control tower (Fig. 5.15).

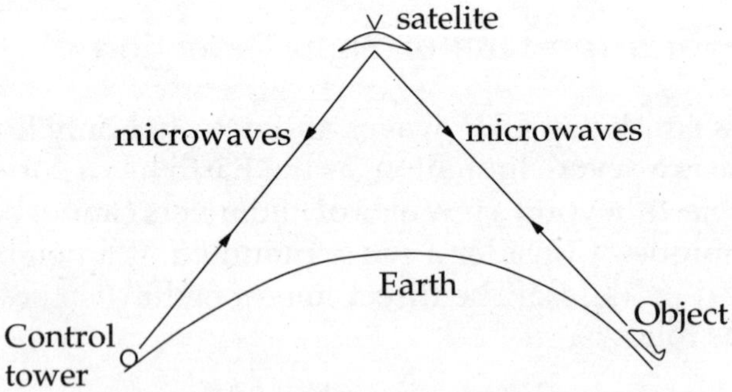

Figure 5.15: Schematic of a beyond-the-line-of-sight radar.

This requires waves, which suffer less diffraction divergence and less scattering in the atmosphere and air. If the radius of the aperture of the parabolic reflector or horn is r_0, then the electromagnetic wave diffraction divergence angle is $\theta_D \approx 1.22\,\lambda/2\,r_0$ and the radius of the beam (after travelling a distance z) becomes

$$r_z \cong r_0 + z\theta_D \approx \frac{z\lambda}{2r_0},$$

where λ is the wavelength of electromagnetic wave. If the intensity of the wave at the aperture is I_0, then after travelling a distance z, it becomes

$$I = I_0\frac{r_0^2}{r_z^2} = 4I_0\frac{r_0^4}{z^2}\lambda^2. \tag{21}$$

Thus, shorter wavelengths would be better suited for radar to reduce the diffraction divergence.

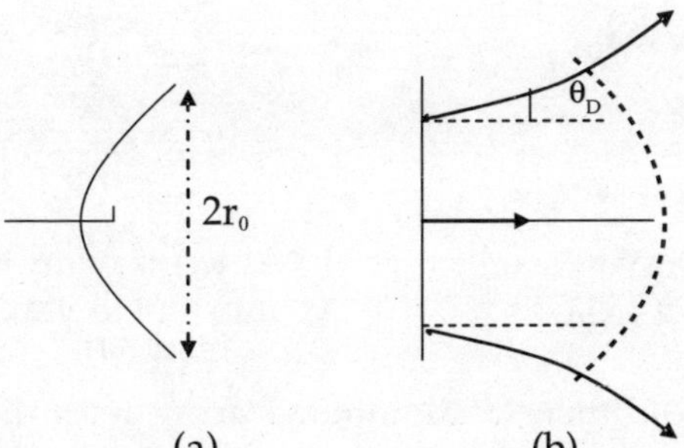

Figure 5.16: (a) A parabolic reflector with dipole antenna at the focus (b) a wave front of finite size, $2r_0$, suffering diffraction divergence at angle $\theta_D \sim 1.22\lambda/2r_0$.

The scattering of microwaves or UHF waves in the troposphere (the lower region of Earth's atmosphere) is larger for shorter wavelengths. The scattered power goes as λ^{-4} (where λ is the wavelength). As a compromise between diffraction and scattering losses, the optimum wavelengths turn out to be in the centimetre and millimetre range. 36 GHz, 96 GHz are specially preferred as the noise at these frequencies is low.

Another important characteristic of a radar is its power to resolve the details of the target surface. For this purpose, shorter wavelengths are preferred. This, in fact, has been a major motivation for the development of *gyrotron* and *free electron laser* that produce millimetre and submillimetre waves.

5.6.4 Phased Array

In order to detect objects all around, one needs to change the orientation of the antenna aperture or horn rapidly. Rotating a horn physically, is a mechanical process and one can hardly do so few times per second. Much faster scanning can be done if the electromagnetic beam is rotated electronically. This is done by using a phased array of antennae. By changing the phases of the currents fed into successive antennae, one can change the orientation of the beam.

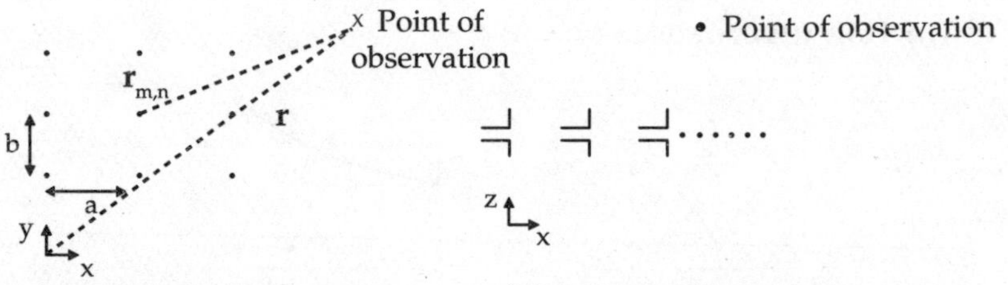

(a) Antenna array as seen normal to z-axis.

(b) Antenna array as seen normal to y-axis.

Figure 5.17: A two-dimensional antenna array with dipole antennae of length $dl\,\hat{z}$, each mounted at (ma, nb, 0), where m = 0,..., M−1, n=0, ..., N−1. $\mathbf{r}_{mn}$ is the distance of point of observation from the (m, n)th element, located at $ma\hat{x} + nb\hat{y}$.

Consider a M × N array of M antenna elements on x-axis and N elements on y-axis. Each antenna has length $dl\hat{z}$. The current in the antenna placed at (ma,nb,0) is $I_0 \exp[-i(\omega t - m\psi_x - n\psi_y)]$. The vector potential at (r, θ, ϕ), due to the antenna at (0,0,0), is

$$\mathbf{A}_{00} = \hat{z}\frac{\mu_0 I_0 dl}{4\pi r} e^{-i(\omega t - \omega r/c)},$$

and the one due to the antenna at (ma, nb, 0) is

$$\mathbf{A}_{m,n} = \hat{z}\frac{\mu_0 I_0 dl}{4\pi r} e^{-i(\omega t - \omega r_{mn}/c - m\psi_x - n\psi_y)}, \tag{22}$$

where $r_{mm} = |\mathbf{r} - ma\,\hat{x} - nb\hat{y}| \approx r - ma\sin\theta\cos\phi - nb\sin\theta\sin\phi$ is the distance of point of observation from the (m,n)th antenna and we have assumed r >> Ma, Nb. The net vector potential at **r**, t is thus

$$\mathbf{A}(\mathbf{r},\ t) = \hat{z}\frac{\mu_0 I_0 dl}{4\pi r} e^{-i(\omega t - \omega r/c)} \sum_{m=0}^{M-1}\sum_{n=0}^{N-1} e^{im\psi_1} e^{in\psi_2}$$
$$= \hat{z}\frac{\mu_0 I_0 dl}{4\pi r}\frac{\sin(M\psi_1/2)}{\sin(\psi_1/2)}\frac{\sin(N\psi_2/2)}{\sin(\psi_2/2)} e^{-i(\omega t - \omega r/c - (M-1)\psi_1/2 - (N-1)\psi_2/2)}, \tag{23}$$

where $\psi_1 = \psi_x - (\omega a/c)\sin\theta\cos\phi$, $\psi_2 = \psi_y - (\omega b/c)\sin\theta\sin\phi$. One may note that the vector potential is strongly enhanced in directions, where $\psi_1 = 2l_1\pi$ and $\psi_2 = 2l_2\pi$ (l_1, l_2 are integers).

The magnetic field at **r**, t is $\mathbf{B} \approx i(\omega/c)\hat{r} \times \mathbf{A}$ and the time average Poynting vector is

$$\mathbf{S}_{\mathbf{av}} = \hat{r}\left(\frac{I_0 dl}{4\pi r}\right)^2 \frac{\sin^2(M\psi_1/2)}{\sin^2(\psi_1/2)}\frac{\sin^2(N\psi_2/2)}{\sin^2(\psi_2/2)}\frac{\omega^2\sin^2\theta}{2c^3\varepsilon_0}. \tag{24}$$

The radiated power density flux is a product of the power flux, radiated by a single antenna, multiplied by a form factor that attains a value of M^2N^2 for $\psi_1 = 0$ and $\psi_2 = 0$ or when ψ_1 and ψ_2 are integral multiples of 2π. This gives a huge enhancement in intensity in the desired direction. For instance, a 10×10 array of 100 antennae would produce an intensity 10^4 times larger than the one produced by a single antenna. The angular half-width of intensity maximum is correspondingly smaller.

PROBLEM 5.3. A linear array of antennae consisting of 100 dipoles, placed on x-axis at regular intervals of 3 cm, has to send a 5 GHz signal in a direction $\theta = \pi/2$, $\phi = \pi/3$. What should be the relative phase difference between currents in successive antennae? Estimate the angular half-width of the maximum intensity lobe in $\theta = \pi/2$ plane.

Solution: In this case, M = 100, N = 0, $\theta = \pi/2$, $\phi = \pi/3$. Hence $\psi_1 = \psi_x - (\omega\, d/c)\sin\theta\cos\phi = \psi_x - \pi/2$. In order to have intensity maximum in this direction, one must choose phase difference between the currents in successive antennae to be $\psi_x = \pi/2$.

The intensity would vanish in a direction $\theta = \pi/2, \phi = \pi/3 + \Delta$, for which $M\psi_1/2 = \pi$ or $\psi_1 = \psi_x - (\omega d/c)\cos(\pi/3 + \Delta) = 2\pi/M, \Delta \cong (4\pi/\sqrt{3}) \times 10^{-2}$ radian.

PROBLEM 5.4. Consider a linear array of 100 identical dipoles, each of length $dl\,\hat{y}$ and placed at (kd, 0, 0) where k = 1, 2, 3...99. The current in the dipole placed at (kd, 0, 0) is $I_0 \exp[-i(\omega t - k\psi_0)]$. Obtain radiation intensity at r, θ, ϕ.

Solution: The vector potential due to the array at a far point is

$$\begin{aligned}\mathbf{A}(\mathbf{r}, t) &= \hat{y}\frac{\mu_0 I_0\, dl}{4\pi r}\sum_{k=0}^{99}\exp(-i(\omega t - \omega r_R/c - k\psi_0)\\ &= \hat{y}\frac{\mu_0 I_0\, dl}{4\pi r}e^{-i(\omega t - \omega r/c)}\sum_k e^{ik\psi_1}\\ &= \hat{y}\frac{\mu_0 I_0\, dl}{4\pi r}\frac{\sin 50\psi_1}{\sin\psi_1/2}e^{-i(\omega t - \omega r/c - 99\psi_1/2)},\end{aligned}$$

where $r_k = |\mathbf{r} - kd\hat{x}| \cong r - kd\sin\theta\cos\phi, \psi_1 = \psi_0 - (\omega d/c)\sin\theta\cos\phi$. The magnetic field at (r, θ, ϕ) is $\mathbf{B} = \nabla \times \mathbf{A} = i(\omega/c)\,\hat{\mathbf{r}} \times \mathbf{A}$, $\hat{r} = \sin\theta\cos\phi\hat{x} + \sin\theta\sin\phi\hat{y} + \cos\theta\hat{z}$. The intensity at (r, θ, ϕ) is

$$I = \mathbf{B}\cdot\mathbf{B}^*/\mu_0^2\varepsilon_0 c = (\cos^2\theta + \sin^2\theta\cos^2\phi)\left(\frac{I_0\, dl}{4\pi r}\right)^2\frac{\omega^2}{2c^3\varepsilon_0}\frac{\sin^2(50\psi_1)}{\sin^2(\psi_1/2)}.$$

5.7 Conclusions

Radio transmission, using medium and short waves, is the most prevalent and economical mode of communication. With frequency modulation, its quality has greatly improved. However, bandwidths of these waves are limited, hence VHF and UHF waves are employed for TV and telephone transmissions. Cellular communication, employing waves around 800-900 MHz, is making rapid advances and quickly taking over landline communication.

The techniques of radio detection and ranging have also undergone a major breakthrough, with the availability of satellites as reflectors and gyrotrons and free electron lasers as the sources of millimetre and submillimetre waves.

UNSOLVED PROBLEMS

1. In the ionosphere, maximum electron density (occurring at a height of 300 km above the Earth) is $10^{12}\,m^{-3}$. Estimate the lowest radio wave frequency that can propagate through it. Ignore collisions and the Earth's magnetic field.

[**ANS.** 5×10^7 rad/s.]

2. An electromagnetic wave of frequency ω and wave vector $\mathbf{k} = (\omega/c)(\cos\theta_i.\hat{z} + \sin\theta_i.\hat{x})$ is incident on a collisionless plasma half-space ($z > 0$) from vacuum, at the angle of incidence θ_i. The plasma frequency of the plasma varies with z as $\omega_p^2 = \omega^2 z^2/L^2$. Using Snell's law ($k_x$ = constant = $\frac{\omega}{c}\sin\theta_i$), obtain the ray trajectory by using the differential equation for the ray $\dfrac{dx}{dz} = \dfrac{v_{gx}}{v_{gz}}$ where $v_g = (c^2/\omega)\vec{k}$ is the group velocity.

[**ANS.** $z = L\cos\theta_i \sin(x/L\sin\theta_i)$.]

3. The plasma frequency in the lower ionosphere varies with height above the ground, h, as

$$\omega_p^2 = \begin{cases} 0 & h < h_0 \\ \omega_{p0}^2 \exp\left[\frac{h-h_0}{L_n}\right] & h > h_0. \end{cases}$$

A short wave of frequency ω is launched at an angle of incidence θ_i into the ionosphere from the ground. Deduce the height of reflection and plot it as a function of ω/ω_{p0} for $\theta_i = 60°$. Neglect the curvature of the Earth and collisions.

[**ANS.** $h_T = h_o + 2L_n \ln(\omega \cos\theta_i/\omega_{p0})$.]

4. A right circularly polarised x-mode of frequency ω is launched vertically into the polar ionosphere, where the ambient magnetic field can be assumed vertical ($\mathbf{B}_s || \hat{z}$) and uniform. The plasma frequency varies as $\omega_p = \omega_{p0} \exp(z/L_n)$, where z is measured from the bottom of the ionosphere. Obtain the height of reflection when $\omega = 2\omega_c = 4\omega_{po}$, where ω_c is the electron cyclotron frequency. Ignore collisions.

[**ANS.** $L_n \ln(8)$.]

5. A radio transmitter launches a 1 MHz surface wave on the surface of the Earth. The Earth has $\sigma = 1$ MKS and $\varepsilon_r = 4$. At some distance, the amplitude of the horizontal component of the wave field on the Earth surface is 10^{-4}V/m. Estimate its value at a depth of 2 m underneath the same location.

[**ANS.** 1.6×10^{-6}V/m.]

6. An array of N identical dipoles is mounted on x- axis with intermediate separation b. Their lengths are parallel to z-axis. If one intends to direct the radiated beam at 30^0 from $\hat{z}$ in the x-z plane, how much phase difference should be given between successive antenna currents of frequency ω to achieve this?

[**ANS.** $\omega b/2c$.]

7. A linear antenna array has N identical dipoles of length $dl\hat{z}$ located on z axis at (0,0,0), (0,0,d) ...The current in the lth dipole is $I_0 \exp[-i(\omega t - l\psi_0)]$. Obtain the modulus of far electric field at (**r**, t) due to the array.

[**ANS.** $(\mu_0 I_0 dl \sin\theta/4\pi r\varepsilon_0 c)(\sin N\psi/2)/(\sin\psi/2)$;
$\psi = \psi_0 - (\omega d/c)\cos\theta$.]

8. A 100 MHz radar signal, when received from a moving object, records a frequency shift of 0.01%. Obtain the speed of the object. If the space between the radar and the object is filled with a plasma of plasma frequency $\omega_p/2\pi = 90\,\text{MHz}$, what change in frequency would you expect?

[**Ans.** $v = 1.5 \times 10^4 \text{m/s}$, frequency shift $= 4.5 \times 10^{-3}\%$.]

9. A 1 GHz electromagnetic wave is launched from a 100 kW antenna of aperture radius 1 metre. Estimate the power intercepted by a similar antenna at a distance of 100 km.

[**Ans.** 1 mW]

10. Power received at a long distance from a transmitter horn at 30 GHz is $1\,\text{mW/cm}^2$. If same power is transmitted at 90 GHz by the same horn, what would be the intensity at the same distance? Neglect scattering losses.

[**Ans.** $9\,\text{mW/cm}^2$.]

References and Suggested Reading

[1] Ginzburg V. L., *Propagation of electromagnetic waves in plasmas*, New York, Gordon and Breach, 1961.

[2] Griffiths J. G, *Radio wave propagation and antennas: an introduction*, Englewood Cliffs, New Jersey, Prentice Hall Intl., 1989.

[3] Peebles P. Z., Jr., *Radar principles*, New York, John Wiley & Sons, Inc., 2004.

[4] Wesolowshi H., *Mobile communication systems*, England, John Wiley & Sons, Inc., 2002.

Unit 6

Satellite Communication

Introduction

The information carrying capacity of an electromagnetic wave increases with the frequency of the wave. Hence, microwaves are widely used for voice, video and data transmission. These waves can be transmitted from one Earth station to another as sky waves, using satellite as an active repeater as they suffer optimal diffraction, scattering and absorption losses in one round trip between the Earth and the satellite. The power loss due to diffraction divergence goes as λ^2 with the wavelength of the carrier electromagnetic wave, whereas scattering loss goes as λ^{-4}. Hence, microwaves (1-30 GHz) and millimetre waves (30-300GHz) are a good trade off for satellite communication over radiowaves and lasers, if one avoids the frequencies corresponding to vibrational excitations of atoms in the troposphere.

The schematic of an end-to-end satellite communication link is shown in fig. 6.1 It comprises terrestrial links, Earth stations and a geosynchronous satellite repeater, that is stationary with respect to the location of the Earth.

6.1 The Terrestrial Link

A terrestrial link processes raw information from the user into a base band signal and transmits it to an Earth station (Fig.6.2). It also performs the reverse role of retrieving information from the signal.

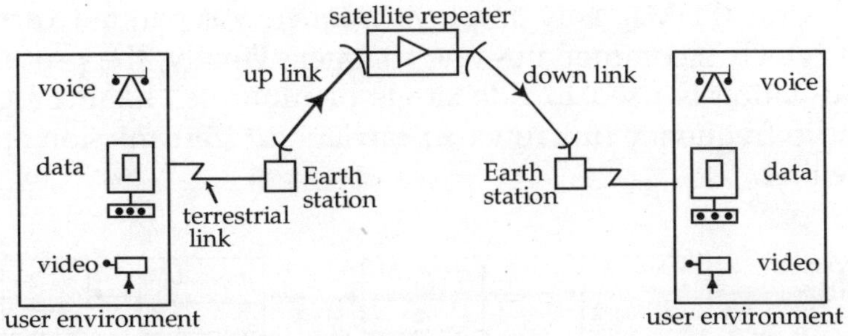

Figure 6.1: Schematic of a satellite communication link.

At the first level of processing, a source signal (voice, video, or data) is coded in digital form or processed in analog form. The information is characterised by bandwidth. Consider, for instance, a telephone speech signal generated by converting acoustic signal into an electrical signal using telephone as the transducer. The telephone speech signal is typically band-limited to the frequency range 300 to 3400 Hz and the bandwidth, assigned to a voice channel, is 4 kHz.

Television signals contain information in electrical form, from which a picture can be recreated. To translate a complete picture into an electrical signal, the picture is scanned in horizontal lines from left to right, starting at the top of the image. When bottom is reached, the process is started again with alternate scanning fields interlaced. The rapid frame rate (25 to 30 per second) creates the illusion of continuous motion, while the field rate (which is twice the frame rate) makes the flickering imperceptible to the human eye. The colour information (chrominance) is conveyed on a sub-carrier, which travels along with the basic colour-independent description (luminance) of the scanned image. The video bandwidth is 4.2 MHz in NTSC-M system and 5.0MHz in PAL-B system. The channel bandwidths, assigned in the two systems, are 6 MHz and 7 MHz respectively.

The next level is multiplexing. In digital transmission, multiple channels are combined into higher level digital signals using time division multiplexing (TDM). This employs separate time slots, each carrying the information from one channel. For analog transmission, individual channels are combined using frequency division multiplexing (FDM). This employs frequency-separated carriers, each of which accommodates one channel. Finally, the process of RF modulation is used to ride single or multiple channel signals onto radio frequency (microwave) carriers for transmission on the satellite link.

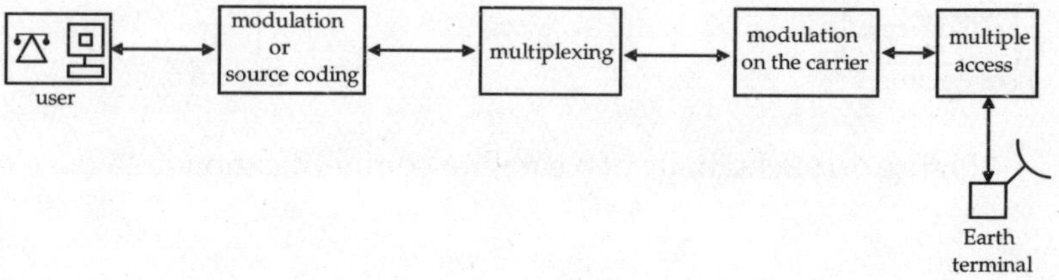

Figure 6.2: Signal processing between user location and satellite Earth station.

6.2 Earth Stations

An Earth station consists of an antenna subsystem, a power amplifier, a low noise receiver and ground communication equipment. It also has power supplies and control, test and monitoring facilities. Antenna diameters range in size from 70 cm (for direct broadcast receive-only applications) to 30 m (for large international gateway stations). The performance of an Earth station is specified by its effective radiated power (ERP)

$$P_{eff} = P_T \, G_T, \tag{1}$$

where P_T is the actual radiated power and G_T is the antenna gain. For a disc antenna of aperture radius r_0, the antenna gain G_T (at wavelength λ) is given by

$$G_T = 4\pi^2 \frac{r_0^2}{\lambda^2}. \tag{2}$$

PROBLEM 6.1. A horn antenna of aperture radius r_0 emits power P_T, at wavelength λ. Assuming the intensity distribution across the aperture to be Gaussian with half-width $r_0' \sim r_0/2$, one may take the diffraction divergence length $R_d = 2\pi r_0'^2/\lambda$ and the beam radius at long distance z to be $r_0'' \simeq r_0' z/R_d$. Obtain the antenna gain.

Solution: At a distance z from the horn, the wave intensity on the axis of the beam would be

$$I = \frac{P_T}{\pi r_0''^2} = \frac{P_T R_d^2}{\pi r_0'^2 z^2}.$$

If the antenna were an isotropic radiator of power P_T, the intensity at z would be $I_0 = P_T/4\pi\, z^2$. Thus, the antenna gain is

$$G = \frac{I}{I_0} = 4\pi^2 \frac{r_0^2}{\lambda^2}.$$

6.3 Satellite Transponders

A satellite may be viewed as a distant microwave repeater, or transponder that receives uplink transmissions and provides filtering, amplification, processing, and frequency translation to the downlink amplifier (Fig.6.3). The uplink and downlink bands are separated in frequency, to prevent oscillations within the satellite amplifiers, while permitting simultaneous transmission and reception at different frequencies.

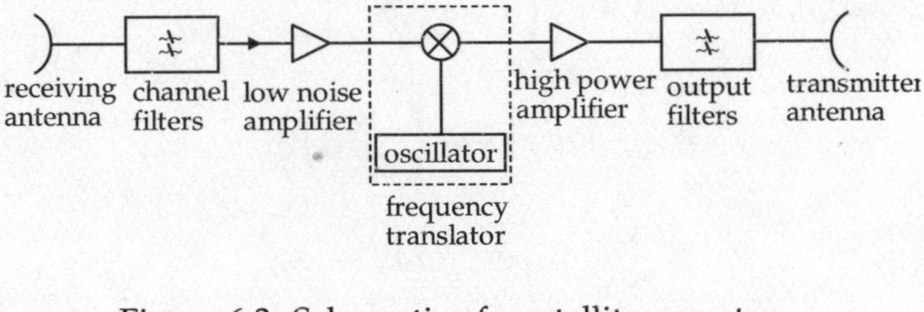

Figure 6.3: Schematic of a satellite repeater.

The signal power, C, intercepted by the receiving antenna, at a distance R from the Earth, is

$$C = \frac{P_T G_T}{4\pi R^2} A_{eff}, \tag{3}$$

where A_{eff} is the effective area of the receive antenna. One may write

$$A_{eff} = \frac{G_R \lambda^2}{4\pi}, \tag{4}$$

where G_R is the antenna gain of the receiving antenna. Since R is large, satellite transponder amplifiers must provide large amplification ($\sim$ 80 to 100 dB), while maintaining relatively low noise operation. Signal to noise ratio is a crucial parameter that determines channel capacity. Noise arises due to several factors, including finite temperature of various subsystems. Following *Nyquist theorem*, the thermal noise power into a matched load is

$$N = k_B T_s B = N_0 B, \tag{5}$$

where k_B is the Boltzmann's constant, T_s is the system temperate and B is the noise bandwidth. Hence, signal to noise ratio (neglecting other factors) is $C/N = C\,/(N_0 B)$,

$$\frac{C}{N_0} = \frac{P_T G_T G_R}{(4\pi R/\lambda)^2 T_s k_B}. \tag{6}$$

Shannon's equation for channel capacity is written as

$$\begin{aligned} H &= B \log_2\left(1 + \frac{C}{N}\right) \\ &= \frac{C}{N_0} \log_2\left(1 + \left(\frac{C}{BN_0}\right)\right)^{BN_0/C} \\ &\simeq \frac{C}{N_0} \log_2 e \quad \text{for } B \to \infty, \end{aligned}$$

wheres H – the 'entropy' of the transmitted information, is measured in bits/s.

PROBLEM 6.2. A horn antenna, of aperture radius 10 metres, emits 10 kW power at 2.4GHz. Estimate the power intercepted by a receiving antenna of aperture radius 4 metre, at a distance of 40,000km.

Solution: The antenna gain of the transmitting antenna is

$$G_T = 4\pi^2 \frac{r_0^2}{\lambda^2} = \omega^2 \frac{r_0^2}{c^2} = 2.5 \times 10^5.$$

Aperture area of the receiving antenna $A_{eff} = 16\pi \ m^2$.
The power intercepted is

$$C = \frac{P_T G_T A_{eff}}{4\pi R^2} \simeq 6.2\,\mu\,W.$$

PROBLEM 6.3. In problem 6.2, if the bandwidth of the signal is 1% and the temperature of subsystems processing the signal is 200^0K, estimate the signal to noise ratio.

Solution: The bandwidth B = 2.4×10^7Hz. Hence the noise is

$$N = k_B T_s B = 6.7 \times 10^{-14} W$$

and the signal to noise ratio is

$$\frac{C}{N} = 0.9 \times 10^8.$$

6.4 Modulation

In analog transmission, the source signal is modulated on a sinusoidal carrier in two ways:

a. amplitude modulation

b. frequency modulation.

6.4.1 Amplitude Modulation

In amplitude modulation, source signal $a_s(t)$ and the carrier $E = a_c \cos \omega_c t$ are fed into a nonlinear device that multiplies the two. The amplitude of the modulated signal follows the source signal (Fig.6.4)

$$E' = a_M \cos \omega_c t, a_M = a_c\, a_s(t). \tag{7}$$

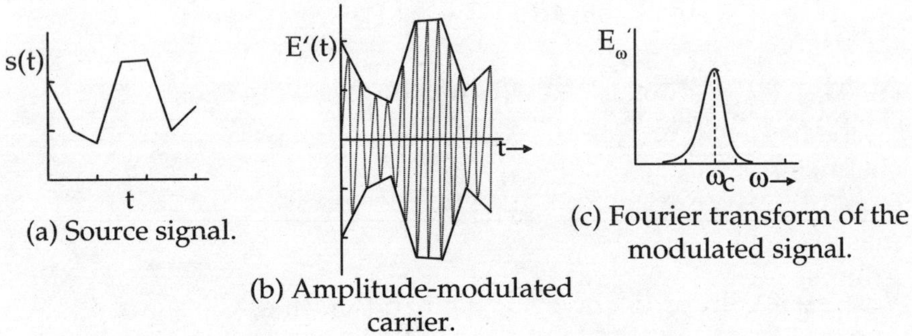

Figure 6.4: Amplitude modulation.

If one Fourier analyses the modulated carrier, it is found to contain lower and upper sidebands around the carrier frequency. The required transmission bandwidth is just the bandwidth of the modulating signal a_s (t).

6.4.1.1 Frequency–Division Multiplexing

Multiple analog channels are combined into higher level composite signals, using frequency division multiplexing (FDM). The FDM signals (assemblies) are generated by the following process. Initially each individual voice band signal is modulated onto a sinusoidal carrier, using an AM modulator. Then, adjacent carrier frequencies are separated by a specific amount (4 kHz for speech transmission). These amplitude modulated signals are then combined, using a summing device, to produce FDM baseband assembly (Fig. 6.5).

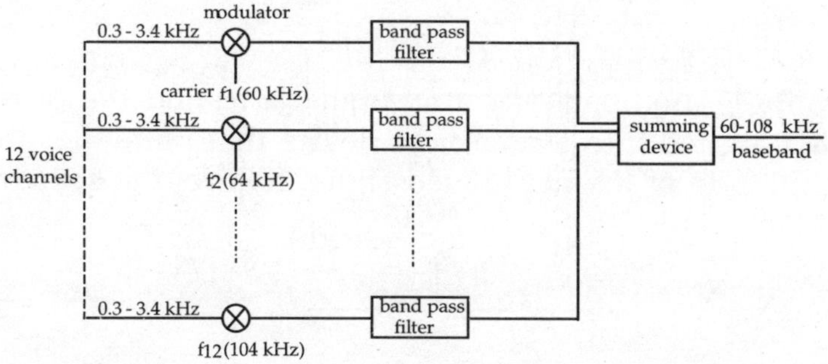

Figure 6.5: Generation of a 12 channel FDM baseband assembly.

The first level in FDM hierarchy is called a *group*, consisting of 12 channels. Several groups can be combined to form a super group (containing say 60 channels). Higher levels in the hierarchy are master groups (600 channels), super master groups and jumbo groups, comprising as many as 6000 channels in a single composite signal. A similar scheme is adopted for video and data transmission.

6.4.2 Frequency Modulation

Frequency modulation (FM) is created by varying the frequency of a sinusoidal carrier E= $a_c \cos \omega_c$ t with the message signal $a_s(t)$,

$$E' = a_c \cos(\phi), \frac{d(\phi)}{dt} = \omega_c(1 + \alpha a_s(t)). \quad (8)$$

The voltage to frequency conversion process results in bandwidth expansion, in the radio frequency (RF) channel (Fig. 6.6). The practical bandwidth for FM signal depends on the FM modulation index $m = \Delta f / f_m$, where Δ f is the peak deviation in frequency and f_m is the highest frequency in the base-band modulating signal.

The multiple FM signals can be combined into higher level complex signals, using frequency-division multiplexing.

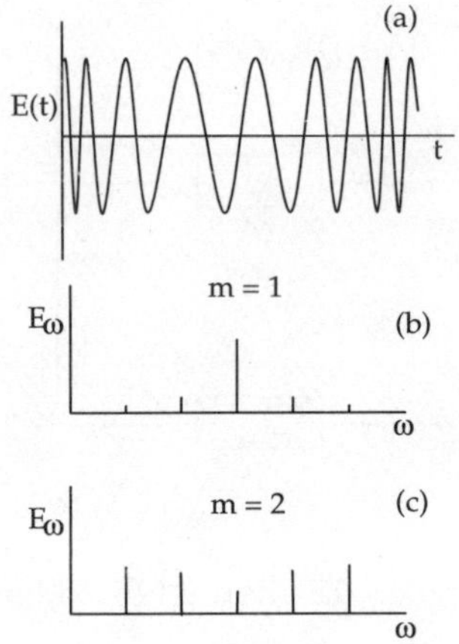

Figure 6.6: A frequency modulated signal (a) and its Fourier spectrum for modulation index m=1 (b) and m=2 (c).

PROBLEM 6.4. A frequency modulated signal is given as

$$E' = A_0 \cos[\omega_0(t + (\alpha/\Omega\sqrt{2})(\sin \Omega t + \cos \Omega t))].$$

Obtain the amplitudes of ω_0, $\omega_0 + \Omega$ and $\omega_0 + 2\Omega$ frequency components.

Solution: Using the Bessel function identity

$$e^{i\alpha \sin\theta} = \sum_{n=-\infty}^{\infty} J_n(\alpha) e^{in\theta},$$

one may write E′ as the real part of

$$\begin{aligned}\psi &= A_0 e^{i(\omega_0 t + \frac{\omega_0\alpha}{\Omega}\sin(\Omega t + \pi/4))} \\ &= A_0 e^{i\omega_0 t} \sum_n J_n\left(\frac{\omega_0\alpha}{\Omega}\right) e^{in(\Omega t + \pi/4)}.\end{aligned}$$

Hence $E' = \text{Re}[\psi] = A_0 \sum_n J_n\left(\frac{\omega_0\alpha}{\Omega}\right) \cos[(\omega_0 + n\Omega)t + n\pi/4]$. The amplitudes of ω_0, $\omega_0 + \Omega$ and $\omega_0 + 2\Omega$ components are respectively, $A_0\, J_0(\omega_0\alpha/\Omega)$, $A_0\, J_1\, (\omega_0\alpha/\Omega)$, $A_0\, J_2\, (\omega_0\alpha/\Omega)$.

6.5 Digital Transmission

Digital source coding for both voice and image signals is rapidly gaining prominence, due to high signal to noise ratio in transmission. *Pulse-code modulation* (PCM) is a conventional waveform coding technique, which directly converts the analog speech waveform into a sequence of multidigit (bit) binary numbers. A voice band signal, occupying a 4 kHz bandwidth, is first applied to a band-pass filler with a bandwidth 300 to 3400 Hz and then applied to a sampler to determine the instantaneous voice signal at regular time intervals. According to Nyquist theorem, as long as the signal is sampled at a rate greater than twice the highest frequency in the band-limited analog signal, $2f_m$, the original signal may be recovered without distortion. Thus, a voice band is sampled at a rate of 8kHz. The output of the sampler is a pulse amplitude modulation (PAM) waveform, with pulse amplitude sample occurring every 125μ s. The amplitude of a pulse equals the instantaneous level of the analog waveform, at the sampling instant.

The PAM signal is then converted into pulse-code modulated signal by applying it to a quantising encoder (Fig.6.7). The range of amplitude occupied by the signal from minimum to maximum, called *dynamic range*, is divided into many quantised levels. In systems used for telephone, approximately $255 = (2^8 - 1)$ levels are used. Each PAM sample amplitude is compared to these levels and each pulse amplitude is assigned the level closest to the sample amplitude, sample by sample. Each of these levels is identified by a number. The 255 level, when expressed in binary system, requires 7 bits. An additional digit is required for +/- sign. Thus, each level is represented by a 8-bit binary code (for details, see unit 7). A voice channel would thus require a data rate of $8\times 2f_m = 64$ kb/s. The code words for successive samples are transmitted in bit-serial form.

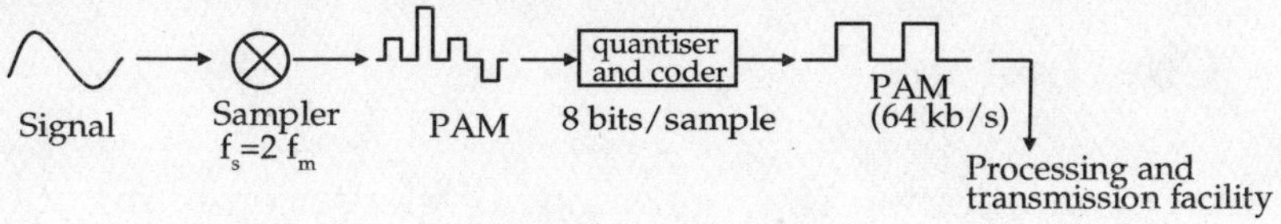

Figure 6.7: Pulse amplitude modulation coder.

6.5.1 Time Division Multiplexing

Multiple digitally-encoded signals are combined into a higher-level composite signal, using time-division multiplexing (TDM). The time interval between successive bits of a voice channel, for instance, is filled with the sequence of other voice channels (Fig.6.8). In DS-(digital signal) hierarchy, DS1 frame (accommodating 24 voice channels) uses 24 8-bit time slots, each reserved for a single PCM encoded voice as digital data signal. Therefore, DS1 frame contains 192 bits (24 × 8) of information, plus a 193rd bit used for framing. The DS1 level has a bit rate of 8000 × 193 = 1.544 Mb/s. Similarly DS4 level, accommodating 4032 voice channels, would have a bit rate of 274.176 Mb/s.

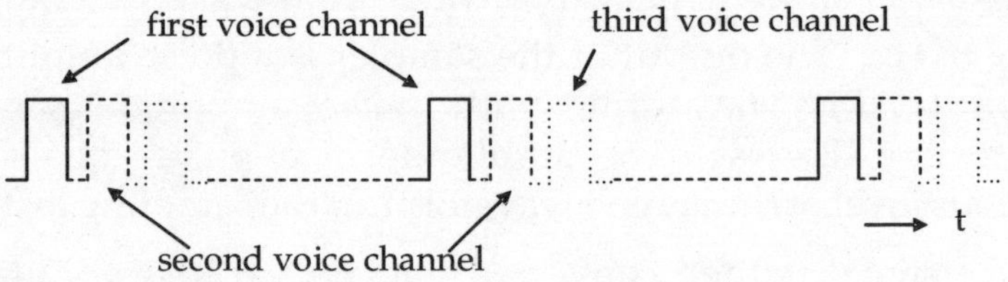

Figure 6.8: Sequencing of different voice channels ——, - - - - - -, ...

At the receiving end, voice channels are separated out. Each successive 8-bit 'word' is converted into a pulse, whose amplitude is specified by the value of the 'word'. The output of the decoder is a PAM pulse sequence, which closely follows the shape of the waveform of the original analog signal. The output of the resampler is passed on to a low-pass reconstruction filter, which produces a replica of the original signal.

6.6 Conclusions

Satellite communication is a vital mode of global communication. With pulse code modulation (digital transmission) one can achieve nearly noise-free communication. The carrier bandwidth is large, hence the transmission rate (b/s) is fast.

UNSOLVED PROBLEMS

1. Estimate the dispersion broadening of an electro magnetic pulse, of carrier frequency 1 GHz and bandwidth 100 MHz, propagating through a distance of 400 km, in the ionosphere of average plasma frequency $\omega_p/2\pi = 5$ MHz and a distance of 40,000 km, in a region having an average $\omega_p/2\pi = 10^4$ Hz. Ignore the effects of Earth's magnetic field and collisions.

   [ANS. 0.03 ns.]

2. A 100 GHz signal, emitted by a 10 kW antenna of aperture radius 30 cm, is received at a distance of 40,000 km. Estimate the electric field amplitude at the receiver.

   [ANS.1.2×10^{-2} V/m.]

3. An antenna, of aperture radius 10m, emits 5 kW power at 10 GHz with a bandwidth of 5%. Estimate the signal to noise ratio, at a distance of 40,000 km when the receiving antenna radius is 4 m and the temperature of the signal processing subsystem is 200°K.

   [ANS.$C/N = 3.5 \times 10^7$.]

REFERENCES AND SUGGESTED READING

[1] Pratt T. and Bostian C. W., *Satellite communications*, New York, John Wiley & Sons, 2001.

[2] Gagliardi R. M., *Satellite communications*, Belmont, CA, Lifetime Learning Publication, 1984.

[3] Feher K., *Digital communication: Satellite / Earth station engineering*, Englewood Cliffs, NJ, Prentice Hall, Inc., 1983.

[4] Pritchard W. L. and Sciulli J. A., *Satellite communication system engineering*, Englewood Cliffs, NJ, Prentice Hall, Inc., 1986.

Unit 7

Laser and Optical Fibre Communication

Introduction

With the advent of semiconductor laser and low-loss ($\leq$ 0.2 db/km) optical fibre, optical communication has emerged as the fastest mode of information transmission at short, long and transoceanic distances. It employs infrared/visible frequencies ($\geq$100 THz) which, within a bandwidth of 1%, can (in principle) transmit data at the rate of 10^{12} bits (pulses) per second or 1Tb/s. A major constraint on bit transmission rate, B, comes from the dispersion broadening and attenuation of the pulsed signals in optical fibres. With the employment of erbium-doped fibre amplifier repeaters, placed at every 60 to 100 km, one can overcome this limitation considerably. Consequently, a 27,300 km fibre optic link around the globe (called FLAG) at a bit rate of 5 Gb/s became commercially operational in 1997. Since then, transoceanic links operating at 100Gb/s have been set up. In this chapter, we will study the basic principles of lasers, with special emphasis on semiconductor laser and wave propogation through step-index and graded-index fibres.

7.1 Laser

Atomic transitions from higher energy states to lower energy states are a common source of radiation. An analytical theory of this process is a subject of quantum electrodynamics and is beyond the scope of this book. However, we shall discuss a very special class of such light sources, called lasers, using *quantum theory of radiation*. Laser is the acronym of **L**ight **A**mplification via **S**timulated **E**mission of **R**adiation.

Atom in any material tends to stay in a state, where total energy of all its orbiting electrons is minimum in accordance with *Pauli's exclusion principle*. When an electron makes a transition from its normal orbit (ground state) to a higher energy orbit (excited state), it stays there only for a short while and then returns back to the ground state, emitting a quantum of radiation called *photon*. If the energy of the excited state is E_2 and the ground sate is E_1 (Fig.7.1), then the frequency of emitted radiation is

$$\omega = \frac{E_2 - E_1}{\hbar}, \tag{1}$$

where $\hbar$ is Planck's constant divided by 2π.

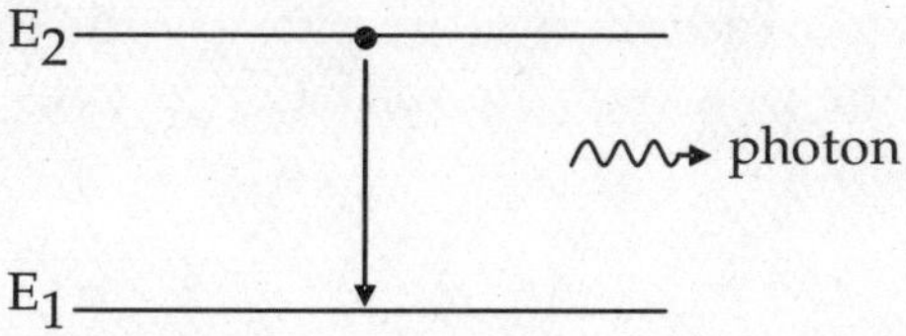

Figure 7.1: Photon generation through the de-excitation of atom from energy level E_2 to E_1.

This natural decay process, inherent to all the excited states of all materials, is referred to as *spontaneous emission*. The radiation produced by different atoms, via spontaneous emission, has random phases with respect to each other and is known as incoherent radiation. Each excited state is characterised by a 'life-time' defined as follows. If the number of atoms per unit volume in an excited state at an instant of time is N, then N decays via spontaneous process (when no other process to populate or depopulate the excited state takes place) at a rate

$$\frac{dN}{dt} = -\frac{N}{\tau},$$

giving

$$N = N_0 e^{-t/\tau}, \tag{2}$$

where N_0 is the density of atoms in the excited state at $t = 0$. This τ is the *e-folding decay time* of density of excited atoms. Typically $\tau \sim 10^{-8}$s. However, certain atoms in specific environments have some excited states with longer life-time. For examples chromium atoms in a ruby rod have a life-time $\tau \sim 1$ ms. Such states are called *metastable states*.

The upward transition from state E_1 to E_2 cannot occur by itself. It requires energy. It can be induced by photons of right frequency ω. More the photon density (number of photons per unit volume), higher the upward transition rate. For this reason, upward transition is a stimulated process.

Einstein argued that the downward transition from E_2 to E_1 could also be stimulated by the presence of photons of frequency $\omega = (E_2 - E_1)/\hbar$, just like the upward process. He termed this process as stimulated emission of radiation. A novel feature of this process is that the generated radiation is in phase with the pre-existing radiation. Thus, the phase of the radiation is preserved and one gets coherent radiation.

7.2 Detailed Balance and Einstein's Coefficients

Consider a system of atoms in thermal equilibrium with radiation. Let the number density of atoms in the ground state E_1 be N_1 and that in the excited state E_2 be N_2. The energy density of photons of frequency $\omega = (E_2 - E_1)/\hbar$ is u_ω. The rate of upward transitions, per unit volume per second, from E_1 to E_2 (stimulated absorption) is

$$R_{abs} = B_{12} u_\omega N_1,$$

where B_{12} is some coefficient.

The rate of downward transitions due to spontaneous process is

$$R_{sp} = A_{21} N_2,$$

where $A_{21} = 1/\tau$ is the inverse of life-time. The rate of downward transitions, due to the stimulated emission process, is

$$R_{st} = B_{21} u_\omega N_2$$

where B_{21} is another constant. In equilibrium, upward transition rate balances the sum of downward transition rates,

$$R_{abs} = R_{sp} + R_{st},$$

$$\Rightarrow B_{12} u_\omega N_1 = A_{21}\, N_2 + B_{21} u_\omega N_2. \tag{3}$$

From statistical mechanics, the ratio N_2/N_1 in thermal equilibrium follows Boltzmann's law

$$\frac{N_2}{N_1} = \frac{g_2}{g_1} e^{-\hbar\omega/k_B T}, \tag{4}$$

where g_1 *and* g_2 are the statistical weights of the two states, k_B is the Boltzmann's constant and T is the absolute temperature. Using eq.(4) in (3), we obtain

$$u_\omega = \frac{\frac{g_2 A_{21}}{g_1 B_{12}}}{e^{\hbar\omega/k_B T} - \frac{B_{21}}{B_{12}}\frac{g_2}{g_1}}. \tag{5}$$

However, from Planck's theory of blackbody radiation

$$u_\omega = \frac{\hbar\omega^3/c^3\pi^2}{e^{\hbar\omega/k_B T - 1}}. \tag{6}$$

Comparing the two expressions, we obtain

$$\frac{A_{21}}{B_{21}} = \frac{\hbar\omega^3}{c^3\pi^2}, \quad B_{21} g_2 = B_{12} g_1. \tag{7}$$

If A_{21}, B_{21} and B_{12} are redefined to absorb the statistical weights, then $B_{21} = B_{12}$. The coefficients A_{21} and B_{12} are called **Einstein's A and B coefficients** respectively. One may note that the stimulated emission will dominate over the spontaneous one, when

$$u_\omega > \frac{A_{21}}{B_{21}}. \tag{8}$$

Further, stimulated emission would dominate over absorption, if $N_2 > N_1$. Such a situation is called *population inversion* and can be created by forcibly pumping electrons from the ground state to some higher energy state E_3, from which they would decay to E_2. If population inversion is achieved, one could obtain coherent radiation via stimulated emission. A device employing this as the working principle is called laser.

A laser comprises a material medium with a metastable energy state, a scheme for pumping and a resonator cavity to trap the radiation to achieve high photon density and hence strong stimulated emission. Before we take up any specific laser system, let us obtain an expression for gain in terms of population inversion density, Einstein's coefficients and cavity parameters.

7.3 Laser Amplification

Consider a system of atoms (gain medium), with ground state E_1 and a metastable state E_2. Let the densities of atoms in these states (under the application of a pumping mechanism) be N_1 and N_2 respectively and the energy density of photons having frequency between ω and $\omega + d\omega$ be $u_\omega d\omega$. The energy widths of states E_1 and E_2 are not zero but finite, due to several factors like Doppler broadening, collisional broadening etc. Let $g_\omega\, d\omega$ be the fraction of atoms participating in the emission/absorption process of photons with frequency between ω and $\omega + d\omega$. The photon energy produced, per unit volume per second (ignoring spontaneous process), is

$$W = \int B_{21} g_\omega u_\omega (N_2 - N_1) \hbar\omega \, d\omega. \tag{9}$$

If the laser amplifier has only a monochromatic radiation of frequency ω_0,

$$u_\omega = \bar{u}_{\omega_0} \delta(\omega - \omega_0),$$

W turns out to be

$$W = B_{21} g_{\omega_0} \bar{u}_{\omega_0} (N_2 - N_1) \hbar\omega_0. \tag{10}$$

$\bar{u}_{\omega_0}$ is related to the intensity of radiation I (the amount of light energy crossing unit area per second) as $I = \bar{u}_{\omega_0} c/\eta$, where η is the refractive index of the medium. Hence

$$W = \alpha I, \tag{11}$$

where $\alpha = B_{21} g_{\omega_0} (N_2 - N_1) \hbar\omega_0 \eta / c$.

Consider the propagation of radiation along z axis (Fig.7.2). Let the intensity at z be I and that at z + dz be I + dI. If we choose a box of unit area of cross-section and length dz, then the light-energy produced in this volume is Wdz. By energy conservation dI = Wdz, i.e.,

$$\frac{dI}{dz} = \alpha I. \tag{12}$$

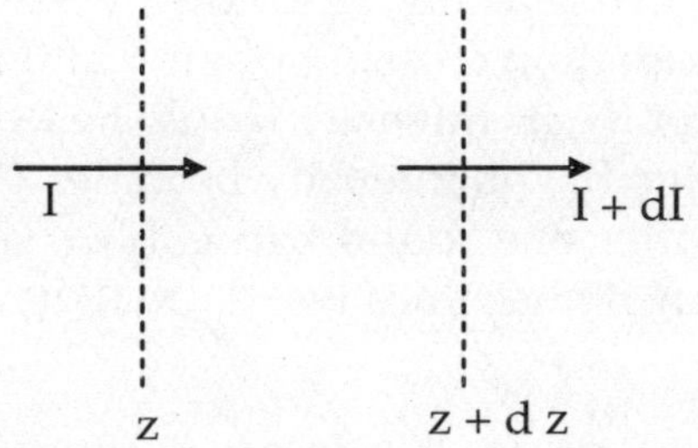

Figure 7.2: Radiation fluxes through unit areas at z and z + dz.

If $N_2 - N_1$ is assumed independent of z, then α is a constant and eq. (12) gives

$$I = I_0 e^{\alpha z}, \tag{13}$$

where I_0 is the intensity at z = 0 – the input end of the amplifier. For $N_2 > N_1$, α is positive and intensity grows exponentially with z. However, at large z when intensity has grown to a high enough level, the downward transitions become brisk and $N_2 - N_1$ falls down rapidly. $N_2 - N_1$ becomes a function of I and I grows at a rate, much less than the exponential rate.

7.4 Laser Oscillator

In order that photon energy density develops with time, one must confine the generated radiation so that the stimulated process quickly takes over the spontaneous process and the photon energy density can acquire a large value. For this reason, the lasing medium is placed inside an open resonator comprising two mirrors of high reflectivities R_1 and R_2 separated by a length L. The mirrors have a finite curvature to offset the diffraction divergence of the beam, when it travels back and forth between the mirrors. The resonator is characterised by a decay time t_c. If it is filled with radiation travelling back and forth along z and there is no generation or absorption of photons inside it, then the radiation energy density u_ω falls off with time due to leakage through the mirrors,

$$u_\omega(t) = u_\omega(o)e^{-t/t_c}. \tag{14}$$

This can be deduced heuristically, ignoring the curvature of mirrors, as follows. Let the radiation crossing point P at time t = 0 be I_0. On reflection from mirror R_2, its intensity would be $R_2 I_0$ (Fig7.3). When it returns from mirror R_1, its intensity becomes $I_0 R_1 R_2$. Thus, the radiation intensity after one round trip is $I_1 = I_0 R_1 R_2$. After two round trips, the intensity becomes $I_2 = I_0(R_1 R_2)^2$, and so on.

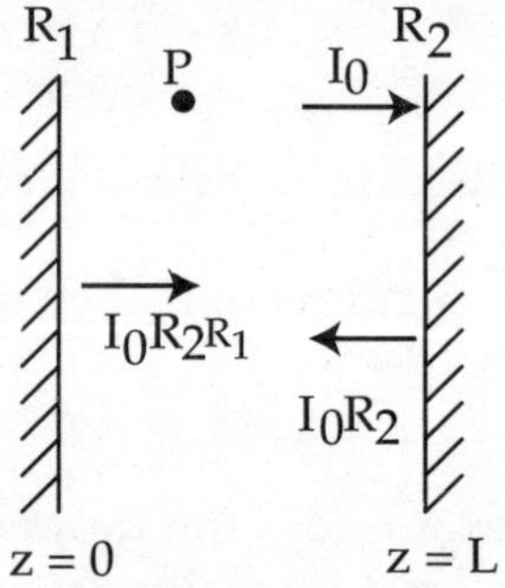

Figure 7.3: Radiation intensity after successive reflections from mirrors of reflectivities R_1 and R_2.

We model the intensity variation (related to that of u_ω by merely a constant) as

$$I = I_0 e^{-t/t_c}. \tag{15}$$

Then, $I_1 = I_0 e^{\frac{-2L\eta}{ct_c}}$ as the time for a complete round trip is $2L\eta/c$. Equating this to $I_0 R_1 R_2$, we obtain

$$t_c = -\frac{2L}{c}\eta\frac{1}{\ln(R_1\,R_2)}. \tag{16}$$

The product of t_c, with frequency ω, is called the **quality factor** Q of the cavity, $Q = \omega\, t_c$.

Now, we include the active role of the lasing medium, characterised by amplification factor α defined above. In this case, the intensity of radiation after successive round trips through the resonator, would be

$$I_1 = I_0 R_1 R_2\, e^{2\alpha L},$$
$$I_2 = I_0 (R_1 R_2)^2 e^{4\alpha L}.$$

We model I as $I = I_0 e^{\gamma t}$. Then

$$I_1 = I_0 e^{\gamma 2L\eta/c}.$$

Comparing the two expressions of I_1, we obtain

$$\gamma = \alpha \frac{c}{\eta} - \frac{1}{t_c}. \tag{17}$$

For $\alpha > \eta/ct_c, \gamma > 0$ i.e. the intensity grows exponentially with time. γ is called the *growth rate*. The threshold population inversion for the growth of intensity or photon energy density (corresponding to $\alpha = \eta/ct_c$) is

$$(N_2 - N_1)_{th} = \frac{1}{(t_c\, g_{\omega_0}\, B_{21}\, \hbar\omega_0)}. \tag{18}$$

One may express α as $\alpha = [(N_2 - N_1)/(N_2 - N_1)_{th}]\eta/t_c c$. Hence,

$$I = I_0 \exp\left[\left(\frac{N_2 - N_1}{(N_2 - N_1)_{th}} - 1\right)\frac{t}{t_c}\right]. \tag{19}$$

This expression, however, is applicable only in the initial stages of growth of photon energy density u_ω. As u_ω acquires large values, $N_2 - N_1$ decreases with u_ω and the growth stabilises. We refer the reader to other texts for a discussion on saturation and information on specific lasers.

PROBLEM 7.1. A parallel plane open resonator of length L has metallic mirrors of complex effective relative permittivity ϵ_m. Obtain the decay time t_c for a resonator mode, using electromagnetic theory.

Solution: Consider a resonator (Fig.7.3) with $z < 0$ and $z > L$ as conducting media of effective relative permittnity ϵ_m and $0 < z < L$ as free space. Taking the time variation of electric and magnetic fields as $\exp(-i\omega t)$ and $\partial/\partial_x = 0$, $\partial/\partial_y = 0$ in the wave equation we obtain,

z < 0

$$\mathbf{E} = \hat{x}A_1\bar{e}^{(\omega t+\omega\eta_m z/c)},$$
$$\mathbf{H} = -\hat{y}A_1(\eta_m/\mu_0 c)e^{-(\omega t+\omega\eta_m z/c)},$$

0 < z < L

$$\mathbf{E} = \hat{x}[A_2 e^{-i(\omega t-\omega z/c)} + A_2' e^{-i(\omega t+\omega z/c)}],$$
$$\mathbf{H} = \hat{y}(1/\mu_0 c)[A_2 e^{-i(\omega t-\omega z/c)} - A_2' e^{-i(\omega t+\omega z/c)}],$$

z > L

$$\mathbf{E} = \hat{x}A_3\, e^{-i(\omega t-\omega\eta_m z/c)},$$
$$\mathbf{H} = \hat{y}A_3(\eta/\mu_0 c)\, e^{-i(\omega t-\omega\eta_m z/c)},$$

where $\eta_m = \sqrt{\epsilon_m}$. The continuity of E_x and H_y at z = 0 and z = L gives

$$A_2 + A_2' = A_1, \quad A_2 - A_2' = -\eta_m A_1$$
$$A_2 e^{i\omega L/c} + A_2' e^{-i\omega L/c} = A_3 e^{i\omega\eta_m L/c}$$
$$A_2 e^{i\omega L/c} - A_2' e^{-i\omega L/c} = \eta_m A_3 e^{i\omega\eta_m L/c}.$$

These equations yield the eigenvalue equation

$$\left(\frac{1-\eta_m}{1+\eta_m}\right)^2 = e^{-2i\omega L/c}.$$

Expressing $\omega = \omega_r - i\Gamma$, where $2\Gamma = 1/t_c$, and taking the modulus on both sides we obtain

$$\left|\frac{1-\eta_m}{1+\eta_m}\right|^2 = e^{-L/ct_c}.$$

One may recognise that the left-hand side is the intensity reflection coefficient of free space-metal interface for normal incidence. Denoting it as R′, we obtain from this equation

$$t_c = -L/(c \ln R').$$

This expression is the same as eq.(16), if one takes $R_1 = R_2 = R'$ and the refractive index of the intermediate region as 1.

The eigenvalue equation gives the eigen-frequencies ω_r also. Expressing $\eta_m = \eta_r + i\eta_i$, $\frac{(1-\eta_m)}{(1+\eta_m)} = (R')^{\frac{1}{2}} \exp(-i\psi)$, $\psi = \tan^{-1}\left(\frac{\eta_i}{(1-\eta_r)}\right) + \tan^{-1}\left(\frac{\eta_i}{(1+\eta_r)}\right)$, the eigenvalue equation gives

$$\omega_r L/c = l\pi + \psi,$$

where ψ is the phase change on reflection and l is an integer. Usually $\omega_r L / c >> \psi$ and one may write the eigen frequency of the lth mode of the resonator as

$$\omega_r = l\pi c/L.$$

PROBLEM 7.2. A laser oscillator with quality factor $Q = 10^4$ at $\omega/2\pi = 6 \times 10^{14}\, s^{-1}$ is maintained at population inversion density $N_2 - N_1 = 4(N_2 - N_1)_{th}$. Estimate the growth rate.

Solution: Radiation decay time for the cavity is

$$t_c = Q/\omega.$$

The growth rate is

$$\gamma = \frac{1}{t_c}\left(\frac{N_2 - N_1}{(N_2 - N_1)_{th}} - 1\right) = 3/t_c = 36\pi \times 10^{10}\, s^{-1}.$$

PROBLEM 7.3. A laser amplifier, at a given population inversion density, has a gain of 10dB when the length of the amplifier is 1m. What will be the gain, if its length is halved? Now the amplifier (1 m long) is converted into an oscillator by placing mirrors of reflectivities $R_1 = R_2 = 0.95$ at the two ends. Obtain the growth rate. Take $\eta = 1$.

Solution: Gain in dB $\simeq (10 / 2.3)\,\alpha L$. Hence $\alpha = 2.3\,\text{m}^{-1}$.

On reducing the length by half, the gain will reduce to 5dB.

When the amplifier is converted into an oscillator, the growth rate (for $\eta = 1$) would be

$$\gamma = \alpha c + \left(\frac{c}{2L}\right) \ln R_1 R_2 \simeq 6.75 \times 10^8\,\text{s}^{-1}.$$

7.5 Semiconductor Laser

So far we have discussed light amplification in media with discrete atomic energy levels. There is another class of materials, called semiconductors, where one encounters energy bands. Two relevant energy bands are designated as *valence* and *conduction* bands separated by a forbidden gap of energy ε_g, ranging from a fraction of an electron volt to a few electron volts (Fig.7.4).

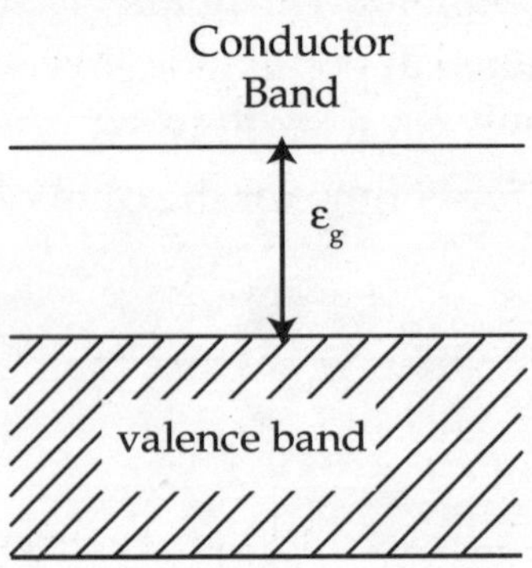

Figure 7.4: Energy bands in a semiconductor.

In an intrinsic semiconductor i.e., the one free form impurities, the concentration of free electrons in the conduction band and holes in the valence band is very low. The situation improves dramatically when one dopes a semiconductor with impurities (Fig. 7.5). A semi conductor doped with donor impurities (n-type semiconductor) has high density of free electrons, whereas the one doped with acceptor impurities (p-type semiconductor) has high density of holes.

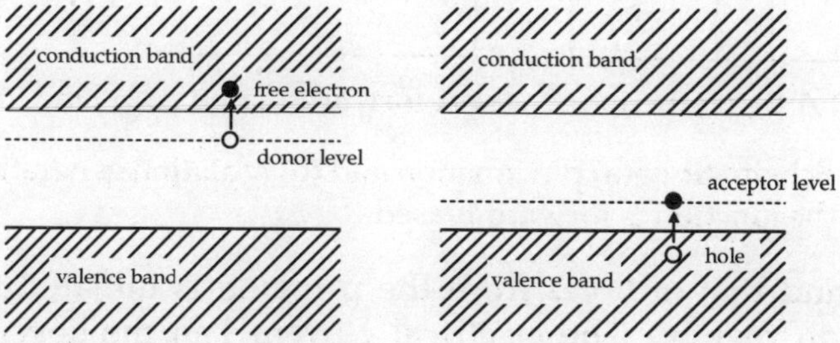

(a) n-type semiconductor. (b) p type semiconductor.

Figure 7.5: Free electron and hole generation in (a) n-type and (b) p-type semiconductors respectively.

A junction of p-type and n-type semiconductors, called a p-n junction (Fig.7.6), has potential to produce laser with great ease of operation and at very low cost. One only needs large concentration of free carriers on both sides of the junction. The width of the transition region from p type to n type region is of the order of fraction of a micron. The free electrons in the n-type region diffuse into the p-type region and the holes in the p-type region diffuse into the n-type region. However, they cannot travel too deep as they are pulled back by the space charge field thus created.

As a result of diffusion, the n-type region has a deficiency of electrons and possesses extra holes from the other side. Hence, it acquires a positive potential. Similarly, the p-type region acquires a negative potential. This potential impedes the further flow of free carriers into the junction region, also called the *depletion region*. However, if we give a forward bias to the p-n junction, i.e. apply a positive potential to p-region and a negative potential to n-region (using a battery of a few volts) the potential barrier is lowered (Fig. 7.7) and a continuous flow of electrons from the

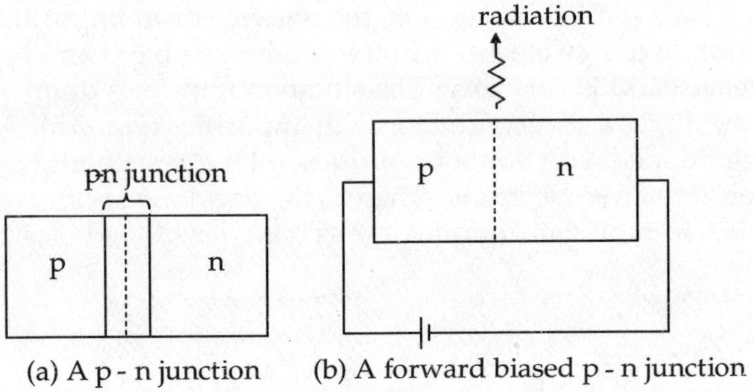

Figure 7.6: Schematic of (a) p-n junction and (b) radiation generation when the junction is forward biased.

n-region and that of holes from the p-region is initiated through the junction region. The electrical current I in the specimen, at temperature T, varies with the applied voltage V as

$$I = I_s(e^{eV/k_B T} - 1),$$

where I_s is called the *saturation current* and it depends on the diffusion coefficients of the free carriers. A stimulated re-combination of electrons and holes in the junction region produces *coherent radiation*.

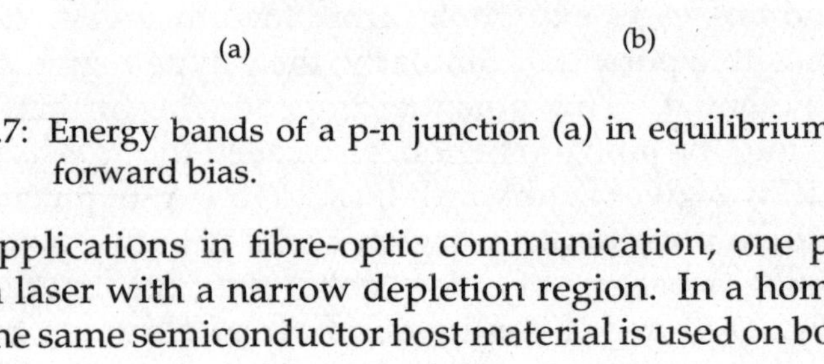

Figure 7.7: Energy bands of a p-n junction (a) in equilibrium, (b) under forward bias.

For applications in fibre-optic communication, one prefers p-n junction laser with a narrow depletion region. In a homojunction, where the same semiconductor host material is used on both sides of the junction, depletion layer width is relatively large. Consequently, the free carrier concentration is low and gain is small. The situation improves considerably in a heterojunction, where a thin layer ($\sim 0.1\mu$m) of a lower band gap material is sandwiched between p- and n-type regions (Fig. 7.8). The discontinuity in band gap helps free-carrier confinement in the narrow region. One can achieve high carrier concentrations and hence high laser gain by using this scheme.

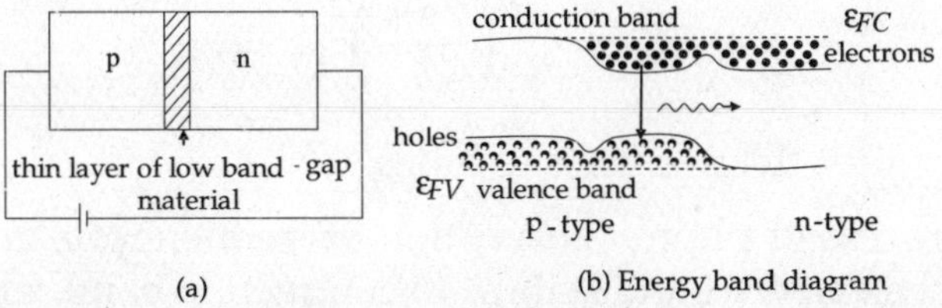

Figure 7.8: Heterojunction under forward bias.

7.5.1 Optical Gain

Consider a forward biased p-n junction (or a heterojunction). In the depletion region, (or the sandwiched region), the free electron concentration is n_{0e} and hole concentration is n_{0h}. Both the free carrier species obey *Fermi-Dirac distribution*. The probability that a state ε_e in the conduction band is occupied is

$$f_c = \frac{1}{e^{(\varepsilon_e - \varepsilon_{Fe})/k_B T} + 1},$$

where ε_{Fe} is the free electron Fermi energy (related to n_{0e} and temperature T) and ε_e is measured from the bottom of the conduction band. The number of energy states per unit volume, between energies ε_e and $\varepsilon_e + d\varepsilon_e$, is $\rho(\varepsilon_e)d\varepsilon_e$, where

$$\rho_e(\varepsilon_e) = 4\pi \left(\frac{2m_e}{\hbar}\right)^{\frac{3}{2}} \sqrt{\varepsilon_e} \tag{20}$$

is the density of states and m_e is the electron effective mass. The probability that a state in the valence band is occupied is

$$f_v = \frac{1}{e^{(\varepsilon - \varepsilon_{Fv})/k_B T} + 1} \tag{21}$$

where ε_{Fv} is related to hole concentration n_{0h} and temperature T. The hole probability distribution function $f_h = 1 - f_v$ can be written as

$$f_h = \frac{1}{e^{(\varepsilon_h - \varepsilon_{Fh})/k_B T} + 1} \tag{22}$$

where ε_h is the hole energy measured downward from the top of the valence band and ε_{Fh} is the hole Fermi energy. The density of states of holes is

$$\rho_h(\mathcal{E}_h) = 4\pi \left(\frac{2m_h}{\hbar}\right)^{\frac{3}{2}} \sqrt{\mathcal{E}_h}, \tag{23}$$

where m_h is the effective mass of holes.

In the process of photon absorption of frequency ω, an electron goes from the valence band state ε_h to conduction band sate state ε_e, where

$$\omega = \frac{\varepsilon_e + \varepsilon_h + \varepsilon_g}{\hbar} \tag{24}$$

and ε_g is the band gap energy. The wave number of the photon is very small as compared to that of the electron or the hole before the recombination. Hence, in a direct band gap semiconductor, we may take

$$\varepsilon_e = \hbar^2 k^2/2m_e \quad \varepsilon_h = \hbar^2 k^2/2m_h. \tag{25}$$

With these expressions, eq.(24) uniquely determines the wave number k of recombining electron and hole and hence ε_e and ε_h in terms of ω. Thus, out of so many free electrons and holes, only those in the vicinity of $\varepsilon_e = (\hbar\omega - \varepsilon_g)/(1 + m_e/m_h)$ and $\varepsilon_h = (\hbar\omega - \varepsilon_g)/(1 + m_h/m_e)$ participate in the stimulated emission of radiation of frequency ω. If the frequency width of radiation is $d\omega$, then the ranges of electron/hole energy levels participating in the emission/absorption process are $d\varepsilon_e = [\hbar/(1 + m_e/m_h)]d\omega$ and $d\varepsilon_h = [\hbar/(1 + m_h/m_e)]d\omega$ respectively.

Let u_ω $d\omega$ be the photon energy density in the frequency interval ω and $\omega +$ $d\omega$. The rate of downward transitions (electron hole recombination) per unit volume per second is proportional to spectral energy density u_ω, density of sates $\rho_e(\varepsilon_e)$, the probability of occupancy $f_c(\varepsilon_e)$ of state ε_e, the probability that the state ε_h is vacant $(1 - f_v(\varepsilon_v))$ and the density of relevant states in the valance band $\rho_h(\varepsilon_h))$.

$$R_{st} = Bu_\omega \rho_e(\varepsilon_e)\rho_h(\varepsilon_h)f_c(\varepsilon_e)(1 - f_v(\varepsilon_v))d\omega, \qquad (26)$$

where $\varepsilon_v = -\varepsilon_g - \varepsilon_h$ is the valence band electron energy measured from the bottom of the conduction band and B is the Einstein's coefficient. Similarly, the rate of upward transitions of electrons from valence band to conduction band, via photon absorption, is

$$R_{abs} = B\, u_\omega \rho_e(\varepsilon_e)\rho_h(\varepsilon_h)f_v(\varepsilon_v)(1 - f_c(\varepsilon_e))d\omega. \qquad (27)$$

The spontaneous emission process may be ignored, if u_ω is substantially high. In order to have net radiation generation, R_{st} must exceed R_{abs}, i.e.,

$$f_c(\varepsilon_e) > f_v(\varepsilon_v) \qquad (28)$$

or

$$\frac{1}{e^{(\varepsilon_e - \varepsilon_{Fc})/k_BT} + 1} > \frac{1}{e^{(\varepsilon_v - \varepsilon_{Fv})/k_BT} + 1},$$

where all energies are measured from the bottom of the conduction band. This condition is satisfied when

$$\varepsilon_e - \varepsilon_{Fc} < \varepsilon_v - \varepsilon_{Fv}$$

$$\text{or } \varepsilon_{Fc} - \varepsilon_{Fv} > \varepsilon_e - \varepsilon_v = \varepsilon_g + \varepsilon_e + \varepsilon_h = \hbar\omega. \qquad (29)$$

The difference between Fermi levels of conduction band and valence band must exceed the band gap and photon energy. This can be achieved at electron hole densities of the order of (or greater than) $10^{23}\,m^{-3}$.

For nearly monochromatic radiation, $u_\omega = \bar{u}_{\omega 0}\delta(\omega - \omega_0)$ and the net radiation energy produced per unit volume per second is

$$W = \int \hbar\omega\, B\bar{u}_{\omega 0}\delta(\omega - \omega_0)\rho_e(\varepsilon_e)\rho_h(\varepsilon_h)(f_c(\varepsilon_e) - f_v(\varepsilon_v))\, d\omega$$
$$= \alpha I, \qquad (30)$$

where $\alpha = \hbar\omega_0\, B\, \eta\, \rho_e(\varepsilon_e)\rho_h(\varepsilon_h)(f_c(\varepsilon_e) - f_v(\varepsilon_v))/c$, $I = \bar{u}_{\omega 0}c/\eta$ is the intensity of radiation and η is the refractive index of the material. Following the treatment of section 7.3, the laser intensity in an amplifier will increase with the distance of propagation as

$$I = I_0 e^{\alpha z}. \qquad (31)$$

However, lengths of p-n junctions are small and single-pass amplification is low. Hence, semiconductor lasers operate as oscillators. One may avoid external mirrors, as the two cleaved (polished) facets of the semiconductor act as mirrors of reflectivity (see Fig. 7.9).

$$R = \left|\frac{\eta - 1}{\eta + 1}\right|^2 . \tag{32}$$

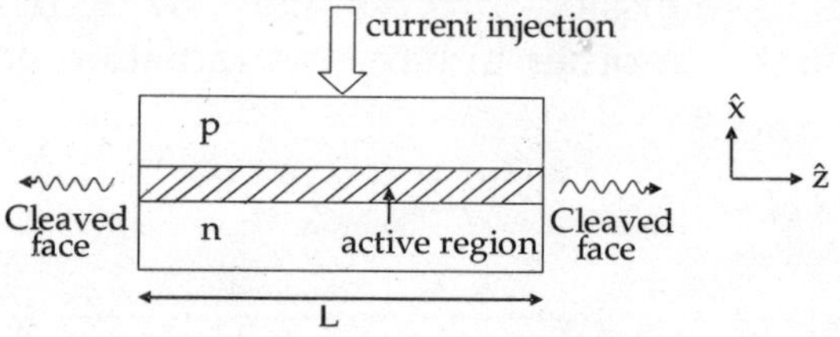

Figure 7.9: A semiconductor laser with cleaved facets.

Typically $\eta \sim 3.5$, hence $R \sim 30\%$. The reflection losses are heavy, yet they can be tolerated when amplification factor α is large. The free carrier density in the active junction region is smaller than that on the two sides, hence it behaves as planar optical waveguide of submicron width, ($\Delta x \sim 0.1\mu m$), supporting only a single mode. The y-width of the laser can be restricted by limiting current injection over a narrow width $\sim 5\mu m$, by depositing a dielectric (SiO_2) layer on top of the p-layer with a central opening for current injection. Such lasers are known as *stripe-geometry semiconductor lasers*. The threshold current is typically 50-100 mA and the emitted radiation is in the form of an elliptic spot of size $\sim 1 \times 5\,\mu m^2$. The spot size, however, is not stable as laser power is increased. For this reason, *index-guided semiconductor lasers* are used, in which active region of dimension $0.1 \times 1\mu m^2$ is buried on four sides by layers of lower refractive index giving an index step $\Delta\eta \sim 0.1$.

7.6 Optical Communication System

A schematic of a fibre-optic communication link is shown in fig. 7.10. The audio, video or data signal is converted into an electrical signal, using transducers. It could be a continuously time-varying waveform, called an analog signal or a discrete signal, e.g., letters on a page, each having its individual entity. The waveform is scanned by a scanner at sample frequency f_s, that is more than twice the highest frequency contained in the waveform. Next, each sampled level is allocated one of the finite numbers of amplitude levels. Typically, one may employ a set of $2^8 - 1 = 255$ levels. Each level can be expressed as a 8-digit binary number, requiring 8-bit allocation. Thus, a sampling frequency f_s would imply $8f_s$ bits/s. In between the bits (pulses), the information bits of other voice, video or data channels are interwoven.

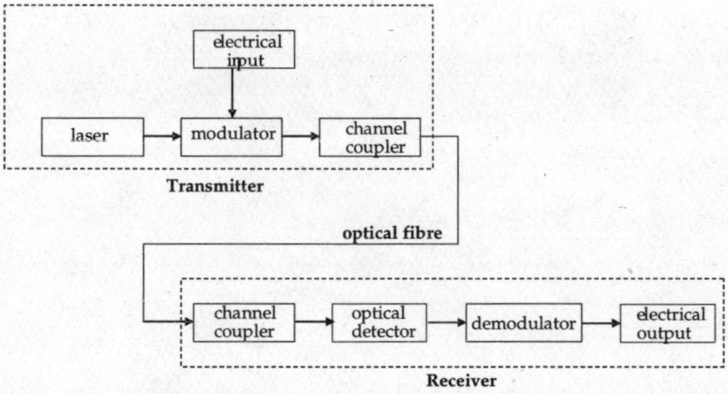

Figure 7.10: Schematic of a fibre-optic communication link.

The pulsed electrical output goes to a modulator–a nonlinear device; where it meets a laser. The output is a succession of short laser pulses, that go into an optical fibre through a channel coupler. The optical path between the transmitter and receiver contains a series of amplifier repeaters at regular intervals. At the receiver, the received signal goes through the channel coupler to optical detector. It is then passed through a demodulator/decoder to produce the electrical output, from which the original signal can be recovered.

We have already discussed the physics of a semiconductor laser. In subsequent sections, we discuss the propagation of the optical signal through an optical fibre.

PROBLEM 7.4. A printed book of 200,000 words contains an average of 5 letters per word. Each letter is represented by seven binary digits, giving ample allowance for upper and lower-case letters, numerals, spaces and punctuations. How many bits is the information content of the book?

Solution:

$$\begin{aligned} \text{Total binary digits per word} &= 35 \\ \text{Total number of words} &= 200,000 \\ \text{Total information content} &= 7\,\text{Mbits.} \end{aligned}$$

PROBLEM 7.5. The bandwidth required for an intelligible speech channel is 3 kHz. Estimate the information rate with sampling frequency 6 kHz and 5 bits per sample.

Solution:

Sample time $T_s = 10^{-3}/6$ second

Bits needed for each sample = 5 bits

Total number of bits per second $= 5/T_s = 30\,\text{kb/s}$

NOTE: In practice digital pulse-code-modulated (PCM) telephone voice channel operates at 64 kb/s. The analog waveform is sampled at $T_s = 125\mu s(f_s = 8kHz)$ and each sample is encoded into 8 bits.

PROBLEM 7.6. According to Shannon's formula, the minimum number of binary digits per second, B, required to transmit a signal of bandwidth Δf(Hz) at a signal to noise ratio A_S / A_N is

$$B = \Delta f \log_2(1 + (A_S/A_N)^2).$$

In a PAL colour television, video signal occupies a bandwidth 5.5 MHz and signal to noise ratio has to be atleast 50 dB. Estimate how many Gbits is required to transmit a 10 minute film.

Solution:

$$\begin{aligned} 10\log_{10}(A_S/A_N)^2 &= 50 \\ (A_s/A_N)^2 &= 10^5 \end{aligned}$$

$$\begin{aligned} B &= \Delta f \log_2(1 + (A_S/A_N)^2) \\ &\approx 5.5 \times 10^6 \times \log_2(10^5) \text{bits/s} \\ &\approx 5.5 \times 10^6 \times 3.32 \times 5 = 91\,\text{Mb/s} \end{aligned}$$

The information transmitted in 10 minutes $= 91 \times 600\,\text{Mb}$
$= 55\,\text{Gbits}.$

PROBLEM 7.7. Convert the analog waveform, given in fig. 7.11(a), into 6-digit binary digital signal by sampling it at period T_s (frequency $f_s = 1/T_s$).

Solution: Sampling the waveform at intervals T_S, we find the signal amplitudes to be 15, 45, 35, 20 and 30. In 6-digit binary, these can be written as

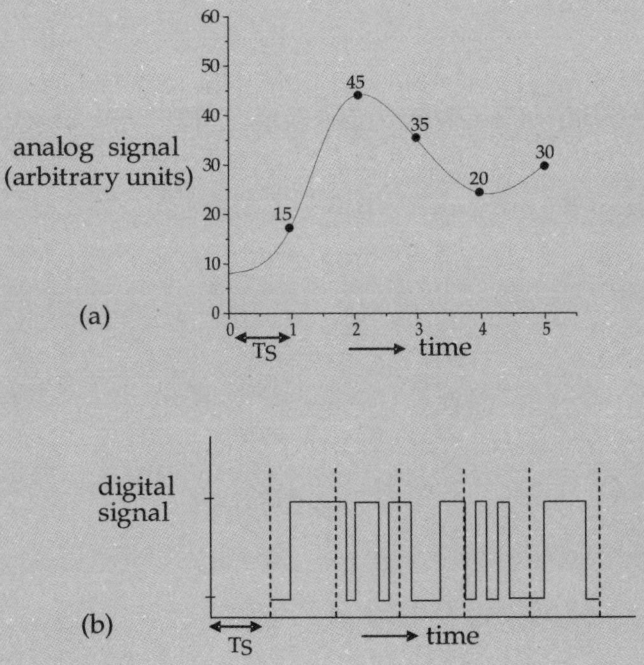

Figure 7.11: (a) An analog signal and (b) corresponding digital signal.

$$
\begin{aligned}
15 &= 0 \times 2^5 + 0 \times 2^4 + 1 \times 2^3 + 1 \times 2^2 + 1 \times 2^1 + 1 \times 2^0 \\
&= 001111 \\
45 &= 1 \times 2^5 + 0 \times 2^4 + 1 \times 2^3 + 1 \times 2^2 + 0 \times 2^1 + 1 \times 2^0 \\
&= 101101 \\
35 &= 100011 \\
20 &= 010100 \\
30 &= 011110
\end{aligned}
$$

If we presume the bit size to be $T_S/6$, the digital signal would be as shown in fig. 7.11(b). For smaller bit size, there would be room to accommodate other signals in between.

7.7 Optical Fibre

An optical fibre is a flexible and thin cylinder of silica with radius 'a' of the order of a few microns (Fig. 5.12). The inner region of a fibre (of radius a), known as core, has refractive index $\eta_0 \approx 1.5$.

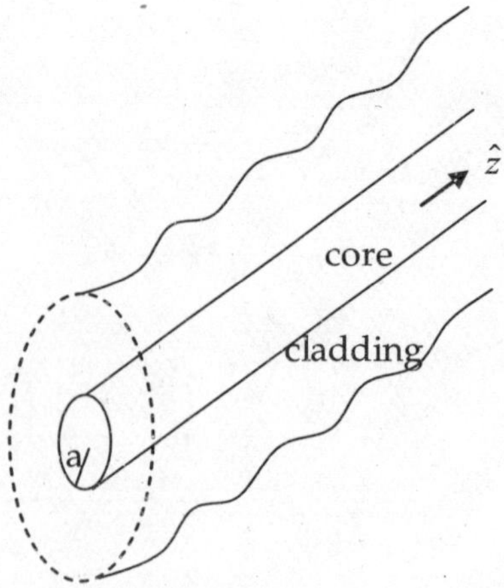

Figure 7.12: Schematic of an optical fibre.

It is surrounded by a cladding of lower refractive index $\eta_1 = \eta_0 - \Delta\eta$. Any ray of light, originating from the fibre axis (r = 0) at an angle α (Fig. 7.13), hits the core cladding interface at an angle of incidence $\theta = \pi/2 - \alpha$. When θ exceeds the critical angle $\theta_c = \sin^{-1}(\eta_1/\eta_0)$, the ray suffers total internal reflection returns back to the axis and remains confined in the fibre. α is related to the angle of incidence i of the ray from free space onto the face of the fibre as

$$\sin \mathrm{i} = \eta_0 \sin \alpha.$$

The maximum value of i, for which the optical rays are trapped in the core, is called the *numerical aperture*, $\mathrm{i_m}$, of the fibre,

$$\begin{aligned} \sin \mathrm{i_m} &= \eta_0 \sin \alpha_\mathrm{m}, \\ \alpha_\mathrm{m} &= \frac{\pi}{2} - \sin^{-1}(\eta_1/\eta_0) \\ \mathrm{i_m} &= \sin^{-1}\sqrt{\eta_0^2 - \eta_1^2} \approx \sqrt{2\eta_0\Delta\eta}. \end{aligned} \tag{33}$$

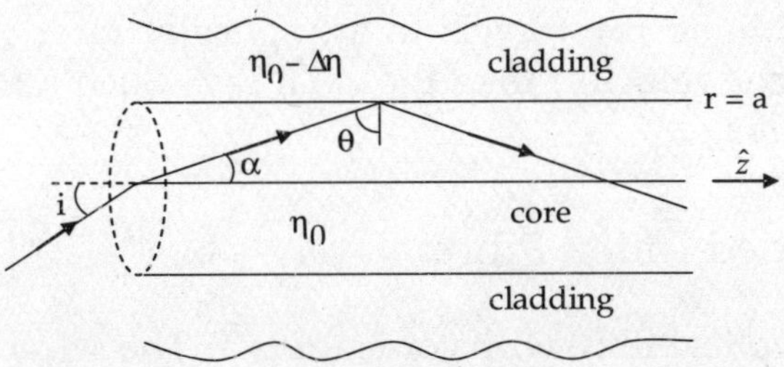

Figure 7.13: Ray guiding in a cladded fibre.

Typically $\Delta\eta \simeq 0.02$, $\eta_0 \simeq 1.5$, hence $\alpha_\mathrm{m} \simeq 15^0$.

In a fibre-optic link, signal is transmitted in short pulses (bits) with small separation in between the pulses, of the order of fraction of a nanosecond. It is necessary that the optical pulses do not suffer too much temporal broadening to cause overlap of two pulses. However, a fibre has an inherent broadening as rays launched into the fibre from a laser source have a range of angles of incidence between 0 and $\mathrm{i_m}$. The ray with i = 0 propagates along the axis of the fibre, whereas the one with $\mathrm{i} = \mathrm{i_m}$ takes an oblique zig-zag path. The former takes a time $\eta_0 L/c$ to travel a length L of the fibre, whereas the latter takes a time $\eta_0 L/c \cos\alpha_m$, leading to a temporal broadening of the pulse by

$$\Delta\tau = \frac{\eta_0 L}{c}\left(\frac{1}{\cos\alpha_m} - 1\right) \\ \cong \Delta\eta L/c.$$

This must be smaller than the separation between the pulses ($= B^{-1}$). Hence

$$BL \le \frac{c}{\Delta\eta}. \tag{34}$$

This restricts the bit rate for a given distance of propagation and vice versa. For $\Delta\eta \simeq 0.02$, $BL \le 15$ MHzkm. This is too restrictive for step index fibre to be of any utility in long distance communication. The main reason is that the velocity of ray propagation, over the straight line axial path and the zig-zag path inside the homogenous core is the same while the path lengths are different–hence the times of travel are different. The problem can be sorted out a great deal, if one has a refractive index (η) variation with radial coordinate r, so that the oblique ray travels in an optically rarer medium (i.e. possesses larger velocity) than the axial ray. Such a fibre is called *graded index fibre.*

7.7.1 Graded Index Fibre

Consider an optical fibre with core radius a. The refractive index is η_0 at r =0 (fibre axis) and decreases monotonically with r to $\eta_0 - \Delta\eta$ at r = a. The core is surrounded by a cladding of index $\eta_0 - \Delta\eta$. A suitable profile that yields minimal broadening of optical pulses is a parabolic profile,

$$\eta^2 = \eta_0^2\left(1 - \frac{2\Delta\eta}{\eta_0 a^2}r^2\right) \qquad \text{for } 0 < r < a \\ = \eta_0^2(1 - 2\Delta\eta/\eta_0) \qquad \text{for } r > a. \tag{35}$$

Consider a ray originating on fibre axis ($r = 0$), at an angle α to z axis. Its local propagation vector has z and r components:

$$k_{0z} = \frac{\omega}{c}\eta_0 \cos\alpha, \quad k_{0r} = \frac{\omega}{c}\eta_0 \sin\alpha$$

where ω is the frequency of the wave. Since the refractive index has no variation with z, according to Snell's law, k_z must remain unchanged as the ray moves deeper into r,

$$k_z = \frac{\omega}{c}\eta_0 \cos\alpha$$
$$k_r = \sqrt{k^2 - k_z^2} = \frac{\omega}{c}\sqrt{\eta^2(r) - \eta_0^2 \cos^2\alpha}. \tag{36}$$

As the ray goes away to higher values of r, k_r decreases and the ray bends towards the axis of the fibre. Total internal reflection occurs at $r = r_c$, where $k_r = 0$, i.e. $\eta(r_c) = \eta_0 \cos\alpha$. The velocity of the ray (ignoring dispersion) is given by

$$\mathbf{v} = \frac{\omega}{k^2}\mathbf{k}.$$

Hence

$$\frac{dz}{dt} = v_z = \frac{\omega k_z}{k^2}, \quad \frac{dr}{dt} = v_r = \frac{\omega k_r}{k^2}.$$

From the last two equations, one obtains

$$\frac{dr}{dz} = \frac{k_r}{k_z} = \left(\frac{\eta^2}{\eta_0^2 \cos^2\alpha} - 1\right)^{\frac{1}{2}}. \tag{37}$$

Squaring both the sides and then differentiating with respect to z, we get

$$\frac{d^2r}{dz^2} = \frac{1}{2\eta_0^2 \cos^2\alpha}\frac{d\eta^2}{dr}.$$

Using eq. (35), this can be written as

$$\frac{d^2r}{dz^2} + \delta^2 r = 0, \tag{38}$$

where $\delta = (2\Delta\eta/\eta_0)^{\frac{1}{2}}/(a \cos\alpha) \simeq (2\Delta\eta/\eta_0)^{\frac{1}{2}}/a$.

This is a harmonic oscillator equation with sinusoidal solution,

$$r = r_m \sin\delta z, \tag{39}$$

where r_m is a constant of integration. At $r = 0$, $z = 0$, the orientation of the ray with respect to z axis is α i.e. $dr/dz = \tan\alpha$. This gives

$$r_m = \frac{1}{\delta} \tan \alpha = a \sin \alpha \left(\frac{\eta_0}{2\Delta\eta}\right)^{\frac{1}{2}}. \tag{40}$$

One may note that every ray follows a sinusoidal trajectory. Rays, with large angle of inclination α, acquire larger amplitude r_m and then return to the axis. The periodicity length for a ray is

$$z_p = 2\pi\delta^{-1} = 2\pi a \cos\alpha \sqrt{\frac{\eta_0}{2\Delta\eta}}, \tag{41}$$

which is quite insensitive to α. In the limit of small α, the rays that originate at $r = 0$, $z = 0$ at different angles α, meet at $z = nz_p$ where $n = 1, 2, 3, \ldots$, and the optical path lengths of different rays upto common meeting points are the same, irrespective of their path. Thus, the broadening of the laser pulse in the limit of $\alpha \to 0$ is zero. For finite α, z_p has a slow variation with α and gives rise to finite (though small) difference in optical path lengths over different paths. The largest permissible value of α for a trapped ray is the one for which total internal reflection occurs at $r = a$, i.e. $r_m = a$ or $\alpha = \alpha_m = \sin^{-1}(2\Delta\eta/\eta_0)^{\frac{1}{2}}$.

An estimate of pulse broadening in a graded index fibre may be obtained, as follows. The time taken by an extreme ray (with $\alpha = \alpha_m$) in travelling a forward distance $z_p/4$ (from $r = 0$ to $r = a$) is

$$\begin{aligned}
\tau_1 &= \int_0^a \frac{dr}{v_r} \\
&= \frac{1}{c} \int_0^a \frac{\eta^2 dr}{\sqrt{(\eta^2 - \eta_0^2 \cos^2 \alpha_m)}} \\
&= \sqrt{\left(\frac{2\Delta\eta}{\eta_0}\right)} \frac{a}{c} \eta_0 \int_0^1 \sqrt{(1 - x^2)} dx + \sqrt{\left(\frac{\eta_0}{2\Delta\eta}\right)} \frac{a}{c} \cos^2 \alpha_m \int_0^1 \frac{dx}{\sqrt{(1 - x^2)}} \\
&= \sqrt{\left(\frac{\eta_0}{2\Delta\eta}\right)} \frac{\eta_0 a\pi}{2c} (1 - \frac{\Delta\eta}{\eta_0}).
\end{aligned} \tag{42}$$

The time taken by the axial ray to travel a distance $z_p/4$ is

$$\tau_0 = \frac{z_p \eta_0}{4c} = \sqrt{\left(\frac{\eta_0}{2\Delta\eta}\right)} \eta_0 \frac{a\pi}{2c} \left[1 - \frac{\Delta\eta}{\eta_0} - \frac{1}{2}\left(\frac{\Delta\eta}{\eta_0}\right)^2\right]. \tag{43}$$

The time difference between the two rays, in travelling a length L, is

$$\Delta\tau = (\tau_1 - \tau_0)\frac{L}{z_p/4} = \left(\frac{\Delta\eta}{\eta_0}\right)^2 \eta_0 \frac{L}{2c} \quad . \tag{44}$$

Since the bit rate B must be slower than $(\Delta\tau)^{-1}$, this gives

$$BL \le 2c\left(\frac{\eta_0}{\Delta\eta}\right)^2 / \eta_0, \tag{45}$$

which is a great improvement over the step index fibre. The skew rays (that do not pass through the axis of the fiber) lead to longer delays and are a matter of concern. In order to improve BL values beyond the limit given above, one goes over to thinner fibres that allow only a single mode of propagation while all other modes are lost to the cladding.

7.8 Modal Analysis

We consider an optical fibre with refractive index $\eta(r)$ in the core $(r \le a)$ and $\eta_0 - \Delta\eta$ in the cladding $(r > a)$. Following the theory of electromagnetic wave propagation in cylindrical waveguides, we assume that the **E** and **H** fields have t, ϕ, z variations as

$$\mathbf{E}, \mathbf{H} \sim e^{-i(\omega t - l\phi - k_z z)}. \tag{46}$$

Hence, we can replace $\partial/\partial t$ by $-i\omega$, $\partial/\partial\phi$ by il and $\partial/\partial z$ by ik_z and write transverse field components E_r, E_ϕ, H_r and H_ϕ in terms of E_z, H_z (see unit 3). The wave equation governing E_z and H_z is

$$\frac{\partial^2 F}{\partial r^2} + \frac{1}{r}\frac{\partial F}{\partial r} + \left(\frac{\omega^2}{c^2} \in - k_z^2 - \frac{l^2}{r^2}\right) F = 0, \tag{47}$$

where F is E_Z or H_Z, $\in = \eta^2(r)$ for $0 < r < a$ and $\in = (\eta_0 - \Delta\eta)^2$ for $r > a$, and we have ignored the $\nabla(\nabla.\mathbf{E})$ term in E_z equation assuming

$$(1/\in)\partial \in /\partial r << k_z$$

(a condition deduced using $\nabla.(\in \mathbf{E}) = 0$ or $\nabla.\mathbf{E} = -(1/\in)\mathbf{E}.\nabla \in$).

Equipped with eq. (47), now we will analyse the cases of step index and graded index fibres separately.

7.8.1 Step Index Fibre

In a step index fiber

$$\begin{aligned} \in &= \eta_0^2 && \text{for } r < a \\ &= (\eta_0 - \Delta\eta_0)^2 && \text{for } r > a. \end{aligned} \tag{48}$$

Eq. (47) becomes a Bessel equation. For guided modes as we shall see below we must have

$$\omega^2(\eta_0 - \Delta\eta)^2/c^2 < k_z^2 < \omega^2\eta_0^2/c^2.$$

Hence we define

$$\alpha_1^2 = \frac{\omega^2}{c^2}\eta_0^2 - k_z^2, \quad \alpha_2^2 = k_z^2 - \frac{\omega^2}{c^2}(\eta_0 - \Delta\eta)^2$$

and write eq. (47) as

$$\frac{\partial^2 F}{\partial r^2} + \frac{1}{r}\frac{\partial F}{\partial r} + \left(\alpha_1^2 - \frac{l^2}{r^2}\right)F = 0 \quad \text{for } r < a, \tag{49}$$

$$\frac{\partial^2 F}{\partial r^2} + \frac{1}{r}\frac{\partial F}{\partial r} + \left(-\alpha_2^2 - \frac{l^2}{r^2}\right)F = 0 \quad \text{for } r > a. \tag{50}$$

The well-behaved solutions of these equations are:

r < a

$$\begin{aligned} E_Z &= A_1 J_l(\alpha_1 r)e^{-i(\omega t - l\phi - k_z z)}, \\ H_Z &= B_1 J_l(\alpha_1 r)e^{-i(\omega t - l\phi - k_z z)}, \end{aligned} \tag{51}$$

r > a

$$\begin{aligned} E_Z &= A_2 K_l(\alpha_2 r)e^{-i(\omega t - l\phi - k_z z)}, \\ H_Z &= B_2 K_l(\alpha_2 r)e^{-i(\omega t - l\phi - k_z z)}. \end{aligned} \tag{52}$$

Using Maxwell's equations, $\nabla \times \mathbf{E} = -\partial \mathbf{B}/\partial t = i\omega\mu_0\mathbf{H}$ and $\nabla \times \mathbf{H} = -i\omega \in_0 \eta^2\mathbf{E}$, one may obtain the transverse field components in terms of E_z and H_z. For instance

$$\begin{aligned} E_\phi &= \frac{i}{p^2}\left(\frac{il}{r}k_z E_z - \mu_0\omega\frac{\partial H_z}{\partial r}\right), \\ H_\phi &= \frac{i}{p^2}\left(\frac{il}{r}k_z H_z + \in_0 \eta^2\omega\frac{\partial E_z}{\partial r}\right), \end{aligned} \tag{53}$$

where $p^2 = \alpha_1^2$ for $r < a$ and $p^2 = -\alpha_2^2$ for $r > a$.

One may note that for α_2 real (i.e. $k_Z > \omega(\eta_0 - \Delta\eta)/c$), E_Z *and* H_Z decay monotonically with r in the cladding region. Using the continuity of E_Z, H_Z, E_ϕ and H_ϕ at $r = a$, one obtains four equations connecting four constants A_1, B_1, A_2, B_2. The condition for obtaining a non-trivial solution gives the dispersion relation

$$\left[\frac{J_l'(\alpha_1 a)}{\alpha_1 J_l(\alpha_1 a)} + \frac{K_l'(\alpha_2 a)}{\alpha_2 K_l(\alpha_2 a)}\right]\left[\frac{J_l'(\alpha_1 a)}{\alpha_1 J_l(\alpha_1 a)} + \frac{(\eta_0 - \Delta\eta)^2}{\eta_0^2}\frac{K_l'(\alpha_2 a)}{\alpha_2 K_l(\alpha_2 a)}\right] = \left(\frac{2lk_z\Delta\eta}{a\alpha_1^2\alpha_2^2}\right)^2, \tag{54}$$

where prime denotes differentiation with respect to the argument. The dispersion relation can be solved graphically or numerically to obtain k_z as a function of ω. It turns out that for given l and ω, more than one values of k_z are permissible. Each k_z gives a specific value of α_1, α_2 and specific radial profiles of the fields. The wave with a specific radial profile is called a **mode of propagation**. In metallic waveguides, we have distinctive TE and TM modes. In a dielectric waveguide, we have purely TE ($E_Z = 0, H_Z \neq 0$) and TM($E_Z \neq 0, H_Z = 0$) modes only for azimuthal mode number $l = 0$, i.e. when fields are azimuthally symmetric. These are denoted as TE_{0n} and TM_{0n} modes, where n refers to the radial mode number (n^{th} root, of the dispersion relation). For other modes, with $l \neq 0$, both $E_Z(1)$ and H_Z are finite. They are called *hybrid modes* and are denoted as $HE_{l,n}$ or $EH_{l,n}$ depending on whether H_Z or E_Z is the dominating term respectively. Each mode has its typical frequency, below which the mode does not propagate.

One may define the effective refractive index of the fibre for a mode as

$$\bar{\eta} = ck_z/\omega \tag{55}$$

and a quantity related to normalised propagation constant as

$$b = \frac{\bar{\eta} - (\eta_0 - \Delta\eta)}{\Delta\eta}. \tag{56}$$

The normalised frequency, called **V parameter**, is defined as

$$V = \frac{\omega a}{c}\sqrt{\eta_0^2 - (\eta_0 - \Delta\eta)^2} \simeq \frac{2\pi a}{\lambda}\eta_0\sqrt{\frac{2\Delta\eta}{\eta_0}}, \tag{57}$$

where $\lambda = 2\pi c/\omega$ is the free space wavelength. Fig. 7.14 shows the variation of b with V for a typical fibre, for a few modes of propagation. The number of modes increases with V. For $V < 2.405$, only one mode exists–the HE_{11} mode. It is a skewed mode having phase variation with ϕ but no cutoff frequency. All other modes have cutoff frequencies or cutoff V parameter. At large V, a fibre supports nearly $V^2/2$ modes and the fractional power carried by the cladding for a mode is about $2\sqrt{2}/(3V)$. Each mode propagates with a different group velocity. Hence in a multimode fibre, broadening of the laser pulse is strong as the same pulse travels in several different modes, taking different times of travel to the repeater.

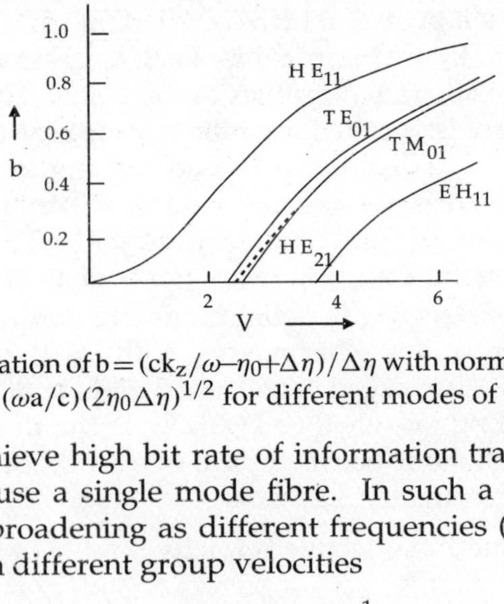

Figure 7.14: Variation of $b = (ck_z/\omega - \eta_0 + \Delta\eta)/\Delta\eta$ with normalised frequency $V = (\omega a/c)(2\eta_0\Delta\eta)^{1/2}$ for different modes of propagation.

In order to achieve high bit rate of information transmission, it is imperative to use a single mode fibre. In such a fibre too, there is laser pulse broadening as different frequencies (in HE_{11} mode) propagate with different group velocities

$$v_g(\omega) = (\partial\omega/\partial k_z) = c\left[\frac{\partial}{\partial\omega}(\bar{\eta}\omega)\right]^{-1} = c/\left(\bar{\eta} - \lambda\frac{\partial\bar{\eta}}{\partial\lambda}\right).$$

For a pulse having wavelengths in the band λ and $\lambda + d\lambda$, the time taken by λ and $\lambda + d\lambda$ in propagating through a distance L is

$$\tau_p = \left(\bar{\eta} - \lambda\frac{\partial\bar{\eta}}{\partial\lambda}\right)\frac{L}{c} \text{ and } \tau_p + \frac{\partial\tau_p}{\partial\lambda}\Delta\lambda.$$

The difference of the two gives the pulse broadening due to mode dispersion,

$$\Delta\tau = \frac{\partial^2 \bar{\eta}}{\partial\lambda^2}\frac{\lambda L}{c}\Delta\lambda. \quad (58)$$

Since the bit transmission rate B must be less than $(\Delta\tau)^{-1}$, one obtains

$$BL < c/\left(\frac{\partial^2 \bar{n}}{\partial\lambda^2}\lambda\Delta\lambda\right). \quad (59)$$

For a semiconductor laser, the spectral width $\Delta\lambda \sim 2-4$ nm and BL product of such fibers can exceed 100(Gb/s) km. In fact, optical links at $\lambda = 1.3\,\mu$m operate at typical bit rate of 2 Gb/s with repeater spacings $\sim 40-50$ km. One can raise the BL product of single mode fibres to 1(Tb/s) km, using single mode lasers with $\Delta\lambda < 1$nm.

The effective refractive index $\bar{\eta}$ of a typical single mode fibre, in the range V = 1.5-2.5, is modelled as

$$\bar{\eta} \simeq \eta_c\left(1 + b\frac{\Delta\eta}{\eta_0}\right), \quad (60)$$

$$b \simeq (1.1428 - 0.9960/V)^2, \quad (61)$$

where $\eta_c = \eta_0 - \Delta\eta$ is the index of the cladding. The dispersion in the fibre depends on $\frac{\partial^2\bar{\eta}}{\partial\lambda^2}$, which arises due to the wavelength dependence of refractive index η_c of the cladding (*material dispersion*) and the nonlinearity of dispersion curves (*waveguide dispersion*).

PROBLEM 7.8. Estimate the highest frequency ω_{max} of radiation, that would propagate in a step index fibre of core radius 3 μm with $\eta_0 = 1.5$, $\Delta\eta = 0.02$ in a single mode. If the frequency is $\omega = 4\omega_{max}$ how many modes would be permitted?

Solution: For only one mode to propagate

$$V = \frac{\omega a}{c}(2\eta_0\Delta\eta)^{1/2} \leq 2.405,$$

$$\omega_{max} = 2.405(c/a)/(2\eta_0\Delta\eta)^{1/2} = 10^{15}\text{rad/s}.$$

At $\omega = 4\omega_{max}$, $V = 4 \times 2.405 = 9.62$ and the maximum number of permissible modes would be

$$N \simeq V^2/2 = 46.$$

7.8.2 Graded Index Fiber

In a parabolic index fibre, with parabolic index profile given by eq. (35), the wave equation for E_z, H_z in the core region takes the form

$$\frac{\partial^2 F}{\partial r^2} + \frac{1}{r}\frac{\partial F}{\partial r} + \left(\alpha_1^2 - \frac{2\omega^2 n_0 \Delta\eta}{c^2 a^2} r^2 - \frac{l^2}{r^2}\right) F = 0, \qquad (62)$$

where $\alpha_1^2 = \omega^2\eta_0^2/c^2 - k_z^2$. Introducing dimensionless quantities, $\xi = r/r_0$, and $\delta = \alpha_1^2 r_0^2$ where $r_0 = (ca/\omega)^{1/2}(2\eta_0\Delta\eta)^{-1/4}$, eq. (62) takes the form

$$\frac{\partial^2 F}{\partial \xi^2} + \frac{1}{\xi}\frac{\partial F}{\partial \xi} + \left(\delta - \xi^2 - \frac{l^2}{\xi^2}\right) F = 0. \qquad (63)$$

If a mode is presumed to be localised well within the core region, then the solution (eigenfunction) of this equation can be written as

$$F = A(r/r_0)^l L_{m-1}^l (r^2/r_0^2) e^{-r^2/2r_0^2} e^{-i(\omega t - k_z z - l\phi)} \qquad (64)$$

with the eigenvalue condition

$$\delta = 2(l + 2m - 1) \qquad (65)$$

where $L_{m-1}^l(r^2/r_0^2)$ are *Associated Laguerre Polynomials*, $l = 0,1,2\ldots$, and $m = 1,2,\ldots$. Eq. (65) gives

$$k_z^2 = \frac{\omega^2}{c^2}\eta_0^2 - \frac{2}{r_0^2}(l + 2m - 1). \qquad (66)$$

The radial extent of lower order modes is of the order of r_0, hence the condition for the validity of this dispersion relation is $r_0^2 << a^2$. For modes extending into the cladding, the TE and TM modes are coupled and the eigenvalue condition is more complicated. We refer the reader to more specialised texts (Ghatak and Thyagarajan)[6].

7.9 Conclusions

Optical communication is making rapid advancements in information carrying capability over transoceanic distances, by overcoming pulse broadening due to various effects and cutting down on losses. Significant promise exists in optical solitons where pulse broadening due to dispersion is compensated, by the self-steepening due to nonlinearity in refractive index. At higher laser intensity, the refractive index of a fibre is an increasing function of intensity and the electromagnetic energy has a tendency to converge towards the maximum refractive index i.e., the intensity maximum. This leads to sharpening and self focussing of the laser pulse. When nonlinear sharpening balances the dispersion broadening, the pulse propagates without distortion.

Besides communications, fibre optics is finding extensive applications in many areas of science and technology, ranging from integrated optics and sensors to medicine.

UNSOLVED PROBLEMS

1. Diamond has a band gap of 5.5 eV. If one could produce adequate density of free electrons and holes in it, then estimate the wavelength of light it would emit.

   [ANS. 0.22 μm.]

2. An open resonator of length L and mirror reflectivities R_1 and R_2 is filled with a weakly ionised gas of relative permittivity $\varepsilon_r = 1$ and conductivity $\sigma = n_e e^2/m(\nu - i\omega)$, where n_e, $-e$, m and ν are the electronic density, charge, mass and collision frequency respectively, ω is the radiation frequency and $n_e e^2/m\varepsilon_0\omega^2 << 1$. Obtain the quality factor of the cavity.

   [HINT: *In one round trip path between the mirrors, the radiation intensity falls down from* I_0 *to* $I_1 = R_1 R_2 \exp(-4 k_i L)$, *where* k_i *is the imaginary part of propagation constant.*]

[ANS. $Q = \frac{2\omega L/c}{(2\nu L/c)\omega_p^2/\omega^2 - \ln R_1 R_2}$, $\omega_p^2 = \frac{n_e{}^2}{m\varepsilon_0}$.]

3. A laser oscillator, operating at $\omega = 4 \times 10^{15}$ rad/s with mirror reflectivities $R_1 = 1$ and $R_2 = 0.95$ and cavity of length L=1 m, has population inversion density twice the threshold value. Estimate the growth rate. Take refractive index $\eta = 2$.

[ANS. $3.75 \times 10^6\ \mathrm{s}^{-1}$.]

4. An Ar-ion laser, of operating wavelength 514 nm and mirror reflectivites $R_1 = 1$, $R_2 = 0.9$, has 10^{12} photons in the cavity in the steady state. Calculate the output power.

[ANS. 5.8 watts.]

5. In a two level laser system–with energy states E_2 and E_1, cavity length L and decay time t_c; the rate equations governing the population densities N_2, N_1 and the photon number density n are:

$$\frac{dN_2}{dt} = -Kn(N_2 - N_1), \qquad \frac{dN_1}{dt} = Kn(N_2 - N_1),$$
$$\frac{dn}{dt} = Kn(N_2 - N_1) - \frac{n}{t_c}.$$

At $t = 0$, $N_2 = N_{20}$, $N_1 = 0$, $n = n_0 << N_{20}$. Solve this set for $t_c = \infty$ to obtain n(t). At time $t = t_s$, when n has attained a large value n_s, t_c is suddenly reduced (the process is called **Q switching**) to $t_c = t_{c0}$ by reducing the reflectivity of one mirror from 1 to R, so that the last term in the last equation dominates and the middle term can be dropped. Obtain the output power as a function of time.

[HINT: *For* $0 < t < t_s$, *you may note* $N_2 - N_1 + 2n$ $=$ *constant*.]

[**Ans.** $P_{out} = n_s h\nu \left(\frac{c}{2\eta}\right) A(1-R)e^{-(t-t_s)/t_{c0}}$
$n_s = n_0 e^{2kt_s/N_{20}}$; ν is the radiation frequency.]

6. A two level laser system of cross section A and length L has developed a photon population density n_0. Now, suddenly the reflectivity of one mirror is reduced from 1 to 0.5 while the other still has a reflectivity 1. Obtain the intensity of light emitted, as a function of time. Neglect atomic transitions during this period and any other form of losses. Take refractive index of the medium as 1.

[**Ans.** $\left(\frac{n_0 \hbar\omega c}{4}\right) e^{-\frac{ct}{4L}}$.]

7. A Gaussian laser beam of frequency ω, initially having (at $z = 0$) a plane wavefront and spot size r_0, acquires a spot size $r_0 f$ and radius of curvature of the wavefront $R_c = f/(df/dz)$ after propagating through a distance z. Here, beam width parameter f is governed by

$$\frac{d^2 f}{d z^2} = \frac{1}{(R_d^2 f^3)}$$

with $R_d = (\omega/c)\, r_0^2$. Show that $f^2 = 1 + z^2/R_d^2$ satisfies it. Obtain the radius of curvature, of the wavefront at $z = 2R_d$.

[**Ans.** $2.5 R_d$.]

8. An open resonator has concave mirrors, each of radius of curvature $R_m = 2L$ and placed at $z = -L/2$ and $z = L/2$. Estimate the spot size r_0 at $z = 0$ for a mode of frequency ω.

[**Hint:** *A mode, of spot size r_0 in the middle of the cavity, must acquire a radius of curvature $R_c = R_m$ at $z = L/2$, so that the reflected rays retrace their paths and the mode is not lost sideways.*]

[**Ans.** $\sqrt{\frac{cL\sqrt{3}}{2\omega}}$.]

9. A 15 page document has 500 words per page, with an average of 6 letters per word. Each letter is represented by 8 binary digits. How much is the information content of the book?

[ANS. 360 kb]

10. An input power of 1 mW decreases to 10 μW, after propagating through a distance of 100 km in a fibre. Obtain the attenuation coefficient of the fibre in dB/km.

[ANS. 0.2 dB/km.]

11. A He-Ne laser operating at $\lambda = 6328\,A^o$ and having oscillator bandwidth of 1500 MHz has cavity length of 0.5 m. How many modes would oscillate?

[ANS. 5 modes.]

12. A step index fibre has $\eta_0 = 1.5$, $\Delta\eta = 0.015$, $a = 5\,\mu$m. How many guided modes can be supported by the fibre, at 0.5 μm wavelength?

[ANS. 88 modes.]

13. A semiconductor laser may be viewed as a planar waveguide with plasma frequency ω_{p0} in the junction region $0 < x < a$ and ω_{p2} for $x < 0$, $x > a$. Taking $E_z = \exp[-i(\omega t - k_z z)]$, $\partial/\partial y = 0$, obtain the dispersion relation for TM modes.

[ANS. $\tan k_\perp a = \frac{2\beta}{(1-\beta^2)}$, $\beta = \frac{\varepsilon_1 \alpha}{\varepsilon_2 k_\perp}$, $\alpha = \sqrt{k_z^2 - \frac{\omega^2 \varepsilon_2}{c^2}}$,
$k_\perp{}^2 = \left(\frac{\omega^2 \varepsilon_1}{c^2} - k_z^2\right)$, $\varepsilon_1 = \varepsilon_r - \omega_{p0}^2/\omega^2$, $\varepsilon_2 = \varepsilon_r - \omega_{p2}^2/\omega^2$.]

14. Consider a cubic metallic cavity of side L. Assuming the wave number quantisation – $k_x = n_1 \pi/L$, $k_y = n_2 \pi/L$, $k_z = n_3 \pi/L$ (n_1, n_2, n_3 being integers) and allowing two independent polarisations for each mode, show that the number of modes in the frequency interval ω and $\omega + d\omega$ inside the cavity is $(L^3 \omega^2 / \pi^2 c^3)\, d\omega$.

References and Suggested Reading

[1] Agarwal Govind P., *Fibre optics communication system*, second edition, New York, John Wiley & Sons, Inc., 2002.
[2] Gowar J. *Optical communication systems*, Englewood Cliffs, NJ, New Delhi, Prentice Hall International, 1984.
[3] Agarwal G. P. and Dutta Niloy K., *Semiconductor lasers*, New York, Van Nostrand Reinhold, 1993.
[4] Thompson G.H.B., *Physics of semiconductor laser devices*, New York , John Wiley& Sons, Inc., 1980.
[5] Shur M. S., *Physics of semiconductor devices*, New Jersey, Prentice Hall, 1990.
[6] Ghatak A. and Thyagarajan K., *Optical Electronics*, Cambridge, Cambridge University Press, 1989.

Unit 8

Geological Seisming and Remote Sensing

Introduction

Underground exploration of oil, minerals and other reserves is an area of high practical interest. The technique of magneto-telluric, employing ultra low frequency (ULF) waves, has emerged as a powerful tool in this endeavour. Extremely low frequency (ELF) waves are useful for undersea exploration and submarine communication. The surface of the Earth, its cloud cover and the atmosphere can be sensed through the techniques of remote sensing, where natural emissions are detected at microwave, millimetre wave and shorter wavelengths using polar and geosynchronous satellites. In this unit, we shall study these fascinating techniques in some detail.

8.1 Magneto-Telluric Method

In this technique, one uses a dipole antenna above the surface of the Earth, to produce ULF/ELF waves, and measures the ratio of horizontal components of electric field E to magnetic field H of the wave (called wave impedance Z = E/H) on the surface of the Earth at different locations. This ratio depends on the conductivity of the Earth, σ, upto a depth $\delta \simeq c(2 \epsilon_0 /\sigma\omega)^{1/2}$, where ω is the frequency of the wave, ϵ_0 is the free space permittivity and c is the velocity of light in free space. By scanning Z with ω, one essentially scans σ with depth (below the surface of the Earth). As one observes Z as a function of horizontal distance on the surface of Earth, anomalies in Z are indicative of the buried ores inside the Earth, just below the location of deviation. For this purpose, one could utilise the naturally occurring electromagnetic disturbance as well.

One may remember that the Earth is an optically denser and absorbing medium for ULF/ELF waves. Hence these waves, generated on the surface of the Earth, travel almost vertically downwards after penetration into the Earth (Fig. 8.1) with

$$k = \frac{\omega}{c}\sqrt{\epsilon_r + i\frac{\sigma}{\omega\,\epsilon_0}} \simeq \left(\frac{1+i}{c}\right)\sqrt{\left(\frac{\sigma\omega}{2\,\epsilon_0}\right)}, \tag{1}$$

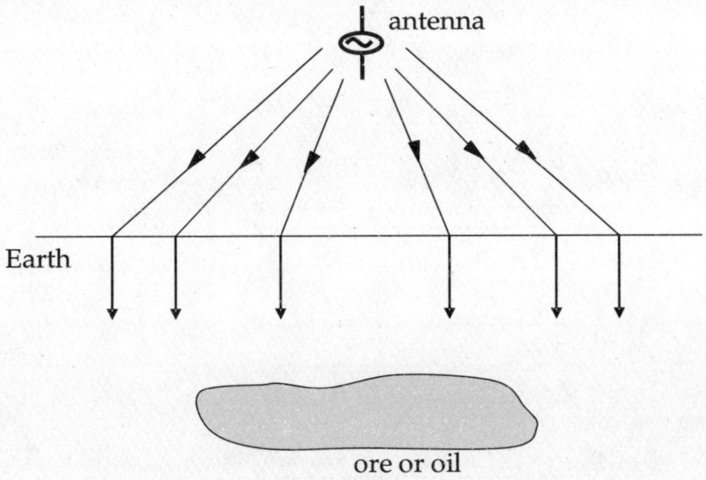

Figure 8.1: Schematic of a magneto-telluric exploration. Arrows indicate that the ULF/ELF wave travels vertically downward inside the Earth.

propagation constant where ϵ_r is the relative permittivity of the Earth. For a wave of 100Hz frequency in the Earth of $\sigma \simeq 1$ mho/ metre, the depth of penetration is $k_i^{-1} \simeq 20$ meters which is suitable for ground water studies. For a conductivity of 0.1 mho/metre and wave period $2\pi/\omega \sim 30$ minutes, the depth of penetration is 67.5 km which corresponds to the depth of the upper mantle.

An insight into the physics of magneto-telluric method can be achieved by considering the normal incidence of an electromagnetic wave, over a multilayered isotropic Earth and estimating planewave impedance.

8.1.1 The Plane Wave Impedance over a Layered Earth

Following *Ward* and *Hohmann*[1], we consider normal incidence of a plane wave from air upon an N-layered Earth. The incident plane wave propagates along positive z, or downward direction (Fig. 8.2.). In air ($z < 0$), its fields are

$$\mathbf{E} = \hat{y} E_0 e^{i(\omega t - \omega z/c)}, \tag{2}$$

$$\mathbf{H} = -\hat{x} \frac{E_0}{c\mu_0} e^{i(\omega t - \omega z/c)}. \tag{3}$$

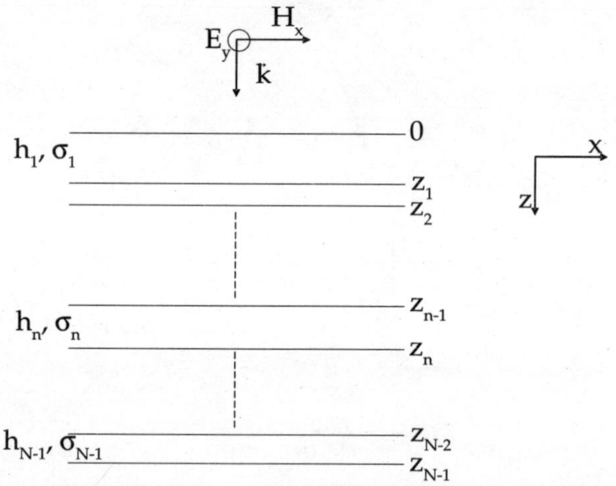

Figure 8.2: Normal incidence of a plane wave, over an N-layered Earth with n^{th} layer, characterised by width $h_n = z_n - z_{n-1}$ and conductivity σ_n.

In the nth layer ($z_{n-1} < z < z_n$) of thickness h_n and conductivity σ_n, the **E, H** fields can be written as

$$E_y = [E_n^+ e^{-i(\omega t - k_n(z - z_n))} + E_n^- e^{-i(\omega t + k_n(z - z_n))}], \tag{4}$$

$$H_x = \frac{-k_n}{\omega\mu_0}[E_n^+ e^{-i(\omega t - k_n(z - z_n))} - E_n^- e^{-i(\omega t + k_n(z - z_n))}], \tag{5}$$

where $k_n \cong \frac{\omega}{c}\left(\frac{\sigma_n}{2\omega \in_0}\right)^{1/2} (1+i)$.

On the plane $z = z_n$,

$$E_{yn} = E_n^+ + E_n^-,$$
$$H_{xn} = \frac{1}{Z_n}(E_n^- - E_n^+), \tag{6}$$

where $Z_n = \omega\mu_0/k_n$ is the intrinsic impedance of the nth layer and $E_{yn} \equiv E_y(z = z_n), H_{xn} = H_x(z = z_n)$. (Note that the phase factor $e^{-i\omega t}$ is implied.)

Eqs. (5) and (6) yield

$$E_n^+ = \tfrac{1}{2}(E_{yn} - Z_n H_{xn}),$$
$$E_n^- = \tfrac{1}{2}(E_{yn} + Z_n H_{xn}). \tag{7}$$

At $z = z_{n-1}$, the continuity of tangential electric and magnetic fields demands

$$E_{y(n-1)} = E_n^+ e^{+ik_n(z_{n-1}-z_n)} + E_n{}^- e^{-ik_n(z_{n-1}-z_n)},$$
$$H_{x(n-1)} = -\frac{1}{Z_n}[E_n^+ e^{+ik_n(z_{n-1}-z_n)} - E_n^- e^{-ik_n(z_{n-1}-z_n)}].$$

Since $z_n - z_{n-1} = h_n$, we get

$$E_{y(n-1)} = E_{yn}\cosh(ik_n h_n) + Z_n H_{xn}\sinh(ik_n h_n),$$
$$H_{x(n-1)} = H_{xn}\cosh(ik_n h_n) + \frac{1}{Z_n}E_{yn}\sinh(ik_n h_n)$$

or

$$\begin{bmatrix} E_{y(n-1)} \\ H_{x(n-1)} \end{bmatrix} = \underline{\underline{T}}_n \cdot \begin{bmatrix} E_{yn} \\ H_{xn} \end{bmatrix}, \tag{8}$$

$$\text{where } \underline{\underline{T}}_n = \begin{bmatrix} \cosh(ik_n h_n) & Z_n \sinh(ik_n h_n) \\ \frac{1}{Z_n}\sinh(ik_n h_n) & \cosh(ik_n h_n) \end{bmatrix} \tag{9}$$

is known as the *transfer matrix of the* n^{th} *layer*. For N-1 layers, we can find a sucession of $\underline{\underline{T}}_n$ matrices, each matrix permitting us to write the fields in one layer in term of the fields in the next layer. Thus, we can readily find the matrix relationship between the fields in zeroth layer (air) and those in the infinite medium below the $(N\text{-}1)^{th}$ layer.

$$\begin{bmatrix} E_{y0} \\ H_{x0} \end{bmatrix} = \underline{\underline{T}}_1 . \underline{\underline{T}}_2 \cdots\cdots \underline{\underline{T}}_{N-1} . \begin{bmatrix} E_{yN} \\ H_{xN} \end{bmatrix}, \tag{10}$$

where E_{yN} and H_{xN} are the fields at the top of the infinite medium.

For an Earth model, consisting of one layer overlying an infinite half space (referred to as a two-layered Earth), one obtains the surface impedance as

$$\bar{Z}_0 = -\frac{E_{y0}}{H_{x0}} \quad = \quad Z_1 \frac{Z_2 - Z_1 \tanh(i\, k_1 h_1)}{Z_1 - Z_2 \tanh(i\, k_1 h_1)}, \tag{11}$$

where Z_1 and Z_2 are the intrinsic impedances of layers 1 and 2. For a homogenous Earth, set $h_1 = \infty$ in eq. (11). Then $\tanh(i\, k_1\, h_1) = -1$ and $\bar{Z}_0 = Z_1$–the impedance of the half plane.

For a two-layered model the effective resistivity of the Earth is defined as

$$\rho_a = \frac{1}{\omega \mu_0} \left|\bar{Z}_0\right|^2 . \tag{12}$$

We have plotted ρ_a/ρ_1 versus $\dfrac{\sqrt{\rho_1 T}}{h_1}$, where $T = 2\pi/\omega$ and $\rho_1 = \dfrac{1}{\sigma_1}$, in fig. 8.3 for a range of resistivity ratios ρ_2/ρ_1 of the two layers. At low values of $\sqrt{(\rho_1 T)}/h_1$, the effective earth resistivity varies slightly with ρ_2/ρ_1. However, at higher $\sqrt{(\rho_1 T)}/h_1$ the contrast is huge as the ULF wave of longer wave period penetrates deeper into the infinite Earth.

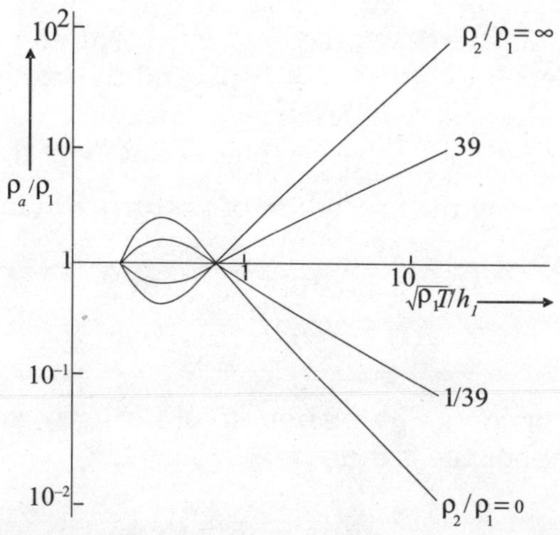

Figure 8.3: Normalised apparent Earth resistivity ρ_a/ρ_1 versus $\sqrt{\rho_1 T}/h_1$ for two-layered Earth.

PROBLEM 8.1. A 100 Hz electromagnetic wave of amplitude 1 V/m is incident normally on Earth of $\sigma = 10$ mho/m. Estimate the E/H ratio at the surface of the Earth. What would be the amplitude of the wave at a depth of 20 m inside the Earth?

Solution: Inside the Earth, effective plasma permittivity, propagation constant of the wave, and intrinsic impedance of the Earth are:

$$\epsilon_{eff} \cong i\sigma/\omega\, \epsilon_0 = 1.8 \times 10^8\, i$$

$$k_1 = \frac{\omega}{c}\sqrt{\epsilon_{eff}} = 2 \times 10^{-2}(1+i)\, m^{-1}$$

$$Z_1 = \omega\mu_0/k_1 = 2 \times 10^{-3}(1-i)\text{ ohm}$$

respectively. The E/H ratio, at the surface of the earth, is $Z_1 = 2 \times 10^{-2}(1-i)$ ohm. The amplitude transmission coefficient is

$$T = 2/(1 + ck_1/\omega) \simeq 10^{-4}(1-i).$$

The amplitude of the electric field at depth d = 10m is

$$A = |T|\, e^{-k_{1i}d} = 7 \times 10^{-5} V/m.$$

PROBLEM 8.2. Consider a two-layered Earth with top layer of $\sigma =$ 10 mho/m, having a width $h_1 = 10$ m, and the bottom half space having $\sigma = 1$ mho/m. A 100 Hz electromagnetic wave is incident on it normally. Estimate the effective resistivity of the earth.

Solution: Following the procedure of problem 8.1, in the top layer

$$k_1 = 6.3 \times 10^{-2}(1 + i)\, m^{-1}$$
$$Z_1 = 6.3 \times 10^{-2}(1 - i)\, \text{ohm}.$$

The intrinsic impedance at the bottom of the Earth is $Z_2 = Z_1\sqrt{10}$. The surface impedance is thus

$$\bar{Z}_0 = Z_1 \frac{\sqrt{10} + \tanh(0.63(i - 1))}{1 - \sqrt{10}\tanh(0.63(i - 1))}$$

and the effective earth resistivity $\rho_a = \left|\bar{Z}_0\right|^2 / \omega\mu_0 \simeq$ **6.4 ohm-m.**

For an N-layered Earth model, we begin with the impedance $\bar{Z}_{N-1}$ at the top of the first layer above the bottom half space,

$$\bar{Z}_{N-1} = Z_{N-1}\left[\frac{Z_N - Z_{N-1}\tanh(i\, k_{N-1}\, h_{N-1})}{Z_{N-1} - Z_N \tanh(i\, k_{N-1}\, h_{N-1})}\right]. \tag{13}$$

The impedance on top of the $(N - 2)^{th}$ layer is

$$\bar{Z}_{N-2} = Z_{N-2}\left[\frac{\bar{Z}_{N-1} - Z_{N-2}\tanh(i\, k_{N-2}\, h_{N-2})}{Z_N - 2 - \bar{Z}_{N-1}\tanh(i\, k_{N-2}\, h_{N-2})}\right]. \tag{14}$$

This way we can move upto the surface of the earth.

In the case when there is a sharp vertical boundary between media of different conductivities, the impedance shows a discontinuity. *Mc Neill* and *Labson*[2] have discussed the problem in detail. We plot the variation of $|E_y/H_x|$ as a function of x/δ_1 in fig. 8.4, for different values of ρ_2/ρ_1 (the resistivity ratio of the two neighboring layers) where δ_1 is the skin-depth in the region of resistivity ρ_1. The sharp change in $|E_y/H_x|$ value indicates the location of the boundary, which is a very valuable information.

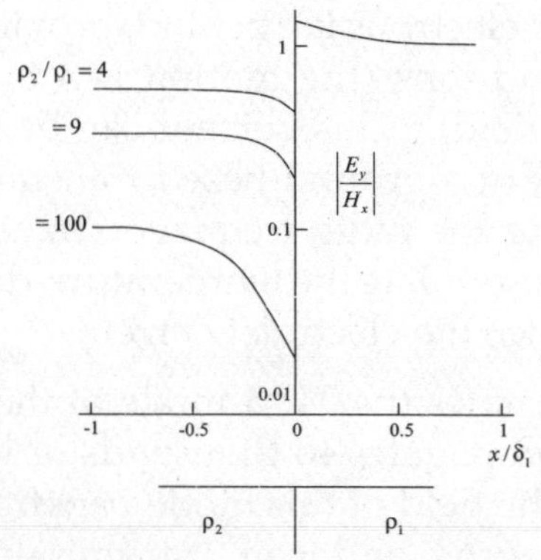

Figure 8.4: Variation of impedance near a vertical contact [after Erceville, I.D. and Kunetz, G., 1962, the effect of fault on the Earth's natural electromagnetic field, Geophys., **27**, 651 (1962).].

8.2 Submarine Communication

Submarines are submerged in sea water, upto a depth of several meters. High frequency waves are not suitable for communicating with them from above the ground, as their depth of penetration

$$\delta_D = c\sqrt{\frac{2 \in_0}{\sigma\omega}}$$

in saline sea water is small. ELF waves are suitable for this purpose. To communicate simultaneously with submarines, widely scattered in the ocean, one essentially needs a global communication system for which a powerful radiation source is required. Wire antennae have low efficiency of radiation at such low frequencies. Moreover, if their lengths are to be comparable to a wavelength ($\sim$ 3000 km), they would extend to hundreds of kilometres. A very attractive alternative is found in the ionosphere. The wavelength of ELF waves in the ionosphere is 2 to 3 orders of magnitude smaller than that in free space, hence size of antenna would be much smaller. Also, a widespread *electrojet current* exists, at a height of about 100 km above the surface of the Earth. This arises due to the rotation of the Earth about its axis in which the Earth's atmosphere and ions and molecules rotate with it due to their large viscosity but the light electrons lag behind, providing a relative drift between electrons and ions (the motion is a bit complicated due to Earth's magnetic field). This current can be modulated at ELF frequency, by means of a ground-based transmitter (Fig. 8.5). By periodically switching the radio frequency transmitter on and off, one can periodically modulate the temperature-dependent electrical conductivity and hence the electrojet current.

The ELF current excites the TEM mode of the Earth-ionosphere waveguide, which propagates to thousands of kilometers without much attenuation. The field of this mode penetrates into sea water, upto a few skin depths and can transmit information to the submarine. The propagation constant of the TEM mode turns out to be

$$\mathrm{k} \approx \frac{\omega}{\mathrm{c}} + \frac{(1+i)}{2\sqrt{2}L}\sqrt{\omega \in_0}\left[\frac{1}{\sqrt{\sigma}} + \frac{1}{\sqrt{\sigma_{\mathrm{u}}}}\right], \tag{15}$$

where σ and σ_u are the conductivities of the Earth and the lower ionosphere respectively and L is the height of the ionosphere above the ground.

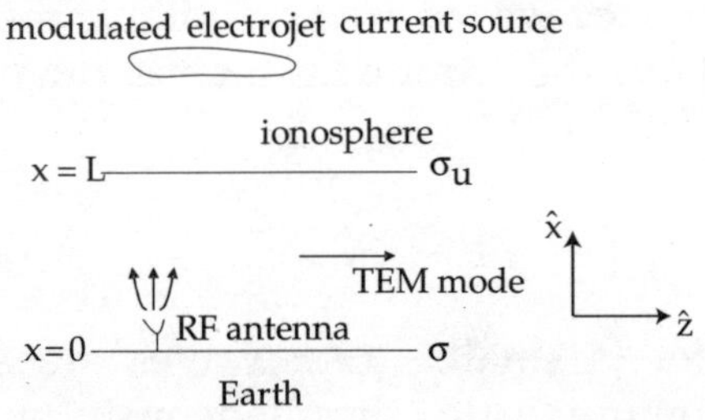

Figure 8.5: Schematic of Earth-ionosphere waveguide. A ground-based radio transmitter (with periodic switching on and off at frequency Ω) modulates the electrojet current in the ionosphere. The modulated current excites the TEM mode of the Earth-ionosphere waveguide at ELF frequency Ω.

PROBLEM 8.3. Consider a parallel plane guiding system (Fig. 8.5) having free space in the region $0 < x < L$ and media of conductivity σ for $x < 0$ and σ_u for $x > L$. Assuming t, z variations of fields as $\exp[-i(\omega t - k_z z)]$ and $\sigma, \sigma_u >> \omega \in_0$, $\partial/\partial_y = 0$, obtain the dispersion relation for the lowest order TM mode ($\mathbf{H} || \hat{y}$).

Solution: The wave equation governing H_y in a medium of effective permittivity $\in_{eff}$ is

$$\nabla^2 H_y + \frac{\omega^2}{c^2} \in_{eff} H_y = 0.$$

Replacing $\partial/\partial z$ by ik_z, this equation takes the form

$$\frac{\partial^2}{\partial x^2} H_y + \alpha^2 H_y = 0,$$

where
$$\begin{aligned} \alpha^2 &= k_x{}^2 = \omega^2/c^2 - k_z{}^2 \text{ for } 0 < x < L \\ &= \alpha_1^2 \approx (\omega^2/c^2) \in_1 \text{ for } x < 0 \\ &= \alpha_2^2 \approx (\omega^2/c^2) \in_2 \text{ for } x > L, \end{aligned}$$

$\in_1 = i\sigma/\omega \in_0, \in_2 = i\sigma_u/\omega \in_0$. The well behaved solutions of the wave equation are

$$\begin{aligned} H_y &= A_1 e^{ik_x x} + A_2 e^{-ik_x x} && \text{for } 0 < x < L \\ &= A' e^{i\alpha_1 x} && \text{for } x < 0 \\ &= A'' e^{-i\alpha_2 x} && \text{for } x > L \end{aligned}$$

where phase factor $\exp[-i(\omega t - k_z z)]$ is implied.
Using $\nabla \times \mathbf{H} = -i\omega \in_0 \in_{eff} \mathbf{E}$, one may write

$$\begin{aligned} E_z &= \frac{-k_x}{\omega \in_0}(A_1 e^{ik_x x} - A_2 e^{-ik_x x}), && 0 < x < L \\ &= \frac{\alpha_1}{\omega \in_0 \in_1} A' e^{-i\alpha_1 x}, && x < 0 \\ &= -\frac{\alpha_2}{\omega \in_0 \in_2} A'' e^{i\alpha_2 x}, && x > L. \end{aligned}$$

Applying the continuity of E_z and H_y at $x = 0$ and $x = L$, we obtain

$$\begin{aligned} &A_1 + A_2 = A', \quad A_1 - A_2 = -(\alpha_1/k_x \in_1)A' \\ &A_1 e^{ik_x L} + A_2 e^{-ik_x L} = A'' e^{\alpha_2 L}, \\ &A_1 e^{ik_x L} - A_2 e^{-ik_x L} = \left(\frac{\alpha_2}{k_x \in_2}\right) A'' e^{i\alpha_2 L}. \end{aligned}$$

These equations give the dispersion relation

$$e^{i2k_xL} = \frac{\left(1+\frac{\alpha_1}{k_x \epsilon_1}\right)\left(1+\frac{\alpha_2}{kx \epsilon_2}\right)}{\left(1-\frac{\alpha_1}{k_x \epsilon_1}\right)\left(1-\frac{\alpha_2}{k_x \epsilon_2}\right)}$$
$$\simeq 1 + \frac{2\alpha_1}{k_x \ \epsilon_1} + \frac{2\alpha_2}{k_x \ \epsilon_2}$$

For $\sigma, \sigma_u \rightarrow \infty$, the right hand side goes to 1, giving $k_x = 0$ i.e., $k_z = \omega/c$ – the dispersion relation for the TEM mode. When σ, σ_u are finite but large, $2k_xL$ is still less than 1 and one may expand the exponential to obtain

$$i2{k_x}^2L = \frac{2\alpha_1}{\epsilon_2} + \frac{2\alpha_2}{\epsilon_2},$$

leading to

$$k_z = \frac{\omega}{c} + \sqrt{(\omega \ \epsilon_0)}\frac{1+i}{2\sqrt{2}L}\left[\frac{1}{\sqrt{\sigma}} + \frac{1}{\sqrt{\sigma_u}}\right].$$

8.3 Remote Sensing

All bodies, at a finite temperature, radiate. The spectrum radiated by a black body obeys Planck's formula

$$u_\lambda = \frac{8\pi hc/\lambda^5}{e^{hc/k_BT\lambda} - 1}, \tag{16}$$

where $u_\lambda d\lambda$ is the energy density of radiation in the wavelength range λ and $\lambda + d\lambda$, T is the temperature of the body, k_B is the Boltzmann's constant, h is the Planck's constant and c is the velocity of light in vacuum. For a body of emisstivity e_λ, the energy density of radiation is $u_\lambda e_\lambda$. e_λ characterises whether the radiating surface is ocean, soil, rock or vegetation. Thus, by observing the spectrum radiated from Earth's surface, one can infer temperature and composition of the surface. Remote sensing employs Earth orbiting satellites (both polar and geostationary) to monitor this radiation from different Earth locations and retrieve valuable information about the Earth, its atmosphere and weather.

The spectral energy density u_λ peaks at a wavelength λ_m, where

$$\lambda_m \, T = 0.3 \, \mathrm{cm}^0 \, \mathrm{K}, \qquad \textit{(Wein's displacement law)}.$$

The temperature on the surface of the Earth varies in the range -40^0 C to 50^0 C (233^0 K to 323^0 K), hence the emission peaks in the far infrared region, and has substantially high levels of microwave and millimetre wavelength components. Fig.8.6 gives a spectrum of radiation from the Earth at 27^0C. We have also plotted the spectrum of solar radiation, received at the surface of the Earth. As the

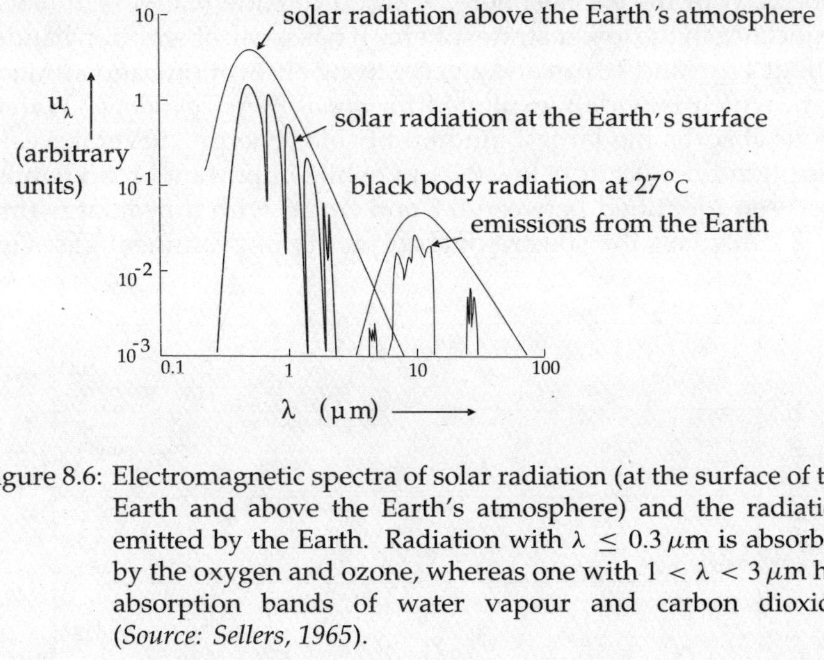

Figure 8.6: Electromagnetic spectra of solar radiation (at the surface of the Earth and above the Earth's atmosphere) and the radiation emitted by the Earth. Radiation with $\lambda \leq 0.3\,\mu$m is absorbed by the oxygen and ozone, whereas one with $1 < \lambda < 3\,\mu$m has absorption bands of water vapour and carbon dioxide. (*Source: Sellers, 1965*).

temperature on the surface of the sun is much higher ($\sim 6000\,^0$C) than that on the Earth's surface, the two spectra are in fairly distinct wavelength domains. The radiation that reaches the Earth is primarily of wavelength $< 4.0\mu$m whereas that emitted by the Earth is principally in the band 4.0 to 40μm. The solar radiation is significantly absorbed by the gases present in Earth's atmosphere. However, there is great variability in absorption from one wavelength to another. Many sharp peaks of absorptivity at some wavelengths are separated by equally abrupt troughs, where the ability of the gases to absorb radiation is very low.

The most important absorbers of radiation in the atmosphere are:

a. oxygen and ozone
b. carbon dioxide
c. water vapour.

Radiation of wavelength $\lambda < 0.1\mu m$ is strongly absorbed by atomic and molecular oxygen (O and O_2) and also by nitrogen (N_2) in the ionosphere. 0.1-0.3μm wavelengths are absorbed efficiently by ozone (O_3) in the ozonosphere. Carbon dioxide (CO_2) is of chief significance in the lower stratosphere. It has weak absorption bands at about 4 μm and 10 μm and a very strong absorption band around 15μm, which is widely exploited for *atmospheric sounding*. Water vapour absorbs the largest amount of solar energy. Several weak absorption lines occur below 0.7 μm, while important broad bands have been identified between 0.7 and 8 μ m with a peak at 6 μm. Fig. 8.7 displays the spectra of absorptivity of prominent gases in

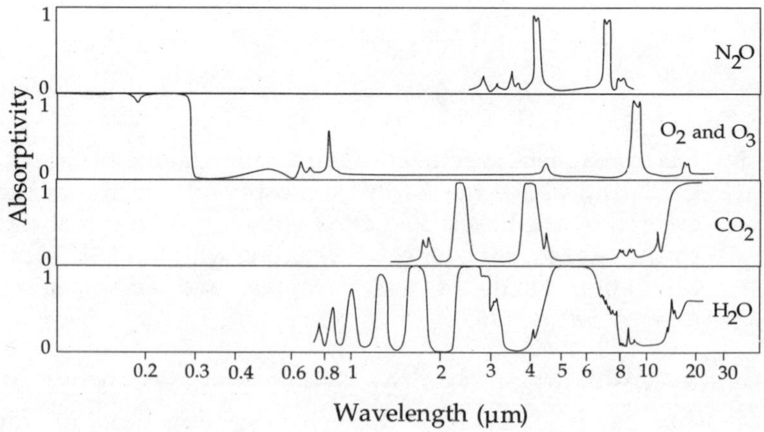

Figure 8.7: Absorptivity versus wavelength for various constituents of the atmosphere (*Source: Fleagle and Businger, 1963*).

the atmosphere. At wavelengths $\lambda < 0.3\mu m$, the atmosphere absorbs almost all the incident radiation. It is transparent in the visible band from $0.3\,\mu m$ to $0.7\,\mu m$. At higher wavelengths, we have sharply defined absorption bands alternating with relatively transparent regions, called *atmospheric windows*.

Atmospheric window wavebands are used for remote sensing of the features on the Earth's surface, especially when they use satellite-borne platforms. The wavebands of maximum absorption may be chosen for some kinds of meteorological and climatological studies. $15\,\mu m$ CO_2-absorption band, for instance, is used for vertical profiling or sounding of the atmosphere in depth. As the molecules of a gas emit radiation of wavelengths at which they absorb them, satellite-borne multichannel radiometers and spectrometers measure the radiant emission from the underlying atmosphere across a series of wavelengths. The wavelengths that are absorbed strongly, are received at the satellite from upper layers of the atmosphere. The wavelengths, which are absorbed to a less extent, originate from deeper layers. When proper weightage is given to the exponential decrease in concentration of gases with height above the surface of the Earth, useful information about different layers of the atmosphere can be obtained.

For remote sensing of the surface of the Earth (and its cloud cover) using satellite, one employs 3.0–$4.5\,\mu m$ and 8.5–$14\,\mu m$ window regions. A noteworthy complication is that in the second window, the clouds often obscure the surface of the Earth. Water vapour is transparent to radiation, but water droplets reflect and scatter incident radiation. These attenuation effects are so strong that even shallow clouds can reduce the transmittance to zero. Therefore, when clouds are present in the atmospheric column, their upper surface acts as the effective radiating surface. An important benefit of this is that the observed radiation (brightness) can be used to map the height of the cloud tops, as temperatures usually decline upwards through the troposphere. The cloud top temperatures can be translated into heights of the cloud tops above the ground. The radiometers used for mapping surface composition measure radiation typically in the 8.4–$9.5\,\mu m$ and 10.2–$11.4\,\mu m$ bands. In case data from the two windows give different brightness temperatures, comparison of measurements from two channels can give clues about types and variations of mineral surfaces.

There is yet another window at microwave and millimetre wavelength ($\lambda > 1$ mm), that is prominently used for remote sensing the Earth. These waves have a clear advantage, that they can perpetrate dense clouds and rain storms. Hence, they have a greater degree of 'all-weather' usability, that other systems lack.

8.3.1 Active Remote Sensing

The schemes of remote sensing, described so far, employ the natural emissions. In active remote sensing, one detects the signals that are artificially produced (e.g., microwaves produced by a Klystron or a travelling wave tube). Radar is a commonly used technique. In this, a signal is sent on to a target and the reflected signal is received. The time delay and the level of the reflected signal provide information about the distance of the target and its surface reflectivity. The radar directs signals in a precise direction and this direction is rapidly changed to monitor a large solid angle. This gives precise information about the terrain, target etc. Sonar (**So**und **Na**vigation and **R**anging) employs acoustical waves for underwater sensing.

8.4 Conclusions

Underground exploration using ULF and ELF waves has made great strides in the last few decades and continues to be at the forefront of research. A significant effort has been put in for understanding and realising the excitation of these waves and accounting for the effects of Earth's curvature, anisotropic ionosphere and other details.

The scheme of ELF generation via the modulation of electrojet current, using ground-based radio transmitters with effective radiated power of ~100 MW (at around 5MHz), has brought out several novel physics concepts, e.g., parametric excitation of Alfven wave and whistler wave, suppression/enhancement of spread F and electrojet instabilities by radio waves etc. Remote sensing continues to be the most powerful tool for atmospheric mapping, short and long-range weather forecasting and exploration of the geological phenomena near the surface of the Earth.

UNSOLVED PROBLEMS

1. A rock of $\sigma = 0.2$ mho/m, $\epsilon_r = 4$ and large thickness is buried at a depth of 100 m inside the earth of conductivity 1 mho/m. What frequencies of electromagnetic wave would you use to infer about the rock?

   [ANS. $\omega/2\pi \sim 30$ Hz.]

2. An electromagnetic wave is incident normally, from a dielectric of relative permittivity $\epsilon_r = 2$ on to a dielectric of $\epsilon_r = 4$. What is the E/H ratio at the interface?

   [ANS. 188.5 Ohm.]

3. An electromagnetic wave of 1 kHz frequency is normally incident from free space, on a dielectric slab of $\epsilon_r = 100, \sigma = 0$ and thickness 10 m, backed by a semi-infinite medium of $\epsilon_r = 4, \sigma = 1$ mho/m. Estimate the surface impedance.

   [ANS. $3.77 \times 10^{-2}(1.6–3.6i)$]]

4. Estimate the wavelength and attenuation length (k_i^{-1}) of 100 Hz ELF wave, propagating in TEM mode in the Earth-ionosphere waveguide. Take Earth's conductivity as $\sigma = 1$ mho/m, ionospheric conductivity $\sigma_u = 2.5 \times 10^{-3}$ mho/m and the height of the ionosphere = 90 km. Ignore the curvature of the Earth.

   [ANS. $\lambda = 3 \times 10^3$ km, $k_i^{-1} = 1.5 \times 10^5$ km.]

References and Suggested Reading

[1] Ward S.H. and Hohmann G.W., *'Electromagnetic theory for geophysical applications'* in Electromagnetic Methods in Applied Geophysics-Theory, Vol. 1, edited by M.N. Nabighian, Society of Exploration Geophysicists, Oklahoma, USA, (1987).

[2] Mc Neill J.D. and Labson V.F., *'Geophysical mapping using radio fields'*, in Electromagnetic Methods in Applied Geophysics- Applications Part A and Part B, edited by M.N. Nabighian, Society of Exploration Geophysicists, Oklahoma, USA, (1991).

[3] Barrett E.C. and Curtis L.F., *Introduction to Environmental Remote Sensing*, Third Edition, Madras, London, Chapman and Hall, (1992).

[4] Wait J. R., *Electromagnetic waves in stratified media*, London, Pergamon Press, (1970).

Unit 9

Relativistic Covariance of Electrodynamics

Introduction

Newton's first law of motion, known as the *law of inertia*, states that a body continues to maintain its state of motion, i.e., its velocity remains constant, unless acted upon by an external force. Conversely, a frame of reference in which the law of inertia holds is called an *inertial frame*. Any frame, moving with constant velocity with respect to an inertial frame, is also an inertial frame. Thus, one may visualise numerous inertial frames, moving with constant velocities with respect to each other.

Newton believed that there is no preferred inertial frame in nature. The laws of nature are the same with respect to all inertial frames.

A major crisis arose, after Maxwell's equations predicted that the velocity of light in free space is $c=1/\sqrt{\mu_0 \epsilon_0}$, where $\mu_0=4\pi \times 10^{-7}$ MKS is the free space permeability and $\epsilon_0= 10^{-9}/(36\pi)$ MKS is the free space permittivity. Since μ_0 and ϵ_0 are the same in all inertial frames, the velocity of light c, must also appear the same in all inertial frames. How could it be that the velocity of light is c, with respect to a ground observer and it is the same, with respect to an observer moving with an arbitrarily large velocity **v** also? This violated the law of addition of velocities. Then, it became imperative to change the very foundation of mechanics. Yet there was a suspicion that there might be a very special frame in which the velocity of light was c and it was different in other frames. This suspicion was conclusively demolished by the Michelson-Morley experiment.

9.1 Michelson-Morley Experiment

A Michelson interferometer was mounted on a table, floating on mercury, so that it could be rotated at will. It comprised a monochromatic broad source of light, two identical parallel-glass plates, two mutually perpendicular mirrors M_1 and M_2 and a lens. The glass plates were inclined at 45^0 to the surface normals of mirrors M_1 and M_2. The rear surface of the first glass plate had a thin silver lining to cause substantial reflection (Fig. 9.1). The light beam, incident on the first plate, splits into two – one goes towards M_1 and the other towards M_2. Lengths OP and OQ are equal to L (however, one can take them to be unequal). The two light beams suffer reflections on the mirrors and combine with each other, after returning to O and reaching the focal plane of the lens.

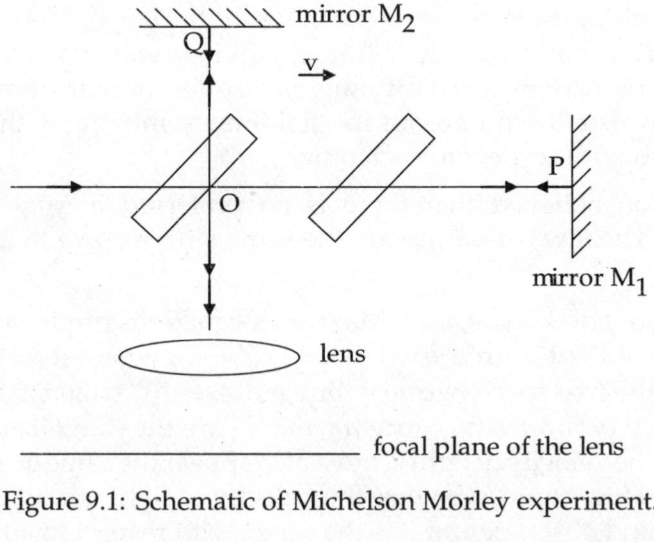

Figure 9.1: Schematic of Michelson Morley experiment.

Let the velocity of light in the special inertial frame (if such a frame exists) be c and that of the Earth (hence of the apparatus) with respect to the special frame be v, as shown in fig. 9.1. The time of flight from O to P is, say t_1 and that from P to O is t_2:

$$ct_1 = L + vt_1, \ ct_2 = L - vt_2$$
$$\Rightarrow t_1 = \frac{L}{c-v}, \ t_2 = \frac{L}{c+v}.$$

Thus the total time for round trip travel from O to P is

$$T_1 = t_1 + t_2 = \frac{2Lc}{c^2 - v^2}.$$

The light that goes towards M_2 hits it at Q′, as M_2 has moved by distance QQ′ during the flight of light (Fig. 9.2). Let the time of flight from O to Q′ be t_1', then $(ct_1')^2 = (vt_1')^2 + L^2$.

$$\therefore, \ t_1' = \frac{L/c}{\sqrt{1 - v^2/c^2}}.$$

The total time for OQO′ travel is

$$T_2 = \frac{2L/c}{\sqrt{1 - v^2/c^2}}.$$

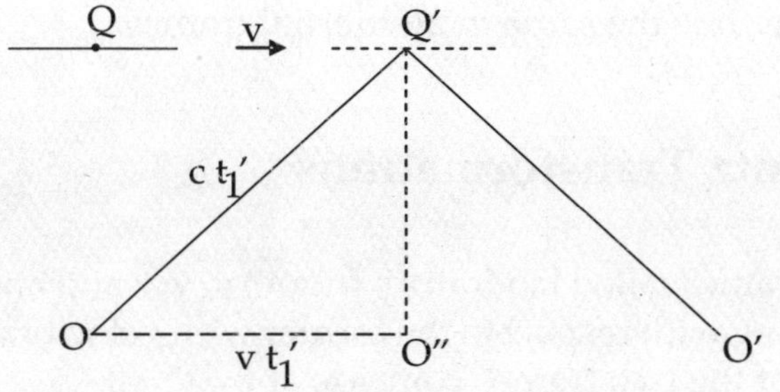

Figure 9.2: Ray path of the vertical ray in the special inertial frame, with respect to which Earth is moving with velocity v.

The phase difference introduced between the two light beams is $\Delta\phi = \omega(T_1 - T_2)$, where ω is the frequency of light in radian/s. For $v^2/c^2 << 1$,

$$\Delta\phi = \frac{\omega L}{c}\frac{v^2}{c^2}.$$

The phase difference will manifest itself in the positioning of the fringes. Next, rotate the entire apparatus by 90^0 so that the time of flight for the ray coming from M_2 becomes T_2 and that for the one coming from M_2 becomes T_1, giving

$$\Delta\phi = -\frac{\omega L}{c}\,\frac{v^2}{c^2}.$$

The fringes must show a shift, corresponding to a phase change of $2\,[(\omega L)/c]\ [v^2/c^2]$. For the parameters of the experiment and v corresponding to the rotational speed of Earth, this phase shift was well within the sensitivity of experiment. However, no phase change was observed. This conclusively proved that there is no special frame of reference and that the velocity of light in all the inertial frames is the same.

9.2 Postulates of Relativity

i. All laws of nature are the same in all inertial frames of reference.
ii. The velocity of light is independent of the motion of the source. It is the same in all inertial frames.

9.3 Lorentz Transformations

Consider a frame (called laboratory frame) x, y, z and another frame x′,y′,z′, moving with respect to the former with velocity $v\hat{z}$ (Fig. 9.3). The origins of the two frames coincide at t = t′ = 0.

A light source placed at O releases a pulse at t = 0. The pulse arrives at P(x,y,z) at time t and triggers a photocell instantly. x,y,z must satisfy

$$x^2 + y^2 + z^2 - c^2t^2 = 0. \qquad (1)$$

To the moving observer, the event of triggering of photocell will appear at x′,y′,z′ at time t′, such that

$$x'^2 + y'^2 + z'^2 - c^2t'^2 = 0. \qquad (2)$$

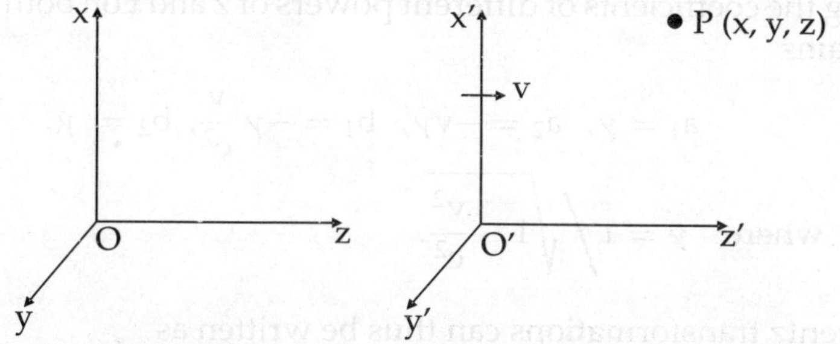

Figure 9.3: Laboratory frame x, y, z and the moving frame x′, y′, z′. The events of release of a light pulse from a source at O and triggering of photocell at P are viewed in these two frames.

The transformation of (x,y,z,t) from one frame to another frame should be such that, the above two equations are simultaneously satisfied. For any other general event $x^2 + y^2 + z^2 - c^2t^2$ need not be zero. However, from eqs. (1) and (2), we get a clue that if the transformation is such that

$$x^2 + y^2 + z^2 - c^2t^2 = x'^2 + y'^2 + z'^2 - c^2t'^2, \tag{3}$$

then eqs. (1) and (2) will automatically be satisfied for the special event of light propagation and triggering of photocell. We further assume that transformation is linear, i.e., (x, y, z, t) depend linearly on (x,y,z,t). It is equivalent to saying that space-time is homogenous and isotropic. Demanding that the motion, parallel to z in the moving frame, must remain so in the lab frame, one gets

$$x' = x, \quad y' = y. \tag{4}$$

We thus write

$$z' = a_1 z + a_2 t, \tag{5}$$
$$t' = b_1 z + b_2 t, \tag{6}$$

where a_1, a_2, b_1 and b_2 are functions of v. As $v \to 0$, one must have $a_1 = 1,\ a_2 = 0,\ b_1 = 0,\ b_2 = 1$. Eqs. (5) and (6) are valid for all z, t. For the motion of O (the origin of the moving frame) $z = vt$, $z' = 0$, and eq.(5) gives $a_2 = -a_1 v$. Using eqs. (5) and (6) in eq. (3) and equating the coefficients of different powers of z and t on both sides, one obtains

$$a_1 = \gamma, \quad a_2 = -v\gamma, \quad b_1 = -\gamma \frac{v}{c^2}, \quad b_2 = \gamma,$$

$$\text{where} \quad \gamma = 1 \bigg/ \sqrt{1 - \frac{v^2}{c^2}}.$$

The Lorentz transformations can thus be written as

$$x' = x, \; y' = y, \; z' = \gamma(z - vt), \; t' = \gamma\left(t - \frac{vz}{c^2}\right). \tag{7}$$

9.4 Space-Time Separation between Events

Let event 1 occur at (x, y, z) at time t and event 2 occur at (x + dx, y + dy, z + dz) at time t + dt. The same events, when viewed from the moving frame, would appear at (x′, y′, z′, t′) and (x′ + dx′, y′ + dy′, z′ + dz′, t′ + dt′) respectively. Using Lorentz transformations (7), one can express primed quantities in terms of unprimed ones and obtains that

$$\begin{aligned}(ds)^2 &= (dx)^2 + (dy)^2 + (dz)^2 - c^2(dt)^2 \\ &= (dx')^2 + (dy')^2 + (dz')^2 - c^2(dt')^2. \end{aligned} \tag{8}$$

ds may be viewed as the space-time separation between the two events, which remains the same in all reference frames. One may visualise a four-dimensional space with four coordinate axes as $x_1 = x, x_2 = y, x_3 = z$ and $x_4 = ict$. An event in this space is represented by a point. From eq.(3).

$$x_1^2 + x_2^2 + x_3^2 + x_4^2 = x_1'^2 + x_2'^2 + x_3'^2 + x_4'^2$$

i.e., the distance of a point in 4-D space from the origin is invariant. Eq. (8) says that the length of any line in 4-D space is invariant under Lorentz transformation. Thus, mathematically, one may view Lorentz transformation like a rotation of coordinate axes in 4-D space.

The Lorentz transformation equations can be written as

$$x'_{\mu} = a_{\mu\nu}\, x_{\nu}, \tag{9}$$

where μ and v take values from 1 to 4, summation over repeated index (v in this case) is implied and $a_{\mu v}$ is the transformation matrix,

$$a_{\mu\nu} = \begin{vmatrix} 1 & 0 & 0 & 0 \\ 0 & 1 & 0 & 0 \\ 0 & 0 & \gamma & i\gamma\beta \\ 0 & 0 & -i\gamma\beta & \gamma \end{vmatrix}; \quad \beta = \frac{v}{c}. \tag{10}$$

Similarly, one may write

$$x_{\mu} = a'_{\mu\nu}\, x'_{\nu}$$

where $a'_{\mu\nu}$ can be deduced from $a_{\mu\nu}$ by replacing β by $-\beta$, i.e., $a'_{\mu\nu} = a_{\mu\nu}$. Thus, $x_{\mu} = x'_{\nu} a_{\nu\mu}$.

Using eq. (9), we get

$$x_{\mu} = a_{\nu\mu} a_{\nu l} x_{l}. \tag{11}$$

Since the RHS must equal x_{μ}, we have

$$a_{\nu\mu} a_{\nu l} = \delta_{\mu l}, \tag{12}$$

where $\delta_{\mu l}$ is a Kronecker delta whose value is 1 when $l = \mu$, otherwise it is zero.

9.5 Proper Time

Consider the motion of a particle, with respect to the laboratory frame. At any instant t, let the particle's velocity be $\mathbf{v} = \mathbf{dr}/\mathbf{dt}$. The time taken by the particle in going from point $P(\mathbf{r})$ to $Q\ (\mathbf{r} + \mathbf{dr})$ is $dt = |\mathbf{dr}|/v$. If one views this interval in a frame moving with velocity **v**, it would be $d\tau$, which from eq. (8) (on taking $dx' = 0,\ dy' = 0,\ dz' = 0$) can be obtained as

$$(d\tau)^2 = (dt)^2 - [(dx)^2 + (dy)^2 + (dz)^2]/c^2$$
$$= (dt)^2\left[1 - \frac{v^2}{c^2}\right].$$

We call this time $d\tau$ as *proper time*

$$d\tau = \frac{dt}{\gamma}, \tag{13}$$

where $\gamma = (1 - v^2/c^2)^{-1/2}$ is called the *Lorentz gamma factor.*

9.6 Four-Vectors

In section 9.4, we have defined a space time 4-vector $x_\mu = (\mathbf{r}, ict)$. One may construct other combinations of a vector and a scalar to have four-vectors. A four-vector A_μ must transform like x_μ, i.e.,

$$A'_\mu = a_{\mu\nu}\, A_\nu.$$

Suppose, we have a scalar ψ (x, y, z, t) that is Lorentz invariant, then

$$\frac{\partial\psi}{\partial x'_\mu} = \frac{\partial\psi}{\partial x_\nu}\,\frac{\partial x_\nu}{\partial x'_\mu} = a_{\mu\nu}\frac{\partial\psi}{\partial x_\nu} \tag{14}$$

where we have used eq. (11), i.e., $x_\nu = x'_\mu a_{\mu\nu}$ and differentiated it. Thus, the differential four-vector operator $\partial/\partial x_\mu$ is a proper four-vector.

One may define four-divergence of a four-vector $\partial A_\mu/\partial x_\mu$, where A_μ is a proper four-vector. One may note that

$$\frac{\partial A'_\mu}{\partial x'_\mu} = a_{\mu\nu}\frac{\partial}{\partial x_\nu}(a_{\mu 1}A_1) = \delta_{\nu 1}\frac{\partial A_1}{\partial x_\nu} = \frac{\partial A_\nu}{\partial x_\nu} = \frac{\partial A_\mu}{\partial x} \tag{15}$$

i.e., $\partial A_\mu/\partial x_\mu$ is **Lorentz invariant.**

9.6.1 Four-dimensional Laplacian Operator

If we substitute $A_\mu = \frac{\partial}{\partial x_\mu}\psi$ in eq. (15), we get

$$\Box'^2\psi = \sum_{\mu=1}^{4}\frac{\partial^2\psi}{\partial x'^2_\mu} = \sum_{\mu=1}^{4}\frac{\partial^2\psi}{\partial x^2_\mu} = \Box^2\psi. \qquad (16)$$

$\Box^2$ is thus Lorentz invariant. When it operates on a four-vector A_μ, the resulting quantity transforms like A_μ.

One may prove that the scalar product of two four-vectors is Lorentz invariant

$$A'_\mu\, B'_\mu = A_\mu\, B_\mu. \qquad (17)$$

Four-vectors are tensors of rank one in a four-dimensional space. One may have a second rank tensor $T_{\mu\nu}$, which will tansform as

$$T'_{\mu\nu} = \sum_{1,m} a_{\mu 1}\, a_{\nu\, m}\, T_{1m}. \qquad (18)$$

9.6.2 Momentum Four Vector

From space-time four-vector, x_μ, one may define a velocity four-vector by differentiating it with respect to proper time.

$$\begin{aligned} v_\mu = \frac{\partial x_\mu}{\partial\tau} &= \left(\frac{\partial\,\mathbf{r}}{\partial\tau}\cdot\frac{\partial t}{\partial\tau},\quad ic\frac{\partial t}{\partial\tau}\right) \\ &= (\gamma\,\mathbf{v}, ic\,\gamma), \end{aligned} \qquad (19)$$

where $\mathbf{v}$ is the actual velocity of the body in the laboratory frame and $\gamma = 1/\sqrt{1-v^2/c^2}$. v_μ transforms as

$$v'_\mu = a_{\mu\nu}\, v_\nu.$$

Multiplying v_μ by the rest mass m_0 of the body, one defines the momentum four-vector

$$p_\mu = (m_0\gamma\,\mathbf{v}, ic\, m_0\,\gamma). \qquad (20)$$

Now, let us examine Newton's second law of motion.

$$\frac{d\mathbf{p}}{dt} = \mathbf{F}. \tag{21}$$

Newton assumed **p** – the momentum of the body, as the product of its mass and velocity and assumed its mass to be constant. Einstein viewed the ($m_0 \gamma \mathbf{v}$) part of the momentum four-vector as momentum

$$\mathbf{p} = m_0\gamma\, \mathbf{v}. \tag{22}$$

In order to visualise the meaning of p_4 – the fourth component of p_μ, consider the work done **F.v** by a force **F** on the body per second. This must equal the rate of change of energy, ε, of the body as

$$\frac{d\varepsilon}{dt} = \mathbf{F.v} = \frac{d\mathbf{p}}{dt}.\mathbf{v}. \tag{23}$$

Multiplying both sides of eq. (23) by $m_0\gamma$,

$$m_0\gamma\frac{d\varepsilon}{dt} = \mathbf{p}.\frac{d\mathbf{p}}{dt} = \frac{d}{dt}\left(\frac{p^2}{2}\right) = \frac{d}{dt}\left(\frac{m_0^2}{2}\gamma^2 v^2\right) = \frac{d}{dt}\left[\frac{m_0^2}{2}c^2(\gamma^2 - 1)\right]$$

$$= m_0^2 c^2 \gamma \frac{d\gamma}{dt}$$

or

$$\frac{d\varepsilon}{dt} = \frac{d}{dt}(m_0 c^2 \gamma)$$

or

$$\varepsilon = m_0\gamma\, c^2. \tag{24}$$

For $v \to 0$, $\gamma = 1$ and $\varepsilon = m_0\, c^2$ is the **rest mass energy**. The kinetic energy of the body is

$$\text{K.E.} = (\gamma - 1)m_0 c^2. \tag{25}$$

For $v/c \ll 1$, this expression reduces to K.E. $= \frac{m_0 v^2}{2}$. At higher velocity, kinetic energy rises more rapidly with velocity and goes to ∞ as $v/c \to 1$. The ratio of momentum to velocity is known as the *relativistic mass* of the body

$$m_R = m_0\gamma. \tag{26}$$

Now onwards, for the sake of brevity, we will drop the subscript zero from the rest mass and denote it by m. Thus

$$\mathbf{p} = m\gamma\,\mathrm{v},\ \varepsilon = m\gamma\,c^2,\ \gamma = \left(1 - \frac{v^2}{c^2}\right)^{1/2} = \left(1 + \frac{p^2}{m^2}c^2\right)^{1/2}. \quad (27)$$

For a charged particle of rest mass m and charge q, moving in electric field **E** and magnetic field **B**, the momentum and energy equations can be written as

$$\frac{d\mathbf{p}}{dt} = q\mathbf{E} + q\mathrm{v} \times \mathbf{B}, \quad (28)$$

$$mc^2\frac{d\gamma}{dt} = q\,\mathbf{E}.\,\mathrm{v}. \quad (29)$$

PROBLEM 9.1. Deduce the Lorentz transformation of velocity.

Solution: Consider a frame moving with velocity $v_f\hat{z}$ w.r.t. laboratory frame. The components of momentum four-vector transform to the moving frame as

$$p_x' = p_x,\ p_y' = p_y,$$
$$p_z' = \gamma_f(p_z - m\gamma\,v_f),$$
$$\gamma' = \gamma_f\gamma(1 - v_f\frac{v_z}{c^2}),$$

where $\gamma_f = 1\sqrt{(1 - v_f^2/c^2)}$ and m is the rest mass of the body. Expressing $\mathbf{p} = mv\gamma$, $\mathbf{p}' = mv'\gamma'$, these equations yield the requisite velocity transformations,

$$v_x' = \frac{v_x/\gamma_f}{1 - v_z v_f/c^2},\ v_y' = \frac{v_y/\gamma_f}{1 - v_z v_f/c^2},\ v_z' = \frac{v_z - v_f}{1 - v_z v_f/c^2}.$$

PROBLEM 9.2. An electron, initially at z = 0 and moving with initial velocity $(c/2)\hat{z}$, is subjected to a dc electric field $\mathbf{E} = -E_0\hat{z}$. Obtain electron energy and velocity as a function of z.

Solution: The energy equation for the electron of rest mass m and charge $-e$, on writing $d/dt = v_z d/dz$, can be written as

$$\frac{d\gamma}{dz} = \frac{eE_0}{mc^2}.$$

It integrates to give

$$v = \gamma\, mc^2 = \gamma_0\, mc^2 + eE_0 z,$$

$$v_z = \sqrt{\left(1 - \frac{1}{\gamma^2}\right)} = c\sqrt{1 - \frac{1}{\left(r_0 + \frac{E_0 z}{mc^2}\right)^2}},$$

where $\gamma_0 = 2/\sqrt{3}$.

PROBLEM 9.3. In free space an electron, initially at rest at $z = 0$, is subjected to an intense laser field $\mathbf{E} = \hat{x}A\cos(\omega t - \omega z/c)$, $\mathbf{B} = \hat{y}(A/c)\cos(\omega t - \omega z/c)$. Show that $\gamma - p_z/mc$ is a constant of motion, hence obtain p_x, p_z and γ.

Solution: The equations governing γ, p_x, p_y and p_z are

$$mc^2\frac{d\gamma}{dt} = -ev_x A\cos\left(\omega t - \frac{\omega z}{c}\right),$$

$$\frac{dp_x}{dt} = -eE_x + ev_z B_y = -e\left(1 - \frac{v_z}{c}\right) A\cos\left(\omega t - \frac{\omega z}{c}\right),$$

$$\frac{dp_y}{dt} = 0,$$

$$\frac{dp_z}{dt} = -e\, v_x\, B_y = -ev_x\left(\frac{A}{c}\right)\cos\left(\omega t - \frac{\omega z}{c}\right).$$

From the first and last equations

$$mc^2\frac{d}{dt}\left(\gamma - \frac{p_z}{mc}\right) = 0,$$

i.e., $\gamma - p_z/mc = \text{constant} = 1$.

The p_x equation gives

$$p_x = -\left(\frac{eA}{\omega}\right)\sin\left(\omega t - \frac{\omega z}{c}\right).$$

The p_y equation gives $p_y = 0$. Using these in $\gamma^2 = 1 + p^2/m^2c^2$, we obtain

$$\gamma = 1 + \left(\frac{a^2}{4}\right)\left(1 - \cos 2\left(\omega t - \frac{\omega z}{c}\right)\right)$$

$$p_z = \left(\frac{a^2}{4}\right) mc\left(1 - \cos 2\left(\omega t - \frac{\omega z}{c}\right)\right),$$

where $a = eA / m\omega c$. One may note that p_z oscillates at frequency 2ω. For $a > 2$, p_z has larger values than p_x. The electron velocity **v** has components at harmonics of laser frequency, producing harmonic current and harmonic radiation.

9.7 Four-Vector Electromagnetic Potential

The Maxwell's equations and the equation of continuity can be written as

$$\nabla.\mathbf{E} = \frac{\rho}{\epsilon_0},$$

$$\nabla.\mathbf{B} = 0,$$

$$\nabla \times \mathbf{E} = -\frac{\partial \mathbf{B}}{\partial t},$$

$$\nabla \times \mathbf{H} = \mathbf{J} + \epsilon_0 \frac{\partial \mathbf{E}}{\partial t}, \tag{30}$$

$$\frac{\partial \rho}{\partial t} + \nabla.\mathbf{J} = 0 \tag{31}$$

where we have assumed $\mathbf{D} = \epsilon_0 \mathbf{E}$ and $\mathbf{B} = \mu_0\mathbf{H}$. The second and third Maxwell's equations suggest that we may write (unit 4)

$$\mathbf{B} = \nabla \times \mathbf{A} \tag{32}$$

$$\mathbf{E} = -\nabla\phi - \frac{\partial \mathbf{A}}{\partial t}. \tag{33}$$

Using eqs. (32) and (33) in the first and fourth Maxwell's equations, we obtain

$$\nabla^2\phi - \frac{1}{c^2}\frac{\partial^2}{\partial t^2}\phi = -\frac{\rho}{\epsilon_0}, \tag{34}$$

$$\nabla^2\mathbf{A} - \frac{1}{c^2}\frac{\partial^2}{\partial t^2}\mathbf{A} = -\mu_0\mathbf{J} \tag{35}$$

under the Lorentz gauge,

$$\frac{1}{c^2}\frac{\partial\phi}{\partial t} + \nabla.\mathbf{A} = 0. \tag{36}$$

One may note from eq. (31), that it can be written as a four-divergence if one defines a four-vector current density as $J_\mu = (\mathbf{J}, ic\rho)$,

$$\frac{\partial}{\partial x_\mu} J_\mu = 0. \tag{37}$$

The Lorentz transformation of $\mathbf{J} = nq\mathbf{v}$ and $\rho = nq$, where n is the number density and q is the charge of each particle, can be alternately deduced from the Lorentz transformation of **r**, t and **v** assuming **q** to be constant (invariant). Due to Lorentz contraction, length $L = L_0/\gamma$. Hence $n = \gamma\ n_0$ (where n_0 is the density w.r.t. a frame where average drift of the particles is zero). Thus, $(\mathbf{J}, ic\rho)$ resemble p_μ with m_0 replaced by n_0q.

Similarly, eq. (36) can be written as a four-divergence of four-vector potential $A_\mu = \left(\mathbf{A}, \frac{i}{c}\phi\right)$,

$$\frac{\partial A_\mu}{\partial x_\mu} = 0. \tag{38}$$

Since four-vector Laplacian operator $\Box^2$ is Lorentz invariant, $\Box^2 A_\mu$ would transform as A_μ. Thus eqs. (34) and (35) can be written as a four-vector equation,

$$\Box^2 A_\mu = -\mu_0 J_\mu. \tag{39}$$

9.8 Covariance of Electrodynamics

We have already demonstrated the relativistic covariance of the equation of continuity and the equations governing **A**, ϕ. Now we would like to put Maxwell's equations in the covariant form, following the Feynman's approach[1].

Let us write the x component of eq. (32).

$$B_x = \frac{\partial A_z}{\partial y} - \frac{\partial A_y}{\partial z}. \tag{40}$$

We may denote the RHS as F_{yz}. Similarly, B_y and B_z may be written as

$$B_y = F_{zx} \equiv \frac{\partial A_x}{\partial z} - \frac{\partial A_z}{\partial x}, \quad B_z = F_{xy} \equiv \frac{\partial A_y}{\partial x} - \frac{\partial A_x}{\partial y}. \tag{41}$$

We may visualise these Fs as the components of a tensor $F_{\mu\nu}$,

$$F_{\mu\nu} = \frac{\partial A_\nu}{\partial x_\mu} - \frac{\partial A_\mu}{\partial x_\nu}. \tag{42}$$

The F_{14}, F_{24}, F_{34} components of $F_{\mu v}$ (realising that $A_4 = i\phi\ /\ c$, $x_4 = ict$) give the electric field (eq. 33).

$$\text{e.g.: } F_{14} = \frac{\partial A_4}{\partial x_1} - \frac{\partial A_1}{\partial x_4} = \frac{i}{c}\frac{\partial \phi}{\partial x} + \frac{i}{c}\frac{\partial A_x}{\partial t} = -\frac{iE_x}{c}. \tag{43}$$

Similarly, one may see that the diagonal terms, $F_{\mu\nu}$ are zero and $F_{\mu\nu} = -F_{\nu\mu}$.

$F_{\nu\mu}$ is the difference of two terms $\partial A_\nu/\partial x_\mu$ and $\partial A_\mu/\partial x_\nu$, each of which transforms like a second rank four-tensor. Hence $F_{\mu\nu}$ is a field four-tensor of second rank. Its components are:

$$F_{\mu\nu} = \begin{vmatrix} 0 & B_z & -B_y & -\frac{i}{c}E_x \\ -B_z & 0 & B_x & -\frac{i}{c}E_y \\ B_y & -B_x & 0 & -\frac{i}{c}E_z \\ \frac{i}{c}E_x & \frac{i}{c}E_y & \frac{i}{c}E_z & 0 \end{vmatrix}. \tag{44}$$

One may note that the first and fourth Maxwell's equations can be written as

$$\frac{\partial}{\partial x_\mu} F_{\mu\nu} = -\mu_0 J_\nu, \tag{45}$$

where summation over μ is implied. For $\nu = 4$ eq. (45) gives the first Maxwell's equation whereas for $\nu = 1, 2, 3$, it gives the (x, y, z) components of the fourth Maxwell's equation.

The remaining two Maxwell's equations, viz., $\nabla \cdot \mathbf{B} = 0$ and $\nabla \times \mathbf{E} = -\partial \mathbf{B}/\partial t$ can be written as the four equations.

$$\frac{\partial}{\partial x_\mu} F_{\nu\lambda} + \frac{\partial}{\partial x_\nu} F_{\lambda\mu} + \frac{\partial}{\partial x_\lambda} F_{\mu\nu} = 0, \tag{46}$$

where λ, μ, ν are any three of the four integers 1,2,3,4. Each term in eq. (46) transforms like a four-tensor of third rank. Hence, eq. (46) is covariant.

9.9 Lorentz Transformation of Fields

The electric and magnetic fields are the components of the field strength four-tensor $F_{\mu\nu}$, which transforms as

$$F'_{\mu\nu} = a_{\mu 1}\, a_{\nu m}\, F_{1m}. \tag{47}$$

On carrying out this transformation, the **E** and **B** fields in the moving frame can be written as

$$\begin{aligned} \mathbf{E}'_\perp &= \gamma(\mathbf{E}_\perp + \mathbf{v} \times \mathbf{B}), \\ \mathbf{B}'_\perp &= \gamma(\mathbf{B}_\perp - \mathbf{v} \times \mathbf{E}/c^2), \\ \mathbf{E}'_\parallel &= E_\parallel,\ B'_\parallel = B_\parallel, \end{aligned} \tag{48}$$

where $\parallel$ and $\perp$ refer to field components parallel and perpendicular to **v**, the velocity of the moving frame respectively.

The phase of a wave $\psi = \omega t - \mathbf{k.r}$ is Lorentz invariant, as this signifies the number of counts of oscillations. $-\partial\psi/\partial x_\mu$ thus is a four-vector $(\mathbf{k}, i\omega/c)$. It transforms as

$$\begin{aligned} \mathbf{k}'_\perp &= \mathbf{k}_\perp, \\ k'_z &= \gamma(k_z - \omega v/c^2), \\ \omega' &= \gamma(\omega - k_z v). \end{aligned} \tag{49}$$

PROBLEM 9.4. A plane electromagnetic wave propagating through a plasma has $\mathbf{E} = \hat{y}\, A\cos(\omega t - \omega x/2c - \omega z/2c)$. What would be its time average Poynting vector, with respect to (i) laboratory observer and (ii) an observer moving with velocity $\mathbf{v}_f = (c/2)\hat{z}$?

Solution: In complex notation

$$\mathbf{E} = \hat{y}\, A e^{-i(\omega t - \mathbf{k}.\mathbf{r})}\, ;\, \mathbf{k} = \frac{\omega}{2c}(\hat{x} + \hat{z}).$$

Using $\nabla \times \mathbf{E} = -\partial \mathbf{B}/\partial\, t$, one may write

$$\mathbf{B} = \mathbf{k} \times \mathbf{E}/\omega.$$

(i) Time average Poynting vector in the laboratory frame is

$$\mathbf{S}_{av} = \frac{1}{2}\mathrm{Re}[\mathbf{E} \times \mathbf{H}^*] = \frac{A^2}{2\omega\mu_0}\mathbf{k}.$$

(ii) In the moving frame

$$\mathbf{E}'_\perp = \gamma_f(\mathbf{E}_\perp + \mathbf{v}_f \times \mathbf{B}_\perp) = \gamma_f(1 - \frac{\mathbf{k}.\mathbf{v}_f}{\omega})\mathbf{E}_\perp$$

$$= \frac{3}{4} y \gamma_f A\, e^{-i(\omega' t' - \mathbf{k}'.\mathbf{r}')}$$

$$\text{Here,} \quad \gamma_f = \frac{1}{\sqrt{1 - \frac{v_f^2}{c^2}}} = \frac{2}{\sqrt{3}}$$

$$\omega' = \gamma_f(\omega - \mathbf{k}.\mathbf{v}_f) = \frac{3}{4}\gamma_f \omega,$$

$$\mathbf{k}' = k_x \hat{x} + \hat{z}\gamma_f(k_z - \omega v_f/c^2) = \frac{\omega}{2c}\hat{x},$$

$$\mathbf{B}' = \mathbf{k}' \times \mathbf{E}'/\omega',$$

$$\mathbf{S}'_{av} = \frac{1}{2}\mathrm{Re}[\mathbf{E}' \times \mathbf{H}'] = \frac{3A^2/4}{2\omega'\mu_0}\mathbf{k}' = \frac{\sqrt{3}A^2}{8\mu_0 c}\hat{x}.$$

One may note that the time average Poynting vector changes in magnitude as well as direction, as one goes from one frame to another.

9.10 Conclusions

The velocity of light in vacuum is the same in all inertial frames of reference. This is an overpowering fundamental law of nature. It revolutionised the concepts of space and time and modified the equations of motion. The Maxwell's equations, however, retained their form in all inertial frames. In fact, it is the constancy of ϵ_0 and μ_0, and hence of c, with respect to all inertial frames that prompted the discovery of *special theory of relativity*. The Lorentz transformations of space and time drove the search for proper four-vectors, leading to fascinating new discoveries like energy mass relation. The electromagnetic potential four-vector and current density four-vector put the Maxwell's equations in covariant form and led to Lorentz transformation of electric and magnetic fields, thereby firmly establishing the 'interdigited' nature of **E** and **B**. A pure electric field in one frame, when viewed by a moving observer in general, has a magnetic field component too and vice versa. The wave vector and frequency also constitute a proper four-vector, with phase of the wave being Lorentz invariant. All four-vectors Lorentz transform in the same way, by the same 4×4 transformation tensor.

The relativity led to intensive explorations in the physical world from the nuclear scale to the dimensions of the universe. Many of the relativistic phenomena of gigantic dimensions are now observable in the laboratory by employing intense short-pulse laser that can impart oscillatory velocity to electrons close to c and accelerate them to GeV energies within a distance of a centimeter. The concept of relativistic Doppler shift has led to the development of free electron laser, a tunable device that produces coherent radiation from submillimetre wavelengths to X rays, by making a relativistic electron beam traverse through *wiggler magnetic field* or colliding it with a counter propagating electromagnetic wave.

UNSOLVED PROBLEMS

1. A particle is executing simple harmonic motion with $z = a \sin \omega t$. Write down its motion, with respect to an observer moving with velocity $v_0\hat{z}$.

$$\left[\textbf{ANS.}\ z' = -v_0 t' + (a/\gamma)\sin \omega'(t' + v_0 z'/c^2), \omega' = \gamma\omega, \gamma = (1 - v_0^2/c^2)^{-1/2}.\right]$$

2. The field of an electrostatic wave in laboratory frame is $\mathbf{E} = \hat{z} A \cos(\omega t - 2\omega z/c)$. To an observer moving with velocity $v_0\hat{z}$, the wave appears as a dc field. Estimate v_0. What would be its wavelength in the moving frame?

$$\left[\textbf{ANS.}\ v_0 = c/2, \lambda' = 2\pi c/\omega\sqrt{3}.\right]$$

3. In a plasma, an electromagnetic wave has $\mathbf{E} = \hat{x} A \exp[-i(\omega t - kz)]$ with $k = (\omega/c)\,(1 - \omega_p^2/\omega^2)^{1/2}$. In what frame would its wave number vanish? What would be the frequency of the wave in this frame?

$$\left[\textbf{ANS.}\ \mathbf{v_0} = \hat{z}c(1 - \omega_p^2/\omega^2)^{1/2}, \omega' = \omega_p.\right]$$

4. The field of a standing wave is $\mathbf{E} = \hat{x} A \cos kz \cos \omega t$. Express it with respect to a frame moving with velocity $v_0\hat{z}$.

$$\left[\textbf{ANS.}\ \mathbf{E}' = \hat{x}(\gamma A/2)[(1 - kv_0/\omega)\cos(\omega' t' - k' z') + (1 + kv_0/\omega)\cos(\omega'' t' + k'' z'],\right.$$
$$\omega' = \gamma(\omega - kv_0), \omega'' = \gamma(\omega + kv_0),$$
$$k' = \gamma(k - \omega v_0/c^2), k'' = \gamma(k + \omega v_0/c^2),$$
$$\left.\gamma = 1/\sqrt{1 - v_0^2/c^2}.\right]$$

5. Estimate the velocity of an electron of rest mass energy $mc^2 = 0.5\,\text{MeV}$, accelerated through a potential difference of 1 MV.

$$\left[\textbf{ANS.}\ v = 2\sqrt{2} \times 10^8\ \text{m/s}.\right]$$

6. An electromagnetic pulse of $\lambda = 1\,\text{mm}$ is incident normally on an object counter moving with velocity $v_f = -0.99c\hat{z}$. Obtain the wavelength of the reflected signal.

[**Ans.** $\lambda_R = \frac{\lambda(1-v_f/c)}{1+v_f/c} = 5\mu\text{m}$.]

7. The ionisation potential of a static atom is 4 eV. The atom is given a velocity $v\,\hat{z}$. A laser (with photon energy in the laboratory frame = 2 eV) impinges on it from the opposite direction. Estimate v, so that the atom could be ionised by the laser through single photon absorption.

[**Ans.** $v = 0.6c$.]

8. A laser with $\mathbf{E} = \hat{y} A \cos(\omega t - \omega z \cos\theta/c - \omega x \sin\theta/c)$ is obliquely incident on a mirror (with $x - y$ plane as the reflecting plane), moving with velocity $v\,\hat{z}$. Obtain the frequency of the reflected light.

[**Ans.** $\omega_R = \gamma^2 \omega[1 - (2v/c)\cos\theta + v^2/c^2]$,
$\gamma = (1 - v^2/c^2)^{-1/2}$.]

9. In the presence of a large amplitude plasma wave $\mathbf{E} = \hat{z} A \cos\psi$, where $\psi = (\omega_p t - \omega_p \eta z/c)$, show that the energy equation for an electron of rest mass m and charge –e having ($v_x = v_y = 0$), takes the form

$$\left(1 - \frac{\gamma/\eta}{(\gamma^2 - 1)^{1/2}}\right) \frac{d\gamma}{d\psi} = (eA/mc\omega_p \eta) \cos\psi .$$

Solve it to obtain the phase space trajectory of the electron. This equation gives trapped electrons for certain initial values of γ, ψ.

[**Hint:** *In the energy equation* $mc^2 d\gamma/dt = -ev_z A \cos\psi$, *replace* d/dt *by* $(d\psi/dt)d/d\psi$.]

10. An electron beam of density n_0 and velocity $0.8\,c\hat{z}$ enters an acceleration region, after which it emerges with velocity $0.99\,c\hat{z}$. Estimate the density of the emergent beam with respect to (i) a ground observer and (ii) an observer moving with the emergent beam. If the acceleration is caused by a uniform dc electric field over a length of 1 metre, estimate the value of the electric field.

[**ANS.** (i) $8n_0/9.9$, (ii) $0.11n_0, E_{dc} = 2.7\,MV/m$.]

11. An electromagnetic wave $\mathbf{E} = \hat{x}A\exp[-i(\omega t - \omega z/c)]$ is incident normally on a medium, moving with velocity $\mathbf{v}_f=(c/2)\hat{z}$. The medium, in a frame where it is at rest, has refractive index $\eta = 1.5$. Obtain the transmitted field and its frequency with respect to laboratory frame.

[**HINT:** *Lorentz transform the incident wave field to the moving frame, obtain the transmitted field in the moving frame and then Lorentz transform the transmitted field to the lab frame.*]

[**ANS.** $\mathbf{E_T} = \hat{x}A_T e^{-i(\omega_T t - k_T z)}$,
$A_T = 2A\gamma_f^2(1+\eta v_f/c)(1-v_f/c)/(1+\eta) = 14A/15$,
$\omega_T = \gamma_f^2(1 - v_f/c)(1 + \eta v_f/c)\omega = 7\omega/6$,
$k_T = \gamma_f^2(1 - v_f/c)(\eta + v_f/c)\omega/c = 4\omega/3c$.]

REFERENCES AND SUGGESTED READING

[1] Feynman R. P., Leighton R. B., and Sands M., *The Feynman lectures on Physics*, vols.1 and 2, New Delhi, Narosa Pub. House, (2003)

[2] Jackson J. D., *Classical Electrodynamics*, NewYork, John Wiley & Sons, Inc., (2005).

[3] Liu C. S. and Tripathi V. K., *The interaction of electromagnetic waves with electron beams and plasmas*, Singapore, World Scientific, (1994).

Unit 10

Radiation from Accelerated Charges

Introduction

A charge, moving with constant velocity in free space, produces electric and magnetic fields. However, to an observer moving with the velocity of the charge, the charge appears static and produces only a static Coulomb field. It does not constitute power flow, i.e., there is no radiation by the charge. Thus, the charges moving with constant velocity in free space do not radiate electromagnetic energy. However, the accelerated charges do radiate. One may view a moving charge as a current source or a small antenna and obtain the vector and scalar potentials and the electric and magnetic fields due to it. In this chapter, we obtain these potentials, called *Lienard-Wiechert* potentials, and deduce the fields and power radiated from accelerated charges.

10.1 Lienard-Wiechert Potentials

Consider a charged particle of charge q and velocity $\mathbf{v_p}(t')$, moving on a trajectory $\mathbf{r_p}(t')$ at time t' (Fig. 10.1). It may be represented by charge and current densities

$$\begin{aligned}\rho(\mathbf{r}',t') &= q\delta(\mathbf{r}' - \mathbf{r_p}(t'))\\ \mathbf{J}(\mathbf{r}',t') &= q\delta(\mathbf{r}' - \mathbf{r_p}(t'))\mathbf{v_p}(t'),\end{aligned} \tag{1}$$

where $\delta(\mathbf{r}' - \mathbf{r_p}(t'))$ denotes the three-dimensional delta function.

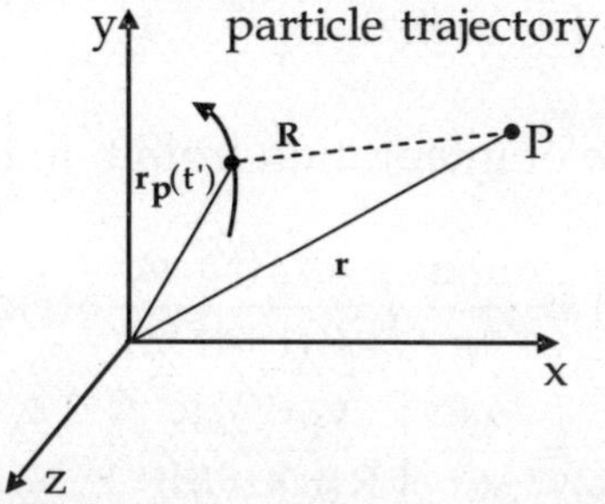

Figure 10.1: The trajectory of a charged particle $\mathbf{r_p}(t')$ and the point of observation P(**r**) at which field due to the particle is observed at time t.

The four-vector current density can be written as

$$J_\mu = \mathbf{J}, i c \rho = (\mathbf{v_p}, i c)\, q\, \delta\, (\mathbf{r}' - \mathbf{r_p}(t')). \tag{2}$$

The vector potential due to a current source is written as

$$\begin{aligned} \mathbf{A}(\mathbf{r}, t) &= \frac{\mu_0}{4\pi} \int \frac{\mathbf{J}(\mathbf{r}', t - R/c)}{R} dV' \\ &= \frac{\mu_0}{4\pi} \int \int \frac{1}{R} \mathbf{J}(\mathbf{r}', t') \delta(t' - (t - R/c)) dV' dt', \end{aligned} \tag{3}$$

where $\mathbf{R} = \mathbf{r} - \mathbf{r}'$ and delta function on t′ ensures that the vector potential at **r**, at time t, depends on the current density at **r**′ at an earlier time $t' = t - R / c$. Similarly, one may write the scalar potential

$$\phi(\mathbf{r}, t) = \frac{1}{4\pi \in_0} \int \int \frac{1}{R} \rho(\mathbf{r}', t') \delta(t' - t + R/c) dV' dt'. \tag{4}$$

Eqs. (3) and (4) can be written as

$$A_\mu = \frac{\mu_0}{4\pi} \int \int \frac{1}{R} J_\mu(\mathbf{r}', t') \delta(t' - t + R/c) dV' dt',$$

where $A_\mu = \mathbf{A}, i\phi/c$ is the four-vector potential. Using eq. (1) and carrying out the volume integral, we get

$$A_\mu = \frac{\mu_0 q}{4\pi} \int \frac{1}{R} (\mathbf{v_p}, i c) \delta(t' - t + R/c) dt', \tag{5}$$

where $R = |\mathbf{r} - \mathbf{r_p}(t')|$. We define

$$f(t)' = t' + \frac{R(t')}{c} - t$$

and change the variable of integration from t′ to f.

$$\begin{aligned} A_\mu(\mathbf{r}, t) &= \frac{\mu_0 q}{4\pi} \int \frac{(\mathbf{v_p}(t'), ic)}{R(t')df/dt'} \delta(f)df \\ &= \frac{\mu_0 q}{4\pi} \left[\frac{\mathbf{v_p}(t'), ic}{R(t')df/dt'} \right]_{ret,} \end{aligned} \tag{6}$$

where ret signifies that the quantity within the bracket has to be evaluated at retarded time t′ = t – R / c. One may express

$$\frac{df}{dt'} = 1 + \frac{1}{c}\frac{dR(t')}{dt'} = 1 - \hat{n}.\boldsymbol{\beta},$$

$$\text{where } \hat{n} = \frac{\mathbf{r} - \mathbf{r_p}(t')}{|\mathbf{r} - \mathbf{r_p}(t')|}, \ \boldsymbol{\beta} = \mathbf{v_p}(t')/c. \tag{7}$$

$\hat{n}$ is the unit vector directed along the point of observation **r** from **r′**– the point from where the signal is released at t′ that arrives at **r** at t. From eq.(6), the scalar and vector potentials, known as **Lienard-Wiechert potentials,** may explicitly be written as

$$\phi(\mathbf{r}, t) = \frac{q}{4\pi \epsilon_0} \left[\frac{1}{KR} \right]_{ret}, \tag{8}$$

$$\mathbf{A}(\mathbf{r}, t) = \frac{\mu_0 c q}{4\pi} \left[\frac{\beta}{KR} \right]_{ret},$$

where $K = 1 - \hat{n}.\beta$.

10.2 Electric and Magnetic Fields

Using the Lienand-Wiechert potentials, one may obtain the electric and magnetic fields. However, we employ the integral form of potential given by eq. (5). Hence the electric field may be written as

$$\mathbf{E}(\mathbf{r},t) = -\nabla\phi - \frac{\partial \mathbf{A}}{\partial t} \tag{9}$$
$$= -\frac{q}{4\pi \in_0}\int \nabla\left(\frac{1}{R}\delta(t'-t+R/c)\right)dt'$$
$$-\frac{\mu_0 q}{4\pi}\int \frac{\mathbf{v_p}(t')}{R}\frac{\partial}{\partial t}\delta(t'-t+R/c)dt'$$
$$= -\frac{q}{4\pi \in_0}\int \hat{n}\left(-\frac{1}{R^2}\delta(t'-t+R/c) + \frac{1}{Rc}\delta'(t'-t+R/c)\right)dt'$$
$$+\frac{\mu_0 q}{4\pi}\int \frac{\mathbf{v_p}(t')}{R}\delta'(t'-t+R/c)dt',$$

where prime over δ function denotes differentiation with respect to argument. Introducing the new variable $f = t' + R(t')/c - t$, and carrying out integration by parts of the terms with δ', we obtain

$$\mathbf{E}(\mathbf{r},t) = \frac{q}{4\pi \in_0 K}\left[\frac{\hat{n}}{R^2} + \frac{d}{dt'}\left(\frac{\hat{n}}{cRK}\right) - \frac{1}{c^2}\frac{d}{dt'}\left(\frac{\mathbf{v_p}(t')}{RK}\right)\right]_{ret}. \tag{10}$$

Using eq. (7),

$$\frac{dR}{dt'} = \frac{-(\mathbf{r}-\mathbf{r_p}(t')).\mathbf{v_p}}{R} = -\hat{n}.\boldsymbol{\beta}c,$$
$$\frac{d\hat{n}}{dt'} = \frac{-\mathbf{v_p}}{R} + \frac{\hat{n}\hat{n}.\mathbf{v_p}}{R},$$

hence eq. (10) can be written as

$$\mathbf{E}(\mathbf{r},t) = \frac{q}{4\pi \in_0}\left[\frac{\hat{n}-\boldsymbol{\beta}}{\gamma^2(1-\boldsymbol{\beta}.\hat{n})^3R^2}\right]_{ret} + \frac{q}{4\pi \in_0 c}\left[\frac{\hat{n}\times(\hat{n}-\hat{\beta})\times\dot{\boldsymbol{\beta}}}{(1-\boldsymbol{\beta}.\hat{n})^3R}\right]_{ret}. \tag{11}$$

where $\dot{\boldsymbol{\beta}} = \dot{\mathbf{v}}_p/c$ is the particle acceleration divided by c and $\gamma = 1\Big/\sqrt{(1-v_p^2/c^2)}$ is the gamma factor.

Following a similar procedure, the magnetic field due to an accelerated charge can be written as

$$\mathbf{B} = (\hat{n}\times\mathbf{E})_{ret}. \tag{12}$$

The $\frac{1}{R^2}$ terms in eqs. (11) and (12) are the induction terms, independent of particle acceleration. These terms do not contribute to radiation emission or energy loss by the particle. The $\frac{1}{R}$ terms are the radiation terms,

$$\mathbf{E}_{rad} = \frac{q}{4\pi \in_0 c}\left[\frac{\hat{n} \times (\hat{n} - \boldsymbol{\beta}) \times \dot{\boldsymbol{\beta}}}{(1 - \boldsymbol{\beta}.\hat{n})^3 R}\right]_{ret}, \tag{13}$$

$$\mathbf{B}_{rad} = \hat{n} \times \mathbf{E}_{rad}. \tag{14}$$

For a non-relativistic charged particle ($\beta << 1$ or $v_p << c$), these fields simplify to give

$$\mathbf{E}_{rad} = -\frac{q}{4\pi \in_0 c^2 R}(\dot{\mathbf{v}}_\mathbf{p} - \hat{n}.\dot{\mathbf{v}}_\mathbf{p}\hat{n}),$$

$$\mathbf{B}_{rad} = \hat{n} \times \mathbf{E}_{rad}/c. \tag{15}$$

10.3 Radiated Power

The Poynting flux is

$$\mathbf{S} = \mathbf{E}_{rad} \times \mathbf{H}_{rad} = \frac{1}{\mu_0 c}\left[\hat{n}\mathbf{E}_{rad}\cdot\mathbf{E}_{rad} - \mathbf{E}_{rad}\hat{n}.\mathbf{E}_{rad}\right].$$

For a non-relativistic particle

$$\mathbf{S} = \frac{\hat{n}}{\mu_0 c}\left(\frac{q}{4\pi \in_0 cR}\right)^2 |\hat{n} \times (\hat{n} \times \dot{\boldsymbol{\beta}})|^2$$

$$= \hat{n}\frac{q^2}{16\pi^2 R^2 \in_0 c^3}\dot{v}_p^2 \sin^2\theta, \tag{16}$$

where θ is the angle between $\hat{n}$ and electron acceleration $\dot{\mathbf{v}}$. The power radiated per unit solid angle is

$$\frac{dP}{d\Omega} = R^2 S = \frac{q^2 \dot{v}_p^2 \sin^2\theta}{16\pi^2 \in_0 c^3}. \tag{17}$$

It has the same θ dependence as that of a short dipole antenna. The total instaneous power, radiated by the particle, can be obtained by multiplying $dP/d\Omega$ with the solid angle element $\sin\theta d\theta d\phi$ and integrating over θ from 0 to π and over ϕ from 0 to 2π,

$$P = \frac{q^2 \left|\dot{\mathbf{v}}_{\mathbf{p}}\right|^2}{6\pi \in_0 c^3}. \tag{18}$$

This is **Larmor's formula** for a non-relativistic charged particle.

The radiated power can be shown to be Lorentz invariant and this fact may be employed to make an intuitive generalisation of Larmor formula for relativistic particle. We may write

$$\left|\dot{\mathbf{v}}_{\mathbf{p}}\right|^2 = \frac{1}{m^2}\frac{d\mathbf{p}}{dt}.\frac{d\mathbf{p}}{dt} \rightarrow \frac{1}{m^2}\frac{dp_\mu}{d\tau}\frac{dp_\mu}{d\tau}, \tag{19}$$

where m is the rest mass of the particle $d\tau = dt/\gamma, p_\mu = (\mathbf{p}, imc\gamma)$ is the momentum four-vector, $\mathbf{p} = m\gamma\mathbf{v}_{\mathbf{p}}$ and $\gamma = 1/\sqrt{(1 - v_p^2/c^2)}$. In the limit $v_p/c \rightarrow 0$, the last expression reduces to $\left|\dot{\mathbf{v}}_{\mathbf{p}}\right|^2$. Using eqs. (18)and (19), the power radiated by a relativistic particle can be written as

$$\begin{aligned} P &= \frac{q^2\ \gamma^2}{6\pi \in_0 c^3 m^2}\left[\frac{d\mathbf{p}}{dt}.\frac{d\mathbf{p}}{dt} - m^2c^2\left(\frac{d\gamma}{dt}\right)^2\right] \\ &= \frac{q^2\gamma^2}{6\pi \in_0 c^3}\left[\left(\mathbf{v}_{\mathbf{p}}\frac{d\gamma}{dt} + \gamma\frac{d\mathbf{v}_{\mathbf{p}}}{dt}\right)^2 - c^2\left(\frac{d\gamma}{dt}\right)^2\right] \\ &= \frac{q^2\gamma^6}{6\pi \in_0 c}\left[\left(\dot{\boldsymbol{\beta}}\right)^2 - (\boldsymbol{\beta} \times \dot{\boldsymbol{\beta}})^2\right]. \end{aligned} \tag{20}$$

The radiated power scales as the square of acceleration. Hence, for a given force, power decreases with particle's rest mass. It has a strong dependence on the relativistic gamma factor, i.e., energy of the particle.

PROBLEM 10.1. In a linear accelerator, an electron is accelerated along $\hat{z}$. Show that the power radiated by the electron is

$$P = \frac{e^2}{6\pi \epsilon_0 m^2 c^3}\left(\frac{d\varepsilon}{dz}\right)^2$$

where $d\varepsilon/dz$ is the rate of change of electron energy per unit distance of propagation.

Solution: For $\mathbf{p} \parallel \hat{z}$, $\gamma^2 = 1 + p_z^2/m^2c^2$.

$$P = \frac{e^2 \gamma^2}{6\pi \epsilon_0 c^3 m^2}\left[\left(\frac{dp_z}{dt}\right)^2 - m^2c^2\left(\frac{d\gamma}{dt}\right)^2\right]$$

$$= \frac{e^2}{6\pi \epsilon_0 c}\left(\frac{\gamma^2}{\gamma^2 - 1}\right)\left(\frac{d\gamma}{dt}\right)^2$$

$$= \frac{e^2 c}{6\pi \epsilon_0}\left(\frac{d\gamma}{dz}\right)^2 = \frac{e^2}{6\pi \epsilon_0 m^2 c^3}\left(\frac{d\varepsilon}{dz}\right)^2$$

where we have used $d/dt = v_z d/dz$, $v_z^2/c^2 = 1 - 1/\gamma^2$.

PROBLEM 10.2. An electron rotates in a circular orbit with

$$\mathbf{v_p} = v_\perp(\hat{x}\cos(\omega_c/\gamma)t) + \hat{y}\sin((\omega_c/\gamma)t)),$$
$$\mathbf{r_p} = (\gamma v_\perp/\omega_c)(\hat{x}\sin(\omega_c/\gamma)t) - \hat{y}\cos((\omega_c/\gamma)t)),$$

where $\gamma = (1 - v_\perp^2/c^2)^{-1/2}$ is a slowly varying function of time, as compared to the phase terms. Estimate the power radiated by the electron.

Solution: $\dot{\mathbf{v}}_\mathbf{p} \simeq -(\omega_c^2/\gamma^2)\mathbf{r_p}$, Using it in eq.(20), we get

$$P = \frac{e^2\omega_c^2\gamma^2 v_\perp^2}{6\pi \epsilon_0 c^3}.$$

This expression holds true for radiation power loss in a synchrotron.

PROBLEM 10.3. In the spring model of an atom, the electron orbiting around the nucleus with frequency ω_0 is equivalent to an electron attached to a spring of spring constant $K = m\omega_0^2$ where m is the electron mass. The electron is subjected to an electric field $\mathbf{E}_0 = \hat{x}A_0 \cos \omega t$. Obtain electron displacement $x_p(t)$ in the steady state, treating the electron motion to be non-relativistic. Obtain the far field at point (d,d,0) of the radiation of frequency ω emitted by the electron. Estimate the time average radiated power.

Solution: The equation of motion for the electron of charge $-e$ and mass m is

$$m\frac{d^2x}{dt^2} = -Kx - eA_0e^{-i\omega t}.$$

In the steady state $x \sim e^{-i\omega t}$, hence replacing $\partial/\partial t$ by $-i\omega$ we obtain

$$x = \frac{eA_0e^{-i\omega t}}{m(\omega^2 - \omega_0^2)},$$

where $\omega_0 = (K/m)^{1/2}$. Electron acceleration can now be written as

$$\dot{\mathbf{v}}_\mathbf{p} = -\hat{x}\frac{e\omega^2 A_0 \cos \omega t}{m(\omega^2 - \omega_0^2)}.$$

Expressing the direction of observation as $\hat{n} = (\hat{x} + \hat{y})/\sqrt{2}$, the far field at (d, d, 0) is

$$\begin{aligned}\mathbf{E}_{rad} &= \frac{e(\hat{n} \times \hat{n} \times \dot{\mathbf{v}}_\mathbf{p})_{t'=t-r/c}}{4\pi \in_0 c^2 r} \\ &= \frac{e^2\omega^2 A_0 \exp[-i(\omega t - \omega d\sqrt{2}/c)}{8\pi \in_0 c^2 r(\omega^2 - \omega_0^2)}(\hat{y} - \hat{x}).\end{aligned}$$

The time average radiated power (see eq. 18) is

$$P_{av} = \frac{e^4\omega^4 A_0^2}{12\pi m^2(\omega^2 - \omega_0^2)^2 \in_0 c^3}.$$

For $\omega < \omega_0$, P_{av} scales as ω^4 or λ^{-4} with radiation wavelength. This is *Rayleigh scattering*.

10.3.1 Frequency Spectrum of Radiation

The electromagnetic fields due to an accelerated charge, in general, are non-monochromatic. We may express them as Fourier integrals (see Appendix A.2). The radiation electric field, for instance, can be written as

$$\mathbf{E}(\mathbf{r},t) = \frac{1}{\sqrt{2\pi}R}\int_{-\infty}^{\infty} \mathbf{g}(\omega)e^{-i\omega t}d\omega, \tag{21}$$

where

$$\begin{aligned}\mathbf{g}(\omega) &= \frac{R}{\sqrt{2\pi}}\int_{-\infty}^{\infty} \mathbf{E}(\mathbf{r},t)e^{i\omega t}dt \\ &= \frac{1}{\sqrt{2\pi}}\frac{q}{4\pi\ \epsilon_0\ c}\int_{-\infty}^{\infty}\left[\frac{\hat{n}\times(\hat{n}-\boldsymbol{\beta})\times\dot{\boldsymbol{\beta}}}{(1-\boldsymbol{\beta}.\hat{n})^3}\right]_{ret} e^{i\omega t}dt.\end{aligned}$$

One may note that $\mathbf{g}(\omega) = \mathbf{g}(-\omega)$. The Poynting flux at $\mathbf{r}$ can be written as

$$\mathbf{S} = \hat{n}\frac{1}{4\pi R^2\mu_0 c}\int_{-\infty}^{\infty}\int_{-\infty}^{\infty} \mathbf{g}(\omega).\mathbf{g}^*(\omega')e^{-i(\omega-\omega')t}d\omega' \tag{22}$$

and the total energy, radiated per unit solid angle, as

$$\frac{dW}{d\Omega} = \frac{1}{4\pi\mu_0 c}\int_{-\infty}^{\infty} dt\int_{-\infty}^{\infty}\int_{-\infty}^{\infty} g(\omega)g^*(\omega')e^{-i(\omega-\omega')t}d\omega d\omega'. \tag{23}$$

Carrying out the time integral first, we note that it merely equals $2\pi\delta(\omega-\omega')$. Hence, ω' integral gives

$$\frac{dW}{d\Omega} = \frac{1}{2\mu_0 c}\int_{-\infty}^{\infty} |g(\omega)|^2 d\omega. \tag{24}$$

The radiated energy per unit frequency interval, per solid angle, is

$$\frac{d^2I}{d\omega d\Omega} = \frac{1}{\mu_0 c}|\mathbf{g}(\omega)|^2. \tag{25}$$

Here, a factor of 2 is included to account for the fact that ω and $-\omega$ represent the same frequency.

Since $t' + R(t')/c = t$, we change the variable of integration in $\mathbf{g}(\omega)$ from t to t′, giving

$$\mathbf{g}(\omega) = \left(\frac{q}{4\pi \in_0 c}\right) \frac{1}{\sqrt{2\pi}} \int_{-\infty}^{\infty} \frac{\hat{n} \times (\hat{n} - \boldsymbol{\beta}) \times \dot{\boldsymbol{\beta}}}{(1 - \boldsymbol{\beta}.\hat{n})^2} \quad e^{+i\omega(t' + R(t')/c} dt'.$$

If the observation point $\mathbf{r}$ is far away from the origin (where an accelerated charge is located), the unit vector $\hat{n}$ is fairly independent of time and one may approximate $R \simeq r - \hat{n}.\mathbf{r_p}(t')$. Thus, $\mathbf{g}(\omega)$ may be simplified to give

$$\frac{d^2 I}{d\omega d\Omega} = \left(\frac{q}{4\pi \in_0 c}\right)^2 \frac{1}{4\pi \mu_0 c} \left| \int_{-\infty}^{\infty} \frac{\hat{n} \times [(\hat{n} - \boldsymbol{\beta}) \times \dot{\boldsymbol{\beta}}]}{(1 - \boldsymbol{\beta}.\hat{n})^2} e^{i(\omega t - \hat{n}.\vec{r}_p(t)/c)} dt \right|^2 , \tag{26}$$

where prime over the variable of integration has been suppressed. For a given particle motion $\mathbf{r_p}(t), \boldsymbol{\beta}$ and $\dot{\boldsymbol{\beta}}$ are known and time integral can be carried out. This is a very profound result, widely used in synchrotron, Bremsstrahlung emission of radiation and numerous other applications.

10.4 Conclusions

Radiation from accelerated charges is one of the most prevalent sources of coherent and incoherent radiation. In a plasma, electrons moving in the vicinity of ions suffer acceleration/retardation due to ion coulomb field, and emit radiation called *Bremsstrahlung radiation*. In the presence of a magnetic field, electrons undergo cyclotron motion and emit radiations at the harmonics of cyclotron frequency. When the electrons have relativistic energies, the radiations of cyclotron harmonics overlap strongly.

Radiation from electron beams constitutes a major source of coherent radiation, e.g., in travelling wave tubes, gyrotrons and free electron lasers – where radiation signal bunches the electron beams in retarding phases, inducing them to emit coherent radiation.

UNSOLVED PROBLEMS

1. The trajectory of an electron is given by $\mathbf{r_p}(t) = \hat{x}a\cos\omega t + \hat{y}a\sin\omega t + \hat{z}v_0 t$. Obtain the retarded time, corresponding to the radiation observed at a distant point $\mathbf{r}(t) = \hat{x}b$ at time t. Assume a<<b.

   [**HINT:** $\left|\mathbf{r} - \mathbf{r_p}(t')\right| = c(t - t')$. *In this equation, approximate* $\mathbf{r_p}(t) = \hat{z}v_0 t'$.]

   [**ANS.** $t' = \gamma_0^2[t - \sqrt{t^2 v_0^2/c^2 + b^2/(\gamma_0^2 c^2)}]$, $\gamma_0 = 1/\sqrt{1 - v_0^2/c^2}$.]

2. The electrostatic field in an ion channel is $\mathbf{E} = \alpha_0 x \hat{x}$. A relativistic electron is launched into it at (a, 0, 0), with velocity $v_0 \hat{z}$. Obtain the trajectory of the electron and the radiation field, $\mathbf{E}(\mathbf{r}, t)$, produced by it at $\mathbf{r} = b\hat{z}$, assuming $v_x << v_0$, a<<b.

$$\left[\begin{array}{l} \textbf{Ans.}\; x(t) = a\cos\omega_b t, z(t) = v_0 t, \omega_b = (e\alpha_0/m\gamma_0)^{1/2}, \\ \mathbf{E}(b\hat{z}, t) = -\hat{x}\dfrac{ea\omega_b^2\cos\omega_b t'}{4\pi\varepsilon_0 c(1-v_0/c)^2\left|b - v_0 t'\right|}, \\ t' = \gamma_0^2[t - bv_0/c^2 - (t^2v_0^2/c^2 + b^2/c^2 - 2bv_0 t/c^2)^{1/2}]. \end{array}\right]$$

> **NOTE.** Ion channel can be created by passing a Gaussian laser beam in a plasma. The laser exerts a transverse *ponderomotive force* on the plasma electrons pushing them away from the laser axis, leaving behind an ion channel.

3. An electron, initially at rest at the origin, is subjected to a dc electric field $\mathbf{E}_0 = -0.5\,\hat{z}$ G V / m. Obtain electron energy as a function of z and plot v_z as a function of z. Estimate the radiation field at $\mathbf{r} = 10\,\hat{x}$ at an instant,

when the electron is at $\mathbf{r_p}(t) = 10^{-4}\hat{z}$ (where $\mathbf{r}$ and $\mathbf{r_p}$ are in metres).

[**ANS.** $\varepsilon = \gamma mc^2$, $\gamma = 1 + 10^3 z$, $v_z = c(1 - 1/\gamma^2)^{1/2}$, radiation field is zero at that instant.]

4. In a free electron laser (FEL), an electron beam moves through a wigglermagnetic field of wave number $k_w\,\hat{z}$ with velocity $v_0\,\hat{z}$. A beam electron emits a photon of energy $\hbar\omega$ and momentum $\hbar\,\mathbf{k}$ by losing energy $\hbar\omega$ and momentum $\hbar\,(k + \mathbf{k}_w)$. Using the energy momentum relation $\varepsilon = mc^2\gamma = mc^2(1 + p^2/m^2c^2)^{1/2}$, $\omega = kc$, show that the operating frequency of the FEL is $\omega_L \cong 2\gamma^2 k_w\,c$, where $\gamma = (1 - v_0^2/c^2)^{-1/2}$ (v_0 is close to c and $\hbar k << mc$).

5. A relativistic electron is passed through a wiggler magnetic field $\mathbf{B_w} = (\hat{x} + i\hat{y})B_0 \exp(ik_w z)$, with an initial velocity $v_0 = 0.99c\,\hat{z}$. Taking $v_x^2, v_y^2 << c^2, z = v_0 t$, obtain the acceleration $\mathbf{a_w}$ and oscillatory velocity $\mathbf{v_w}$ of the electron.

[**ANS.** $\mathbf{a_w} = -iev_0\mathbf{B_w}/m\gamma_0$, $\mathbf{v_w} = -e\mathbf{B_w}/m\gamma_0 k_w$, $\gamma_0 = 7.1$.]

6. A non-relativistic electron, gyrating in a magnetic field, $B_s\hat{z}$ has $x_p = a\cos\omega_c t$, $y_p = a\sin\omega_c t$. Obtain $\mathbf{S_{av}}$.

[**ANS.** $\hat{r}\left(\frac{ea\omega_c^2}{4\pi r}\right)^2 \frac{1+\cos^2\theta}{4\varepsilon_0 c^3}$, θ is the angle between $\hat{r}$ and $\hat{z}$.]

References and Suggested Reading

[1] Landau L. D. and Lifshitz E. M., *Classical theory of fields*, translated by Hamermesh M., Reading, MA, Pergamon Press, Oxford and Addison-Wesley, 1987; **ibid**, *Electrodynamics of continuous media*, Addison-Wesley, MA, Reading, 1984.

[2] Jackson J. D., *Classical Electrodynamics*, New York, John Wiley & Sons, Inc., 2005.
[3] Marshall T. C., *Free electrons lasers*, New York, Mc Millan Pub. Co., 1985.

Appendix A
Fourier Series and Fourier Integral

A.1 Fourier Series

Often we encounter functions that repeat themselves after an interval of time

$$f(t+T) = f(t). \tag{A.1}$$

Here, T is called the **time period**. Such functions can be expressed as a sum of sine and cosine functions, each of which repeats itself after time T. Consider sin $(2\pi nt/T)$ and cos $(2\pi nt/T)$, where n is an integer. These functions repeat themselves, after a time interval T, e.g., $\sin n\,\omega\,(t+T) = \sin n\omega\, t$, where $\omega = 2\,\pi\,/\,T$. We express f(t) as

$$f(t) = a_0 + \sum_{n=1}^{\infty} a_n \cos\, n\,\omega\, t + \sum_{n=1}^{\infty} b_n \sin\, n\,\omega\, t, \tag{A.2}$$

where a_0, a_n and b_n are constants to be chosen, such that the right-hand side correctly represents the function on the left at all times. We employ the orthonormality property of sine, cosine functions to evaluate them.

$$\frac{1}{T}\int_{t_0}^{t_0+T} \sin(m\,\omega t)\sin(n\,\omega t)\,dt = \frac{1}{T}\int_{t_0}^{t_0+T} \cos(m\,\omega t)\cos(n\,\omega t)\,dt = \delta_{mn}/2,$$

$$\frac{1}{T}\int_{t_0}^{t_0+T} \cos(m\,\omega t)\sin(n\,\omega t)\,dt = 0, \tag{A.3}$$

where δ_{mn} is the Kroneker delta with $\delta_{mn} = 1$ for $m = n$ and zero otherwise. Multiplying eq. (A.2) successively by dt / T, $(dt/T)\cos(r\omega t)$, $(dt/T)\sin(r\omega t)$ and integrating over t from 0 to T, we obtain

$$a_0 = \frac{1}{T}\int_{t_0}^{t_0+T} f(t)\,dt,$$

$$a_r = \frac{2}{T}\int_{t_0}^{t_0+T} f(t)\cos(r\omega\, t)dt,$$

$$b_r = \frac{2}{T}\int_{t_0}^{t_0+T} f(t)\sin(r\omega\, t)dt. \tag{A.4}$$

As an example, consider the function (Fig. A.1)

$$f(t) = 1 \quad \text{for } 0 < t < T_1$$
$$= 0 \quad \text{for } T_1 < t < T$$

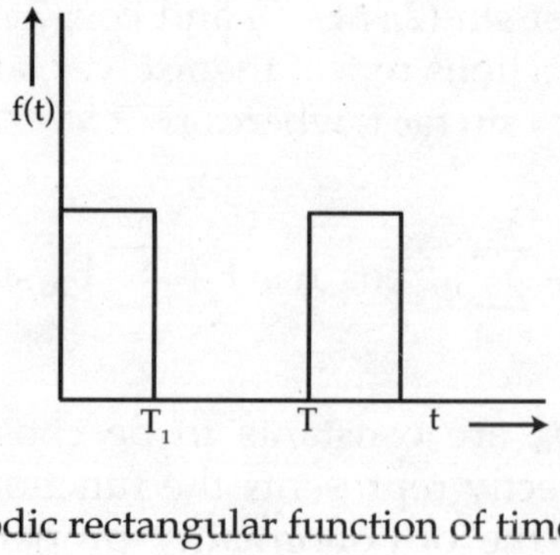

Fig. A.1. Periodic rectangular function of time with period T.

with $f(t + T) = f(t)$. Using $f(t)$ in eqs. (A.4), we have $a_0 = T_1/T$, $a_r = (1/\pi r)\sin(r\omega T_1)$, $b_r = (1/\pi r)(1 - \cos(r\omega T_1))$.

A.2 Fourier Integral

When the function is non-periodic but localised in time, we can treat t as a function of very large time period $T \to \infty$. In such a case $\omega \to 0$. Further, since f (t) is localised in time,

a. $\int_{-\infty}^{\infty} f(t)dt$ is finite

b. a_0, from eq.(A.6), goes to 0 as $T \to \infty$.
Denoting ω by $\Delta\omega$ and using eqs. (A.4) in the Fourier series (with $t_0 = -T/2$, $t_0 + T = T/2$, $T \to \infty$ as the limits), we write

$$f(t) = \sum_n \frac{\Delta\omega}{\pi} \int_{-\infty}^{\infty} f(t') \cos(n\Delta\omega t')\, dt' \cos(n\Delta\omega t)$$

$$+ \sum_n \frac{\Delta\omega}{\pi} \int_{-\infty}^{\infty} f(t') \sin(n\Delta\omega t')\, dt' \cos(n\Delta\omega t)$$

$$= \sum_n \frac{\Delta\omega}{\pi} \int_{-\infty}^{\infty} f(t') \cos(n\Delta\omega(t - t'))\, dt'. \qquad \text{(A.5)}$$

The summation over n may be replaced by integration using $\Delta\omega = ds$ and $n\Delta\omega = s$,

$$f(t) = \frac{1}{\pi} \int_0^{\infty} \left[\int_{-\infty}^{\infty} f(t') \cos[s(t - t')]\, dt' \right] ds$$

$$= \frac{1}{2\pi} \int_{-\infty}^{\infty} ds \int_{-\infty}^{\infty} f(t') \cos[s(t - t')]\, dt'. \qquad \text{(A.6)}$$

as cos [s (t−t′)] is an even function of s. Further, sin [s (t−t′)] is an odd function of s. Hence

$$0 = \frac{1}{2\pi}\int_{-\infty}^{\infty} ds \int_{-\infty}^{\infty} f(t')\sin[s(t-t')]\,dt'. \quad (A.7)$$

Multiplying eq. (A.7) by i, subtracting it from eq. (A.6) and changing the variable of integration from s to ω, we get

$$f(t) = \frac{1}{\sqrt{2\pi}}\int_{-\infty}^{\infty} g(\omega)\,e^{-i\omega t}d\omega, \quad (A.8)$$

$$g(\omega) = \frac{1}{\sqrt{2\pi}}\int_{-\infty}^{\infty} f(t')\,e^{i\omega t'}dt'.$$

g (ω) is known as the Fourier transform of f (t) and integral representation of f as the Fourier integral.

PROBLEM A.1. Obtain the Fourier transform of Gaussian function $f(t) = \exp(-t^2/\tau^2)$.

Solution : The Fourier transform is

$$g(\omega) = \frac{1}{\sqrt{2\pi}}\int_{-\infty}^{\infty} e^{-t^2/\tau^2}e^{-i\omega t'}dt' \left[= \frac{1}{\sqrt{2\pi}}\tau\int_{-\infty}^{\infty} e^{-(\xi - i\omega\tau/2)^2}e^{-\omega^2\tau^2/4}d\xi\right]$$

$$= \left(\tau/\sqrt{2}\right)e^{-\omega^2\tau^2/4}.$$

The Fourier transform peaks at ω=0 and has a frequency width $\sim 2/\tau$. Thus, a sharply peaked function of time has large frequency width.

Appendix B
Vector Identities and Derivatives

Cartesian Coordinates

$$\nabla f \equiv \hat{i}\frac{\partial f}{\partial x} + \hat{j}\frac{\partial f}{\partial y} + \hat{k}\frac{\partial f}{\partial z}$$

$$\nabla.\mathbf{A} = \frac{\partial A_x}{\partial x} + \frac{\partial A_y}{\partial y} + \frac{\partial A_z}{\partial z}$$

$$\nabla \times \mathbf{A} = \hat{i}\left(\frac{\partial A_z}{\partial y} - \frac{\partial A_y}{\partial z}\right) + \hat{j}\left(\frac{\partial A_x}{\partial z} - \frac{\partial A_z}{\partial x}\right) + \hat{k}\left(\frac{\partial A_y}{\partial x} - \frac{\partial A_x}{\partial y}\right)$$

$$\nabla^2 f = \frac{\partial^2 f}{\partial x^2} + \frac{\partial^2 f}{\partial y^2} + \frac{\partial^2 f}{\partial z^2}$$

Cylindrical Coordinates

$$\nabla f \equiv \hat{r}\frac{\partial f}{\partial r} + \hat{\phi}\frac{1}{r}\frac{\partial f}{\partial \phi} + \hat{z}\frac{\partial f}{\partial z}$$

$$\nabla.\mathbf{A} = \frac{1}{r}\frac{\partial}{\partial r}(rA_r) + \frac{1}{r}\frac{\partial A_\phi}{\partial \phi} + \frac{\partial A_z}{\partial z}$$

$$\nabla \times \mathbf{A} = \hat{r}\left(\frac{1}{r}\frac{\partial A_z}{\partial \phi} - \frac{\partial A_\phi}{\partial z}\right) + \hat{\phi}\left(\frac{\partial A_r}{\partial z} - \frac{\partial A_z}{\partial r}\right) + \hat{z}\left(\frac{1}{r}\frac{\partial}{\partial r}(rA_\phi) - \frac{1}{r}\frac{\partial A_r}{\partial \phi}\right)$$

$$\nabla^2 f = \frac{\partial^2 f}{\partial r^2} + \frac{1}{r}\frac{\partial f}{\partial r} + \frac{1}{r^2}\frac{\partial^2 f}{\partial \phi^2} + \frac{\partial^2 f}{\partial z^2}$$

Spherical Coordinates

$$\nabla f \equiv \hat{r}\frac{\partial f}{\partial r} + \hat{\theta}\frac{1}{r}\frac{\partial f}{\partial \theta} + \hat{\phi}\frac{1}{r\sin\theta}\frac{\partial f}{\partial \phi}$$

$$\nabla.\mathbf{A} = \frac{1}{r^2}\frac{\partial}{\partial r}\left(r^2 A_r\right) + \frac{1}{r\sin\theta}\frac{\partial}{\partial \theta}(\sin\theta\, A_\theta) + \frac{1}{r\sin\theta}\frac{\partial}{\partial \phi}A_\phi$$

$$\nabla \times \mathbf{A} = \hat{r}\frac{1}{r\sin\theta}\left(\frac{\partial}{\partial \theta}(\sin\theta A_\phi) - \frac{\partial A_\theta}{\partial \phi}\right)$$

$$+\hat{\theta}\frac{1}{r}\left(\sin\theta\frac{\partial A_r}{\partial \phi} - \frac{\partial}{\partial r}(rA_\phi)\right) + \hat{\phi}\frac{1}{r}\left(\frac{\partial}{\partial r}(rA_\theta) - \frac{\partial A_r}{\partial \theta}\right)$$

$$\nabla^2 f = \frac{1}{r^2}\frac{\partial}{\partial r}\left(r^2\frac{\partial f}{\partial r}\right) + \frac{1}{r^2\sin\theta}\frac{\partial}{\partial \theta}\left(\sin\theta\frac{\partial f}{\partial \theta}\right) + \frac{1}{r^2\sin^2\theta}\frac{\partial^2 f}{\partial \phi^2}.$$

Triple Products

$$\mathbf{A}.(\mathbf{B} \times \mathbf{C}) = \mathbf{B}.(\mathbf{C} \times \mathbf{A}) = \mathbf{C}.(\mathbf{A} \times \mathbf{B})$$

$$\mathbf{A} \times (\mathbf{B} \times \mathbf{C}) = \mathbf{B}(\mathbf{A}.\mathbf{C}) - \mathbf{C}(\mathbf{A}.\mathbf{B})$$

$$\nabla.\nabla \times \mathbf{A} = 0$$

$$\nabla \times \nabla f = 0$$

$$\nabla \times (\nabla \times \mathbf{A}) = \nabla(\nabla.\mathbf{A}) - \nabla^2\mathbf{A}$$

Product Rules

$$\nabla(fg) = f\nabla g + g\nabla f$$

$$\nabla(\mathbf{A}.\mathbf{B}) = \mathbf{A} \times (\nabla \times \mathbf{B}) + \mathbf{B} \times (\nabla \times \mathbf{A}) + (\mathbf{A}.\nabla)\mathbf{B} + (\mathbf{B}.\nabla)\mathbf{A}$$

$$\nabla.(f\mathbf{A}) = (\nabla f).\mathbf{A} + f\nabla.\mathbf{A}$$

$$\nabla \times (f\mathbf{A}) = (\nabla f) \times \mathbf{A} + f\nabla \times \mathbf{A}$$

$$\nabla.(\mathbf{A} \times \mathbf{B}) = \mathbf{B}.\nabla \times \mathbf{A} - \mathbf{A}.\nabla \times \mathbf{B}$$

$$\nabla \times (\mathbf{A} \times \mathbf{B}) = (\mathbf{B}.\nabla)\mathbf{A} - (\mathbf{A}.\nabla)\mathbf{B} + \mathbf{A}(\nabla.\mathbf{B}) - \mathbf{B}(\nabla.\mathbf{A})$$

Fundamental Theorems

Gradient Theorem: $\int_A^B \nabla\phi.\mathbf{dl} = \phi\,(B) - \phi\,(A)$

Gauss's Theorem: $\int_V \nabla \cdot \mathbf{A}dV = \oint_s \mathbf{A.ds}$

Stoke's Theorem: $\int_s \nabla \times \mathbf{A.ds} = \oint_{line} \mathbf{A.dl}.$

Index

E

F

G

H

I

K

L

M

N

O

P

Q

R

S

T